ENERGY RISK ASSESSMENT

ENERGY RISK ASSESSMENT

Herbert Inhaber

Atomic Energy Control Board of Canada
now at Oak Ridge National Laboratory

GORDON AND BREACH SCIENCE PUBLISHERS
New York London Paris

Gordon and Breach, Science Publishers, Inc.
One Park Avenue
New York, NY 10016

Gordon and Breach Science Publishers Ltd.
42 William IV Street
London WC2N 4DE

Gordon & Breach
58, rue Lhomond
75005 Paris

Library of Congress Cataloging in Publication Data

Inhaber, Herbert, 1941–
Energy risk assessment.

Includes bibliographical references and index.
1. Energy development–Environmental aspects.
2. Environmental impact analysis. 3. Risk. I. Title.
TD195.E49I53 333.79'1 82-3060
ISBN 0-677-05980-9 AACR2

ISBN 0-677-059809.
Printed in the United States of America.

For my sister, Marsha, and
brothers, Phillip and Douglas

Contents

PAGE

List of Figures

List of Tables

Foreword

Historically, public acceptance of technology was based almost entirely on economics. If two technologies were in competition, the less costly of the two was inevitably chosen with little attention paid to health and environmental problems. This strong emphasis on economics was natural in a society where few could afford to delay employment long enough to go to high school, where nearly everyone had to work eighty hours per week to avoid starvation and maintain standards of clothing and housing that we now associate with abject proverty, where there was little time or facilities for recreation, and where one was surrounded by imminent threats of disability or death from disease. With economic considerations constantly forcing people into life-threatening and life-shortening situations, improved economics was the first priority.

As living conditions improved early in this century, attention was turned to noneconomic problems. Workers fought for, and eventually won, safer working conditions. Public health measures were greatly improved, and medical research blossomed. With the prosperity following World War II came an awareness of other types of problems, and cities began to clean up the obvious pollution that was detracting from their beauty and the comfort of their population.

Activities of this sort were generally inexpensive and had little impact on the cost of living. They also yielded quick and obvious benefits and thus were almost universally approved. In the late 1960's, however, there was a sharp escalation as the environment suddenly became a hot public issue. Congress passed landmark legislation on air and water pollution abatement and the Environmental Protection Agency was set up and rapidly became a major factor on the Washington scene. Citizen groups and student organizations sprang up and received strong support from public and private foundations, and political careers were built and flourished, all under the banner of "environmentalism." The mass media and its individual stars strongly supported the movement. It was considered the "in" thing to do, and everyone agreed the environment should be cleaned up even if doing so was in conflict with economic considerations. To some extent, this was a reasonable extension of economics. As pollution is an unpaid cost of industrial activity so its costs should be included in the economic equation in some way.

The environmental movement had some quick and easy targets, and rapidly overwhelmed them. Where raw sewage was being dumped into streams, treatment plants were built. Where an industrial plant was polluting a large area, it was forced to clean up its emissions. The worst abuses of strip-mining were corrected. After these more easily recognized problems had been rapidly solved, momentum carried the movement on. The next step was to look into some of the tougher and less obvious environmental problems, which often involved larger economic penalties. The large black stuff in air pollution that soils clothing and makes everything dirty was obviously bad and, fortunately, was easy to remove. However, there are thousands of other components of air pollution that we can't see, smell, or taste, and some of them are undoubtedly harmful. What should be done about them? Some of the chemicals that industry was dumping into rivers were obviously bad to anyone who could smell, taste, or count dead fish, but what about some of the other chemicals that don't contribute to those effects but may still have subtle health effects? What about chemicals that are buried in the ground; they can't be doing much harm now, but will they come back to haunt us?

The answers to these questions and others like them clearly require a great deal of research, and research takes time. Waiting for research results is not an acceptable mode for some of the organizations in the environmental movement. One easy alternative is to concentrate on the research results now available, and act as though no others will be forthcoming, assuming that "what you don't know won't hurt you." This is obviously illogical, and penalizes industries that have acted responsibly in carefully studying relevant environmental impacts, specifically, the nuclear power industry. Yet, this is the course taken by a large segment of the environmental establishment, including its media supporters.

Historically, technologies have been developed by private enterprise, which had little to gain by worrying about environmental impacts and therefore never did so. In contrast, though, a rare exception to this rule is the case of nuclear energy. It was developed by the U.S. government as a contribution to the future well being of the nation and the world and was sold to private industry only with great effort and "hard-sell" techniques. As part of the original development, the government did what is by far the most thorough scientific investigation of environmental impacts ever undertaken by man. In this effort thousands of scientists were recruited and billions of dollars were spent, and fortunately the scientific and technical problems proved to be quite tractable: Biological effects of radiation stem basically from a far simpler phenomenon than health effects of chemicals, radiation is much easier to measure, detection sensitivities are thousands of times higher, and radiation science has now had forty years with generous research support to develop in an orderly fashion. There has also been time and money for elaborate experimental and computer studies of the elements of accident scenarios and for development of sophisticated risk assessments, including estimates of probabilities and health consequences for various potential events.

It is this fund of information that has been taken out of context, distorted, and used as ammunition by the antinuclear movement. This industry has taken great pains to study and evaluate its environmental impacts and has been severely penalized for its efforts, and may well die as a result of them.

But what is a rational course of action under the present circumstances? One possibility is to wait for the research in other areas to come in, and do nothing, that is, give little or no consideration to nonobvious environmental impacts in choosing between competing technologies. That approach was always used until ten or fifteen years ago, and it served us well, constantly improving our standard of living and increasing our life expectancy. Mistakes were made that are obvious in retrospect: exposure to asbestos, vinyl chloride, and some other chemicals. However, their effect was overwhelmed by the favorable economics of the system. Wealth is a very important contributor to health—well-to-do people live an average of four years longer than poor people in the United States, and people in wealthy nations live ten to thirty years longer than those in poor nations, whereas loss of life expectancy due to ***all*** work-related accidents combined reduces life expectancy by only 0.2 years.† Therefore, it might be acceptable to tolerate a technology that increases, or saves, wealth even if it also causes some adverse health effects.

However, ignoring environmental considerations does not seem to be

---

†B. L. Cohen and I. S. Lee, "A catalog of risks," Health Physics **36**, 707 (1979).

wise at this point in time. Environmental issues have important impact and they must be considered. Moreover, there is now accumulating a body of research in many environmental areas, and this should not be ignored.

Under these circumstances, it seems most reasonable to make the best estimates we can with the knowledge now available. That is the science of risk assessment, a large and growing discipline permeating all areas of technology. But in no area is it more important than in comparisons among alternative energy technologies. Choices must now be made every day between these technologies, and if environmental aspects are to be given weight in these choices, we must have comparative risk assessments for them. That is the task Dr. Inhaber has undertaken in this book, the culmination of his many years of work on this problem.

Since these risk assessment studies require the use of health effects estimates that are still quite uncertain, anyone undertaking this task is subject to criticism. Fortunately, however, a number of these estimates have been made by previous researchers; Dr. Inhaber needed only to average among them. This may not satisfy everyone, but it would be difficult to suggest a more acceptable alternative. Surely it is better than assuming there are *no* effects, as is often done.

Such a wide ranging risk assessment involving many hundreds of information sources and estimates could easily contain errors or misinterpretations, but this study has been through several revisions with exposure to the scientific community for each. Thus, the probability is high that any such errors have by now been found and corrected.

A study of this magnitude requires many choices to be made: How do you compare deaths with injuries? Do you include backup energy sources? And so forth. One may not like Dr. Inhaber's choices on these and/or similar issues but he clearly spells them out, making it easy in many cases to deduce the result that would be obtained had other choices been made.

Regardless of personal bias on matters of this sort, everyone must agree that Dr. Inhaber has made a valiant effort on this very important problem, and his assessment is based on at least as firm a set of foundations as any other. It would be difficult to understand how anyone could justify not taking it seriously.

Another valuable aspect of this book is its clear demonstration of what goes into a risk assessment. Members of the public generally form opinions about risks of various technologies on the basis of fewer than ten "facts," many of them obtained from highly unreliable sources such as newspaper stories or TV programs. They are sure to recognize the reliability of the hundreds of important sources used in this book, and by the scientific skill employed in interrelating them. It is hoped that they will realize that risk assessment of technologies is too complex a task to be assessed only by the man in the street or a TV personality. Perhaps they will even accept the results of Dr. Inhaber's study as a valuable guide to the path to the truth rather than just use their own intuition.

In any case, we must all be extremely grateful to Dr. Inhaber for assembling this vast collection of information on energy risk assessment.

Bernard L. Cohen
University of Pittsburgh

Preface

Every form of human activity involves risk of accident or disease resulting in injury or death. Generation of energy is no exception. Although such risk has previously been considered for conventional systems (coal, oil, and nuclear), a similar analysis for the so-called alternative or nonconventional systems (solar, wind, ocean thermal, and methanol) has been lacking. This work presents an evaluation of the risk, both occupational and to the public, of these nonconventional energy systems. They are considered both in absolute terms and in relation to conventional systems.

This book is based, in part, on a document that was issued in 1978 (with two subsequent editions), *Risk of Energy Production*, Atomic Energy Control Board, Ottawa, Canada, Report AECB 1119. Since that time, the original data have been revised, and new data added. What you read is an improved study, with considerably more documentation behind its statements. For example, many of the tables have been revised, and the present work has over 600 references, where AECB 1119 had less than 200. However, the general principles and overall conclusions remain substantially the same.

Because these conclusions proved surprising to some people, the original report produced controversy. It was thought it would prove interesting to readers to see what was said, and what the rejoinders were; some of these letters, memos, and articles appear in Appendix Q. It should be remembered that many of the comments and replies have been overtaken by the new data and interpretations presented in this book. Most of the material in this appendix is at least somewhat critical of the original document. However, it should be noted that much of the feedback was favorable. As they were too numerous, these favorable comments are generally not reproduced here. The assertions of Appendix Q often bear out the remark of Benjamin Disraeli in the 19th century, "It is much easier to be critical than to be correct."

I would like to thank Alice Pagé, Sylvie Brisson, Connie Monk, Doris Hawkins, Nicole Smits, Nicole Coté, and Denise Fortin for typing a long and difficult manuscript. Paul Hamel supplied me with encouragement. Dave Head and Frank Campbell spent many hours going over the manuscript. I profited from extended conversations with Jim Elks while Peter Dyne, Phil Cockshutt, and K. Tupper made valuable suggestions. Rein Lemberg, of Lemberg Consultants Limited, Oakville, Ont., did an analysis of an earlier draft. Others who helped were Harold Stocker, Dave Smythe, Alex Danilov, Bill Bush, Bill Gummer, John Beare, Jon Jennekens, Bob Atchison, Hugh Spence, Paolo Ricci, Geoff Knight, Mike White, Richard Caputo, Steve McReynolds, Maryl Weatherburn, Michael Stock, André Potworowski, Dave Rogers, A. L. Hamilton, Peter Musgrove, Archie Robertson, and Richard Wilson. Bob Lidstone probably spent more time going over the study than anyone else, and I have appreciated his perception and helpfulness.

Herbert Inhaber

1. Introduction

To be useful, therefore, a technology assessment must go far beyond conventional engineering and cost studies to look at what else may happen in achieving an immediate goal, to the total range of social costs . . . (1)

But it is always easy to make a theoretical solution look better than the current situation (278)

"There is no free lunch." This restatement of the second law of thermodynamics applies to all forms of human activity. In particular, for each type of energy production, there is a risk–defined here as the collective chance or probability of accidents and disease resulting in injury or death. This risk (or, more exactly, consequences of risk) is part of the social costs of energy production, which include air and water pollution, land abuse, depletion of natural resources, and other factors.

None of the above is remarkable. There have been many studies on comparative risk of so-called conventional energy sources, such as coal, oil, nuclear, and natural gas. However, in the past few years there has been an upsurge of interest in "nonconventional" or "renewable" energy sources, such as solar, wind, geothermal, ocean thermal gradient, and biomass. An indication of this interest is shown by the 7½ pages of abstracts (approximately 700 in all) on solar energy alone in a recent annual cumulation of energy abstracts (2). Nonconventional sources–defined here as those which are not now producing large amounts of energy in Canada—are sometimes characterized as "benign" or "soft" (3, 128). The unstated implication is that they are risk-free. The object of this book is to evaluate and compare risks arising from major existing or proposed energy sources, both conventional and nonconventional.

This information will permit a better understanding and assessment of the relative risk of energy systems. In effect, another assessment tool is being added to those already in use, such as economic or environmental assessments.

When risk is considered, it might be concluded that it is mostly due to operation of an energy facility. (It is assumed that sources, systems, and facilities are synonymous.) For example, risk of nuclear accidents or risk to the health of the general public due to air pollution from coal-based electricity plants may be considered. However, to be complete, risk should be evaluated for the entire energy cycle, not merely the end the public sees. This is what this report attempts to do.

While risk to human health is an important consideration in evaluating energy systems, it cannot be the only factor. Economics, resource depletion, and conservation all play large roles in discussions of energy. While these and related factors are vital, they are not dealt with here.

A. Energy Use in Canada

This book considers eleven methods of generating electricity or energy. Five are conventional sources: coal, oil, natural gas, nuclear, and hydroelectricity. Six are nonconventional. Of the six, three either are or will be used in Canada, and three others are probably more applicable to regions outside Canada.

Most of the energy systems considered in this report can be used to generate electricity. Electricity is the form of energy with perhaps the widest variety of end uses. In this study, all final units of energy produced are deemed equivalent.

In order to put risk due to electricity production into perspective, present and projected energy usage in Canada can be noted. According to "An Energy Strategy for Canada" (133) the percentage of all energy use in 1975 was as follows: electricity, 26.1%; coal, 8.8%; natural gas, 19.4%; and oil, 45.7%. A small amount of energy (of the order of 0.1%) was contributed by biomass in the form of heat from wood. Other biomass energy used in forestry is apparently not included. The energy contributed by nonconventional systems like solar and wind was negligible. Almost two-thirds of the electricity produced was hydroelectricity.

By 1990, the proportion of all energy use is expected to be as follows: electricity, 31.3–31.7%; coal, 8.6%; natural gas, 20.1–20.3%; and oil, 39.4–40.0%. Of course, some of the electricity is produced from the other three sources. Nonconventional technologies are expected to have only a minor impact on energy supply, although this may change beyond 1990.

The 1975 data for electricity production (134) showed the following capacities, in thousands of megawatts: hydroelectric, 36.8; nuclear, 2.66; and others (coal, oil, and natural gas), 20.08. By 1990, generating capacity in thousands of megawatts is expected to be as follows: hydroelectric, 66.2–72.5; nuclear, 22.2–29.5; and others, 36.8–49.8. Some of these values may have decreased since the publication of Ref. 134. Perhaps more important are the changes expected to occur in generating capacity between 1975 and 1990. Using the "high-growth" scenario, the capacity changes in thousands of megawatts, taking into account replacement of present facilities, are expected to be as follows: hydroelectric, 35; nuclear, 27; and others, 29. In other words, the expected additions to electricity capacity will be about equally divided among those three major sources.

Nonconventional electricity sources are expected to be negligible in 1990. The situation may change after this date.

B. Organization

In order to simplify the presentation as much as possible, the detailed calculations for each energy system are shown in the appendices. The methodology common to each is outlined in the next section. Conclusions are discussed in Section 3, although there are a brief analyses of the results within each appendix.

Among the appendices, most conventional technologies–coal, oil, natural gas, and nuclear*–are grouped at the beginning and are followed by nonconventional sources. There has already been considerable risk analysis done for conventional energy sources, and so the appendix devoted to each of these is relatively short. Because risk analysis of nonconventional technologies is new, the appendices on these systems are

*Because of data availability, data on nuclear power generally refer to light-water reactors, rather than the Canadian CANDU system. Details are given in Appendix D.

detailed, with assumptions spelled out. Hydroelectricity follows the nonconventional sources.

At least three of the nonconventional technologies analyzed–ocean thermal, solar thermal electric, and solar photovoltaic–will not be used in Canada for the foreseeable future, primarily because of climate. Since the analyses for the nonconventional systems are all tied together mathematically, the three are included here. It was felt that their inclusion would be of interest to some readers.

In order to emphasize those nonconventional systems that are more appropriate to Canadian conditions, the three sources mentioned above are separated graphically from the others in the concluding diagrams. In this way, the reader can concentrate, if so desired, on those energy sources deemed to be applicable to Canada.

C. Catastrophic and Noncatastrophic Risk

Public attention to risk of energy systems often concentrates on catastrophes, past or potential, as opposed to the routine, day-to-day risk incurred in the factory or generating station. Radioactive releases from nuclear reactors, oil or gas pipeline failures, hydroelectric dam bursts–these are what capture headlines. One event that kills 100 people is naturally noticed more than 100 events that each kill one person.

Catastrophes *do* take place. The actual or estimated risk to the public of dam failure and accidents at reactors, while very low, can never be zero. However, as shown below, the largest proportion of risk to human health from all the energy systems considered is either from industrial and occupational risk or pollution effects. While death due to an industrial accident or respiratory ailment is indeed a catastrophe to the individual involved, it does not have the same impact on the public as if hundreds were stricken.

Not all of the energy systems considered are subject to catastrophes. The nonconventional systems, while they can have substantial overall risk, generally do not have those which affect large numbers of the public. On the other hand, conventional systems like oil and natural gas do have pipeline or tanker explosions, nuclear reactors can also have severe accidents, and hydro dams can fail resulting in large loss of life.

In the calculation of overall risk, that resulting from catastrophes, which can be roughly defined as those in which five or more people are killed (623), can be added to that of an noncatastrophic origin. In one sense, apples are being added to oranges. In another sense, the same thing is being added, since the cost to society, as measured by the number of deaths, is the same. The appendices indicate risk of both catastrophic and noncatastrophic sources, so the two may be segregated if desired.

D. Conventional and Nonconventional Energy Sources

Nonconventional energy sources are far from risk-free. However, this does not necessarily imply they should not be used. Even the most ardent advocates of conventional technologies, such as coal or nuclear, recognize and indeed emphasize that nonconventional technologies, such as solar or wind, will be increasingly used in the future. The questions about this latter group have traditionally been When? How much? and At what economic cost? This book proposes an additional question: At what risk?

All forms of energy production should be as risk-free as possible. Considerable work has gone into reducing risk from conventional sources. There has been comparatively little thought given to reducing risk from nonconventional sources, primarily because it was thought to be zero or negligible. A fundamental purpose of this book is to stimulate interest in reducing accidents and disease from all energy systems, regardless of the particular mix used in the future.

E. Time Period of Assumptions

An important assumption in the calculations is that present-day technology, models, and systems, with their corresponding risk, are used. Generally speaking, energy systems either in present use or likely to be used in the near future were analyzed. In essence, more-established technologies are compared with less-established ones, an unavoidable requirement. However, there is no universal relationship between the length of time an energy system has existed and its degree of risk.

Reliance on present-day technology avoids assuming anything about the future. For example, risk for some conventional energy sources is to a large degree dependent on the effects of the pollution they release. Standards of allowable pollution may be made more stringent in the future, lowering this risk. Breakthroughs may be made in wind or solar technology, reducing the amount of steel, glass, and other material required. Since risk is partly dependent on the materials used, it may be diminished. For this report only those standards, material requirements, and other specifications in present use are employed. Breakthroughs for some technologies and not for others, as has sometimes been assumed in other energy analyses, will not be implied.

On the other hand, risk estimates could increase in the future, due to (a) discovery of new risks from old pollutants; (b) technical breakthroughs requiring new materials with higher risk; or (c) extraction of leaner ores for metal production or less productive fields for fuels, leading to greater risk per unit product.

F. Efficiencies and Load Factors

Making comparisons between energy systems involves knowledge of their relative efficiencies. About 35% of the energy produced at a thermal power station (such as a coal-burning plant) can be delivered to the consumer as usable power. Almost all of the power generated by solar electricity plants would be deliverable to the consumer, not counting transmission losses. Subsequent calculations will take these various efficiencies into account.

Knowledge of the load factor of an energy system also plays an important part in determining its risk. Every energy system has a "design capacity," or maximum ability to deliver energy. Due to factors like maintenance, accidents, testing, or lack of sunlight and wind, the actual energy delivered may not equal the energy that could be produced if the system operated continuously. The ratio of the actual to the continuous production level may be roughly defined as the load factor. This factor is vital in determining the material requirements of a system. For each system, the assumed load factor is noted.

Solar load factors can vary according to the amount of sunlight available. For solar thermal electric and solar photovoltaic, these factors have been adjusted to take account of Canadian conditions. However, the original data relating to the U.S. Southwest can also be used.

G. Centralization of Energy Sources

Among energy experts, there has been considerable discussion in the last few years of the relative advantages of centralization and decentralization of energy sources. Some commentators, such as Lovins (3), have suggested inherent positive features of decentralized systems like solar space heating. These features were stated to include lower cost, greater reliability, and less dependence on political and economic authority.

This report does not address these questions. However, the analyses in the appendices show that low risk is not inherent in decentralized systems. Highly centralized systems such as natural gas and nuclear seem to have lower risks than decentralized systems like solar space heating. As well, the risk for decentralized systems is calculated assuming centralized or mass production. The decentralization of these systems refers to their deployment, not their production. Decentralized production could well increase their risk estimate. While decentralized systems may offer political and economic benefits, an inherently low degree of risk to human health is not always one of their advantages.

H. Units

There are many units used in energy calculations. For consistency, energy units in this report are in terms of megawatt-years over the lifetime of the system. In effect, this is 1 megawatt of power (or capacity) operating continuously for 1 year. To save space, this is occasionally referred to as "unit energy."

As an example of the meaning of this term, consider an energy system that produces 1000 megawatts of power over the course of a year. Let the lifetime of the system be 30 years. By "lifetime" is meant the length of time, on the average, that the system lasts before replacement is necessary. The total energy produced over the system's lifetime is then 30,000 megawatt-years. If the number of deaths from all sources from this system is R, the risk averaged over the system's lifetime is then $R/30{,}000$ deaths per megawatt-year.

By way of illustration of energy units used, "An Energy Strategy for Canada" (125) notes that 262,000 megawatt-years (7867 trillion British Thermal Units) of all forms of energy were used in Canada in 1975. This includes residential, industrial, transportation, and other end-uses. Assuming a population of about 22 million, this is a use of about 262,000/22,000,000 = 0.012 megawatt-year per Canadian. Put another way, 1 megawatt-year will supply all the annual energy requirements for 1/0.012 = 83 Canadians. For perspective, 1 megawatt-year is the equivalent of 10,000 100-watt light bulbs burning for a year.

This amount of energy is much greater than the electrical energy used. Reference 275 indicates that 30,400 megawatt-years of electricity were produced in 1974 in Canada. This corresponds to 1 megawatt-year of electricity in that year for about 22,000,000/30,400 = 720 Canadians.

The risk calculated here is based on a "life-cycle" concept. Every energy system has capital equipment that eventually is replaced. This capital equipment is more directly related to the rated capacity (or power) of the system, rather than the rate of energy production. The risk attributable to this capital is amortized over or divided by the estimated total energy production over its lifetime. In addition, some energy systems have a continuous need for fuel. The risk attributable to this fuel is added to the "capital" risk.

2. Methodology

Risk evaluation, the methodology used in this paper, is similar in many ways to energy accounting. In the latter method, all the energy requirements of a system are summed, instead of their costs. In risk evaluation, all of the risks of accidents, disease, and death incurred in producing a unit of energy are added together. We can then compare the risks from various forms of energy production to see which has the lowest value.

This section provides a brief description of risk evaluation. A more extended discussion is given elsewhere (4).

A. Components of Risk

Any form of evaluation must include all of the items under consideration. While it cannot be proved that all sources of risk have been accounted for, the seven major ways shown in Figure 1 probably comprise almost all the risk in energy production. These seven are: material and fuel production, component fabrication, plant construction, operation and maintenance, public health, transportation, and waste disposition.

To illustrate the reasoning used, consider two technologies: solar space heating and coal-fired electricity plants. In terms of raw material and fuel production, solar requires the mining of copper for tubing, while the coal-fired plant requires the mining of coal as fuel, iron ore for building turbines, and so on. While some energy technologies do not require fuel in the ordinary sense, all need raw materials. Each type of raw material production incurs risk.

The components are then fabricated: copper tubing, steam turbines, cooling towers, and all other parts of each system. This produces further risk.

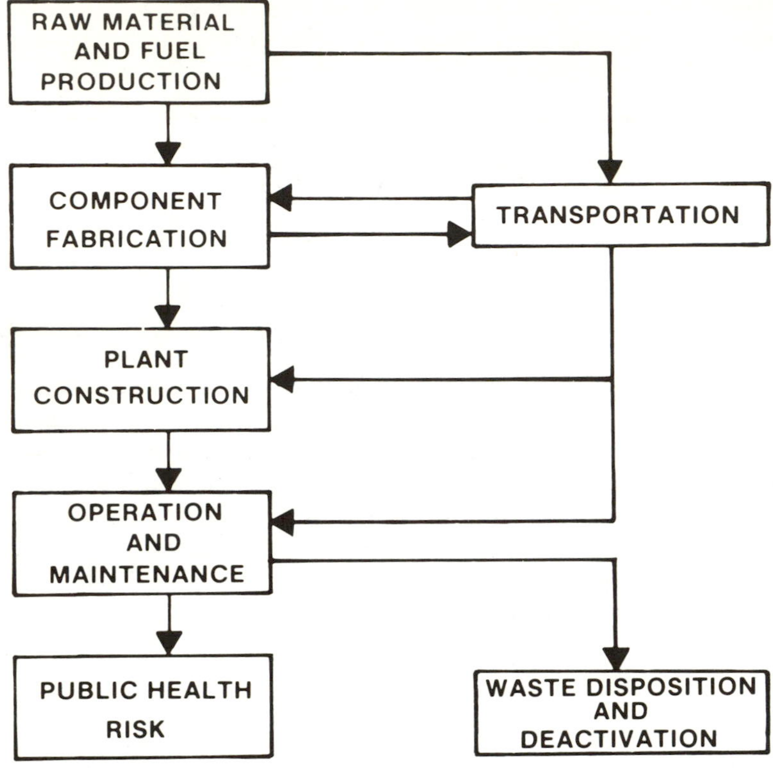

Figure 1. *Sources of risk in energy production. The relative importance of each of the components depends on the energy system. For example, there will be no fuel requirement for some nonconventional systems. Public health risk due to coal will be primarily from air pollution, and due to nuclear from the possibility of reactor accidents and waste management. Transportation plays a crucial part in many components.*

Most raw materials must be moved to an appropriate destination for fabrication into components. The components are then moved to the building site where they are assembled into a power plant. Maintenance also requires transportation for replacement of components. Since transportation plays such a significant part, it is shown as interacting with four aspects: raw materials, component fabrication, plant construction, and operation and maintenance.

After materials have been transported, the energy plants are then constructed, incurring further risk. Some systems may be highly centralized, and others decentralized, as in certain solar technologies.

Operation and maintenance of a system are often overlooked in risk analysis. It will be shown later that this can be a significant fraction of overall risk.

Public health risk is produced by all energy systems. Coal and oil generate air pollution. There are also indirect effects from most systems, such as the air pollutants generated from producing metals. In addition, some of the transportation risk mentioned above can affect the public. Nuclear power produces radioactive effluents both in normal operation and in the case of a catastrophic accident. Although it will be shown that the best available estimates of the last-named risk indicate that it is probably very low, some people are more concerned about this kind of risk than any other type. This danger can be put into context by considering its relationship to the overall risk of many energy systems.

Although not shown in Figure 1, all energy systems have risk associated with decommissioning or dismantling. This is not considered here, except for nuclear power (see Appendix D), due to insufficient data. As a crude approximation, it is probably proportional to construction risk for most systems.

Finally, there is risk inherent in waste disposition. Most public attention to this aspect has focused on nuclear wastes, although there have been disasters associated with coal wastes in the United Kingdom. While there is risk associated with industrial wastes from material production and manufacturing, as exemplified by recent incidents at the Love Canal in New York State, this risk is not evaluated here due to lack of data. The magnitude of this risk is probably roughly proportional to the amount of materials used. Wastes–carcinogenic, mutagenic, or otherwise harmful–must be managed in all energy systems.

Considerable work has previously been done in risk analysis. The many references used in this report are testimony to this. One of the pioneering works was WASH-1224 in 1974 (14). Other works have been Comar and Sagan (23), Hamilton (26), Jet Propulsion Laboratory (6), Hittman Associates (31), and Gotchy (458).

B. Risk Calculation—Raw Materials, Components and Construction

Much of the detailed risk calculations in this paper are centered on three of the items in Figure 1: material and fuel production, component fabrication, and plant construction. While the other four items are also important, the risk attributable to them is estimated by different and less complicated methods.

For the three sources, the calculation proceeds as follows. The amount of materials required to produce a component is determined. The number of man-hours* required to produce this material is then found. If construction, as opposed to material acquisition, is being considered, then the time required to install or build a component is estimated.

For each type of industrial activity, labor statistics are available which show the number of deaths, injuries, or time lost due to disease per million hours worked. From an occupational standpoint, some industries such as coal mining or logging are much more dangerous than others. The number of man-hours required per operation is then multiplied by the deaths, accidents, or disease per man-hour to produce the occupational risks.

As an example, suppose mining X tons of coal require Y man-days. If the number of man-hours lost per day of work is Z, then the number of man-hours lost per ton of coal is YZ/X. The risk associated with each part of the system is found in the same way and added to determine the total.

There are at least four sets of data needed for these calculations:

(a) Industrial production statistics, so estimates can be made of the times required per unit of production;
(b) Rates of industrial illnesses, accidents, and deaths;
(c) Times required for construction;
(d) Raw material requirements for industrial processes.

Much of these data are summarized elsewhere (16). Brief examples of the first three are shown in Tables 1, 2, and 3. More complete data are shown in Appendix M.

---

*Both men and women produce the materials required for an incur the risk produced by energy systems. For simplicity, the terms man-hours and man-days are used, although they should, strictly speaking, be person-hours and person-days.

Table 1 shows the relationship between production, in metric tons, and average manpower for a few industries. From data of this type the number of man-hours required per metric ton can be determined.

Table 1. Production Per Unit Time in Selected Industries (f)

	1973 Production, (millions of metric tons) (a)	*Manpower*	*Man-Hours (per metric ton)* (e)
Steel	120 (b)	603,000	10
Cement	77 (c)	32,600	0.84
Iron ore	68 (d)	15,000	0.44

(a) U.S. statistics.
(b) From Ref. 17.
(c) From Ref. 18.
(d) From Ref. 177.
(e) 2000 man-hours are assumed to equal 1 man-year.
(f) May involve substantial approximations in some industries. See Appendix M.

Table 2 shows more occupational health and safety statistics for certain industries. Trades like coal mining, roofing, and sheet metal have substantial risk. How can this table be used? Consider making one ton of steel. If the materials required, such as iron ore and coal, and the times to produce them, taken from Table 1, are known, Table 2 can be used to calculate the total deaths and man-days lost through accident and illness. This illness is occupational, not ordinary disease.

The above procedure applies only to material and equipment acquisition. To calculate construction and installation risk, Table 3, which lists the man-hours required per metric ton of construction materials, is used. Risk attributable to the demolition and removal of energy structures after the expiry of their lifetimes is considered only for the case of nuclear power, but it is expected to be small for all systems.

Table 2. Selected Occupational Health and Safety Statistics, per 100 man-years (15), (a), (f)

Industry	*Deaths* (b)	*Man-Days Lost, Occupational Illness*	*Man-Days Lost, Accidents*
Hard coal mining (d)	0.120 (e)	5.2	199.8
Roofing and sheet metal	0.056	2.4	172.3
Metal mining (d)	0.047	2.0	112
Fabricated metal products	0.009	2.7	89.3
Flat glass (c)	0.007	1.7	62.0

(a) Data available in Canada are presently not so finely subdivided. As a result, U.S. data were used. However, a consideration of industrial practices in the two countries indicates that the accident rates were probably not substantially different.
(b) Limited data were available on death rates. To estimate them, the following ratios of deaths to total man-days lost due to nonfatal accidents and illnesses were used: mining, 0.00059; contract construction, 0.00032; manufacturing, 0.00011 (60).
(c) Statistics generally based on the stone, clay, and glass products industry.
(d) Ratio of accident and illness determined from cement industry.
(e) Death rate in Norwegian coal mining is 0.18; in Poland, 0.028; both in the same units. The rate will depend on the grade and type of coal, safety precautions, and other variables (242).
(f) Data are generally from 1973–74. It might be expected that future rates may decrease. For example, the death rate in all U.S. industries fell by 30% from 1966 to 1976, for an average drop of 2.7% per year (554). However, accident rates can rise as well as fall. Of 23 industries reporting injury frequency rates in the United States for 1971 and 1976, 18 showed an increase between these two years (207). Another report (550) indicates that while the injuries per man-year in the U.S. private sector fell about 12% from 1973 to 1977, the number of work-days lost rose by about 18%. In Sweden, from 1964 to 1970, industrial accidents rose from 117,000 to 120,000 (279). These are numbers of accidents, not rates, but the rates probably remained about constant. Pochin (235) found accidental death rates in 13 British industries to be relatively stable from 1960 to 1972, falling only by about 0.5% per year. *Miners' Voice* in Canada states (344) "And the rate of death on the job in mining hasn't come down in over 25 years. Ontario's 1976 fatality rate, for instance, was 0.93 per million hours worked–the same as in 1951."

Table 3. Man-Hours Required per Metric Ton of Material in Construction (19), (a)

Material	*Man-Hours/Metric Ton*
Structural	162
Pipes	132 (b)
Major equipment	114
Concrete	0.33

(a) See footnote (e), Table M-1, for discussion of these values.
(b) See footnote (i), Table E-2.

The analysis is shown graphically in Figures 2 and 3. The former figure shows how risk is calculated as a function of unit weights in material acquisition, and the latter figure shows how it is calculated for unit times in construction.

The final materials required for system construction often require intermediate and raw materials. For example, steel requires iron ore, coal, and other basic materials. For this case, it is assumed that 1.67 kg of iron ore and 1.5 kg of coal are required to produce 1 kg of steel (20). This is a broad simplification, but the same assumption is made for each of the energy systems considered here.

To summarize the methodology, the number of man-hours needed to produce both the basic materials and to construct the energy facilities are calculated from Tables 1 and 3. These values are then multiplied by the rates of Table 2, yielding fatalities, accidents, and illnesses.

C. Risk Calculation–Transportation, Operation and Maintenance, Public Health, and Waste Disposition

Because of the widely varying nature of the four categories in this section's title, and their dissimilarity from material acquisition, their risks were estimated in different ways. The principles of this estimation are contained in this section; details are given in appropriate appendices.

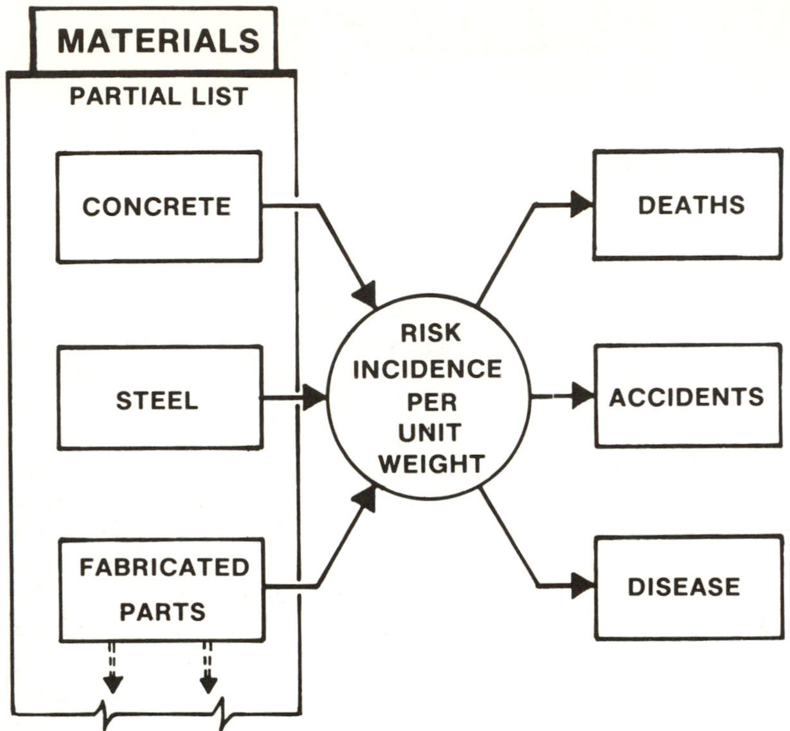

Figure 2. *Risk from material acquisition. Each of the raw materials that goes into an energy system has an associated calculable risk. The risk depends on the accident, illness, and death rate per unit weight produced in the appropriate industry. The dashes indicate that other materials are used.*

For transportation, estimates were available for risk incurred for conventional energy systems, such as coal. This risk could be transformed into risk per unit weight of material transported. From Table A-2, total coal transportation deaths, both occupational and public, range between 0.0024 and 0.0069; injuries between 0.029 and 0.064; and total man-days lost between 15 and 47, for each megawatt-year net output. The weight of construction materials from Table A-1 is 11.2 metric tons per unit energy. The weight of coal transported is 3500 metric tons per unit energy, for a total of 3511 metric tons. The transportation risk for nonsand and cement materials is assumed to be proportional to the coal risk per unit weight. In effect, the risk due to transporting sand and aggregate is excluded.

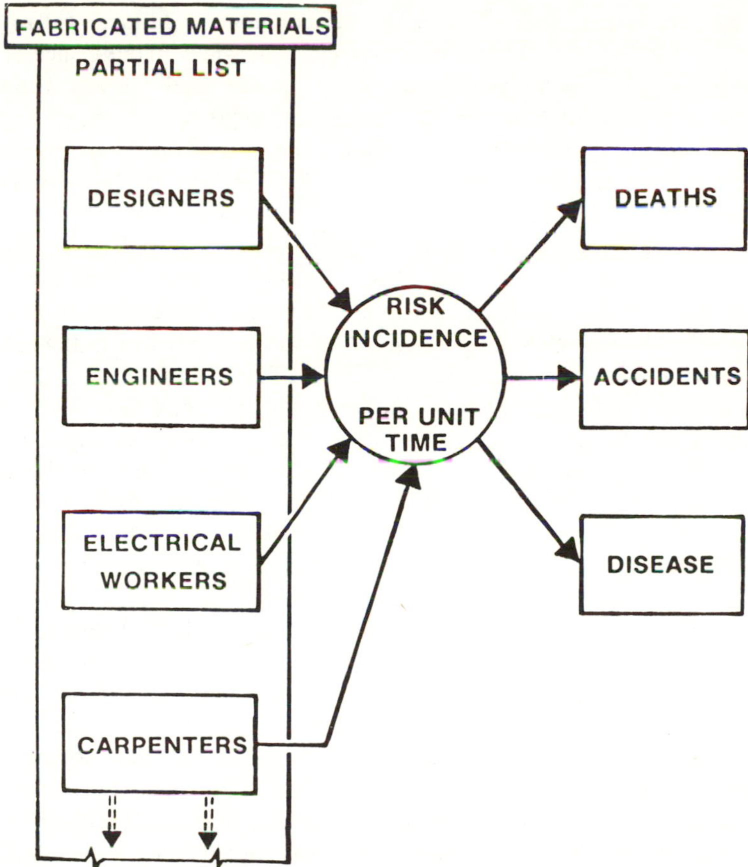

Figure 3. *Risk from construction. Each of the labor trades that are used in the construction of an energy system has an associated calculable risk. The dashes indicate that other trades are used.*

For example, if a system has a total nonsand and cement weight 0.1 that of coal, transportation man-days lost is 1.5–4.7. This implies that, on the average, materials for all energy systems are transported about the same distance to their destination. This is not always true, and research is proceeding to determine how these distances vary. The mode of transport–rail, truck, water, or other–also affects the risk.

Operation and maintenance risks have been previously estimated for the conventional technologies (7) and for certain nonconventional systems like solar thermal electricity. Other systems had their maintenance requirements estimated in analogy to those already well known. These requirements were of the form "*M* man-hours required per year." This time was then assigned to the appropriate trade or industrial activity, and the risk calculated from occupational statistics. Risk due to maintenance materials is not included.

Public health risk fell into two categories. The first, and by far the largest, is risk due to air pollution. Many air pollutants can damage human health. Because the statistical relationship between health and concentration of only sulfur oxides in air is at all understood, this is the sole pollutant considered in this report. There are few quantitative estimates of health effects from pollutants other than sulfur oxides. This group of pollutants is produced from coal and oil burning, and so these two technologies have a public health risk. However, nonconventional technologies often require coal and oil to produce their fabricated materials. For example, solar photovoltaic systems require steel, which in turn requires coal. As a consequence, nonconventional systems, as well as conventional ones, often have public health risk.

Sulfur oxides affecting public health can also be emitted during the processing of oil, natural gas, copper, cement, and material transport and during natural gas burning. This level of complexity is not considered here.

A second aspect of public health risk lies in potential catastrophic accidents. Nuclear power is generally acknowledged to have this potential; oil fires and natural gas explosions have occurred. Hydroelectric power can have catastrophic public risks due to dam failure. The subject is discussed more fully in Section 1C (dealing with catastrophes) and Appendix D (dealing with nuclear power).

Risk of waste disposition has been calculated for nuclear power. These wastes are defined as excluding those of air pollution from coal, oil, and gas burning. However, public health risk from gathering and handling coal (see Appendix A) is probably from burning culm (coal waste) banks. Other energy sources were assumed to have little or no risk from this source.

D. Types of Risk

On the basis of the two previous sections (2B and 2C), there are clearly many sources and types of risk. For some systems, like coal, much of the occupational risk is produced from coal mining (operation and maintenance) and comparatively little from construction. For systems like solar space heating, much of the risk is from construction, and none from fuel gathering. The appendices detail the magnitude of each source and type of risk. A later figure outlines the major types of risk for each energy system.

For clarity, the risk data are divided into occupational and public risk. Occupational risk can be defined as that which is produced on the job. Public risk is incurred by people living close to the site of an energy system, or at a distance in the case of air pollution.

E. Lifetimes

No energy system lasts forever. Parts wear out and eventually the entire system must be replaced. Since risk per unit energy over a system's span of production is evaluated here, its lifetime is an important value. For most of the systems considered in this book, a lifetime of 30 years was assumed. Exceptions are noted in the appendices. As noted in Section 1H, dealing with units, the value assigned to lifetimes plays a key part in calculations.

It is sometimes difficult to tell from the literature whether the lifetime referred to is the actual physical lifetime, or the amortization period over which the investment in money is "written off." As a result, some uncertainty should be attached to the values used for lifetimes.

Since some nonconventional systems discussed here are not in widespread use, their lifetimes are not accurately known. It is implicitly assumed they will be built with the same durability as systems that are more familiar.

F. Combining Deaths, Accidents, and Disease

How can deaths and less severe health problems, such as accidents and disease-related disabilities, be compared? While the impact of death on an individual is absolute, fatalities have different consequences for each group in society. The premature death, by a month, of a 90 year-old person will have less impact on society than that of a 20 year-old just starting a productive career. While there is no simple method for assessing the impact of a death, at least five studies have equated it to 6000 man-days lost (7, 31, 70, 136, 485). (On the other hand, Smith, Weyant, and Holdren (8) assume 100 man-days lost per nonfatal cancer.) This is the equivalent of about 24 working years at 250 days per year ($24 \times 250 = 6000$), or 16.5

calendar years, an approximation to the time lost due to occupational deaths. This simplifying assumption will be used in this book to compare deaths with accidents and disease. The sensitivity of final results to this assumption and whether deaths, nonfatal diseases, and accidents can be combined at all is discussed in Appendix O.

As an example of calculations, consider the upper left-hand corner of Table A-2. The accidental death rate for occupational risk is 0.0007 to 0.00150. Multiplied by 6000, this produces between 4.2 and 9 man-days lost. The injury rate is between 0.04 and 0.07. An estimate of the average number of days lost per injury is 93 (31) in this case, so the man-days lost to injuries are between 93 X 0.04 = 3.7 and 93 X 0.07 = 6.5. The total of man-days lost is then between 4.2 + 3.7 = 7.9 and 9 + 6.5 = 15.5. (The number of man-days lost per accident is generally taken as 50 (8), unless otherwise specified.)

The age at death from chronic ailments (disease) is probably higher than that caused by industrial accidents. While this will influence the total number of man-days lost per death, its effect could not be calculated from the available data.

The number of man-days lost per disability or disease can vary, depending on how the disabling occurs. In the appendices, published data are used wherever possible. Notations are made to the appropriate tables to indicate the values employed and their source.

G. Nonconventional Energy Systems

Generally speaking, the risk from nonconventional systems was calculated in the same way as for conventional systems. Because the former group is built and operated differently from the latter group, the risk components were rearranged to bring out the differences more clearly. However, the basis of the methodology is the same for all systems.

There are at least six readily identifiable risk components for nonconventional systems. First, there is the risk incurred in material acquisition and the fabrication and construction of the energy-capturing equipment, such as solar panels and windmills. The appendices show that nonconventional systems often have greater material needs per unit energy output than conventional systems, at least in their present stage of development.

Second, operation and maintenance produce risks, just as in conventional systems. Some nonconventional systems require considerable maintenance.

Third, emissions produced in the course of acquiring construction materials will incur public health risk. It is small for conventional energy sources. Consider the steel used in building a coal-fired plant. Coal is used in steel smelting (and much is coked), producing public health risk. Far more coal is used to generate electricity for 30 years than to smelt steel. However, the public health risk in producing materials may be large for nonconventional systems. The category can be called "pre-emission" risk, since the risk occurs before the energy system goes into operation, rather than after.

The fourth and fifth components of risk are those due to energy storage and back-up. Because of the importance of these concepts, they are discussed separately in the next section.

Sixth, the materials required must be moved to their destination. This produces transportation risk.

This listing by no means exhausts all the possibilities of risk in nonconventional technologies. For example, gathering of wood and production of fuel in the methanol system produces risk. For simplicity, this is assigned to "operation and maintenance." Other possibilities of risk, such as those listed in Reference 73, are generally unquantified, and probably insignificant. They are not considered here.

Figure 4 shows the six categories of risk schematically. The importance of each category will depend on the system being considered.

Computations of risk from nonconventional sources, as shown in Appendices E through J, are more detailed than those for conventional sources for two reasons. First, the calculations concern components like storage and back-up which do not play a large part in conventional systems. Second, some of the computations have apparently not been done elsewhere.

H. Storage and Back-up

All energy systems have storage systems. Some are less obvious than others. A slight increase in water level behind hydroelectric dams, the small volume required for storage of fuel bundles in nuclear power–neither are noticeable. Piles of coal or oil tanks are more common in public experience.

Regardless of their degree of visibility, storage for conventional energy systems have one factor in common. The amount of risk involved in constructing, fabricating, installing, and operating them is generally small. However, some nonconventional energy sources, such as solar and wind,

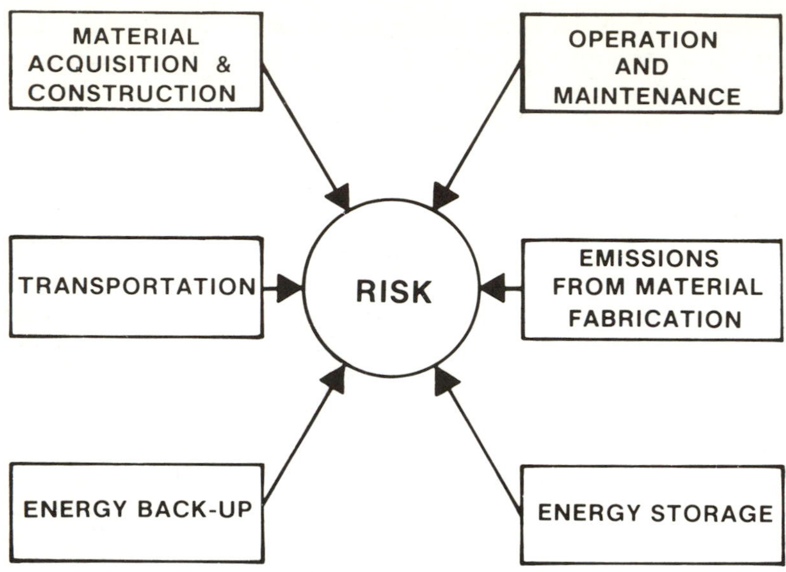

Figure 4. *Risk of nonconventional energy systems. Nonconventional systems have a different distribution of risk compared to conventional systems like coal, oil, or nuclear. While the calculations are made using the same principles, this diagram shows which factors are included. For example, many nonconventional systems have no fuel requirements but may need back-up and storage.*

require comparatively large storage systems if one is assuming "baseload" reliability of energy supply. The sun doesn't always shine and the wind doesn't always blow, but the consumer always wants reliable heat and power. The construction and operation of these storage systems should be taken into account when computing risk. In equalizing energy systems with this method, the philosophy of Lovins (5) is followed:

> . . . compare the total cost (capital and life cycle) of the solar system with the total cost of the other complete systems that would otherwise have to be used in the long run . . .

Only by considering storage and back-up can it be ensured that the Lovins philosophy is carried out.

A recent comprehensive work (569) noted the need for back-up:

> Early in the present study, it was believed that chemical energy storage would make autonomous or 100 percent solar power plants economically attractive. As the systems studies proceeded, however, it became clear that the economics of autonomous solar thermal electric plants were not favorable and that this fact could be satisfactorily established by systems studies of autonomous and hybrid plants with continuous constant demand profiles. . . .

In this reference, "hybrid" refers to the use of back-up.

Caputo (24) notes:

> Occasionally (a ground solar plant) will be down for one or more days due to adverse weather. This reduces the reliability of a stand-alone solar plant compared to a conventional plant operating at the same annual load factor. Because of this, it is necessary to install extra margin (backup) capacity and use some form of backup energy to increase the reliability to that of a conventional plant. A valid economic comparison should include these extra margin requirements for a solar plant.

This point is reinforced in a recent book by Schurr *et al.* (433). In a discussion of the economics of solar space heating, they state:

> To facilitate comparisons, half of the capital, operation, and maintenance costs of a back-up system are also assumed to be included in these (capital) estimates . . . this makes marginal costs of back-up fossil energy equal to marginal costs of purely fossil systems.

It is not clear why half rather than the entire cost of back-up systems is included in their evaluation, but it is clear that some portion of the risk attributable to back-up should be included for those systems that require it.

Storage can be in the form of rocks, liquid, pumped air, or other systems. The appendices detail the particular mode assumed for each system. Some nonconventional sources require additional back-up energy in the form of conventional energy, such as coal, oil, and nuclear. Not all nonconventional technologies require this back-up capacity. Ocean thermal and methanol will probably dispense with back-up energy.

What is the relationship between the amounts of storage and back-up required? If the reliability aimed for in "baseload" electrical distribution systems (usually one hour of power loss in a year) is taken as a guide, the amount of energy storage required to meet this guideline will be huge. To keep the storage system to a reasonable size, back-up capacity is employed. Some combination of storage and back-up is the optimum in terms of minimum energy used consistent with meeting reliability guidelines. Appendix E, Section v, discusses this in detail.

The type of back-up will influence the risk. What energy source should be used? Herrera (64) says:

> It is unlikely that a nuclear plant will be used for a solar plant back-up since a nuclear plant is unsuitable for this use. It is more likely that oil or gas would be used for peaking back-up. These fuels will tend to be unavailable toward the end of the century. Therefore, the extra back-up margin is based on using coal in a manner similar to the reference coal plant.

In order to give nonconventional systems the benefit of the doubt in the direction of low risk, relatively high-risk coal will not be assumed as back-up. Instead, a weighted average of the risk attributable to all the electricity now being generated, including some relatively low-risk sources, will be assumed as back-up. This ensures that nonconventional energy systems are treated fairly in terms of risk. Calculations are shown in Appendix N.

A schematic diagram of how energy is transferred into and out of a system by storage is shown in Figure 5. By judicious use of storage and back-up, material requirements for a nonconventional system can be reduced.

In principle, any energy system can be used as a back-up. While nuclear energy would be wasted if it were used as back-up for nonconventional systems, it could in theory be employed. Risk calculations using relatively low-risk nuclear or natural gas are also discussed in Appendix N. Mathematically, these results are close to what would be found with no back-up. While the absolute value of risk for the three systems assumed to require back-up (solar thermal electric, solar photovoltaic, and wind) is reduced when nuclear or natural gas is used as back-up (or no back-up at all is assumed), the relative rankings of the eleven systems in terms of total man-days lost per unit energy remain fairly similar. It can be concluded that the type of back-up used does not strongly influence the final results.

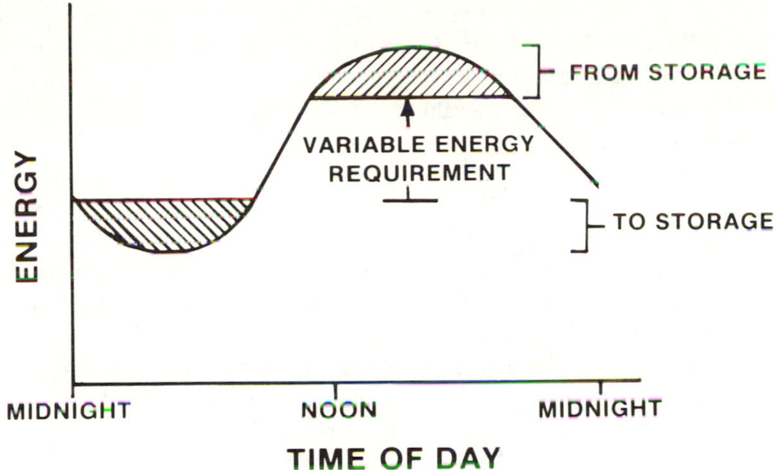

Figure 5. *Matching energy requirements to energy available by using storage. When there is little demand, energy can be sent to storage. When demand outstrips capacity, energy can be taken from storage. This applies only to those systems which require energy storage. The diagram is only schematic. For some nonconventional energy systems, there is no energy input at certain times of day, and the output will approximate the energy demand, which may have a different shape from that shown here. The object of the storage is to allow the system to meet baseload demands.*

I. Emissions

Davidson *et al.* (76) state:

> Associated with the production of solar energy equipment will be certain amounts of pollution resulting from industrial activity. This is an important factor since the material requirements in many solar technologies is immense compared to conventional technologies.

Emissions used in their acquisition can have significant effects on human health. The main impacts are aggravated heart and lung disease, chronic respiratory disease, and asthma. Health effects depend on whether emissions occur in urban or rural settings.

Only oxides of sulfur are considered, although it is now known that other pollutants play a part in health effects. These include carbon monoxide, trace metals, polycyclic aromatics, nitrogen oxides, etc. Also, only Canadian emission values are evaluated. The methodology can be extended to the emission values for other countries.

About $(1.6–4.7) \times 10^{-4}$ deaths, 0.9–2.8 illnesses, and 5.8–18 days lost are associated with one metric ton of airborne sulfur oxides in the presence of particulates (23). This is based on assuming (22) that chronic respiratory disease, asthma attacks, and respiratory diseases in children correspond to 5, 1, and 1 days lost, respectively. Values may be a factor of 2 too low or 10 too high (23).

Canadian sulfur oxides emissions in 1974 from iron and steel, including primary production, iron ore roasting and benefication, pyrrhotite roasting, and metallurgical coke production were 134,000 metric tons. Emissions for aluminum production were 49,600 metric tons (220). Part of this may be due to fossil fuels used for making aluminum. Canadian steel production in 1974 was 13.6 million metric tons (61); aluminum production in 1974 was 1.01 million metric tons (121). The ratio of sulfur oxides emissions to metal weight is then $(1.34 \times 10^5)/(1.36 \times 10^7) = 0.010$ and $(4.96 \times 10^4)/(1.01 \times 10^6) = 0.049$ for steel and aluminum, respectively. Then one metric ton of steel corresponds to $0.010 \times (5.8–18) = 0.058–0.18$ man-days lost. Corresponding values for aluminum are $0.049 \times (5.8–18) = 0.3–0.9$ man-days lost. There is reason to believe that the last range of values is low. Reference 233 notes that about 5.4 metric tons of coal are required to produce 1 metric ton of aluminum. Using footnote (x) of Table A-2, about 3500 metric tons of coal are associated with a total of 610–1900 man-days lost. Assuming that coal used in producing aluminum has about the same health effects as that used in producing electricity, one metric ton of aluminum corresponds to (5.4/3500) (610–1900) = 0.9–2.9 man-days lost.

The emission factors chosen are realistic, rather than theoretical (96). If a pollution control device is designed to remove 50% of a given pollutant but generally removes only 25%, the latter factor is used in Reference 220. As a result, the calculations show the actual picture in terms of sulfur oxides.

It should be noted that these relationships are based on data applicable to the U.S. Northeast. It is not clear if they apply to all parts of Canada.

J. Data Limitations

While some data used in this paper are known to a high degree of certainty, other information is not. This is especially true for nonconventional technologies, where the lack of operating system experience means that many quantities are only estimates. In each of the appendices, only what seemed to be the most reasonable values were used. To aid the reader in evaluation, other estimates and sources of data are noted. As more is learned about nonconventional systems, it is likely that some of the figures in this report will have to be altered.

Because some data are not known to great precision, many entries in the appendix tables are given as a range of values rather than one number. For example, a risk may be calculated to be 10 man-days to within a factor of 2. The range for this quantity is then $(10/2)–(10 \times 2) = 5–20$ units. It should not be assumed, however, that where a range is not shown that there is universal agreement on a value. A detailed discussion of all possible uncertainties in material and fuel requirements, industrial productivity coefficients, health risk data, etc. is beyond the scope of this report. Readers are referred to the references cited in the text for further information.

Occupational statistics, as in Table 2, refer to average values, not the best or worst. It clearly would be unfair to compare the best of some industries to the worst of others. The incidence of fatal accidents in coal mining, or any other industry, can vary widely between the best and worst companies in the sample (126). For example, the most dangerous Ontario mine in 1976 had an accident rate 10 times that of the best one (344).

K. Other Energy Systems

In the past few years, many new forms of nonconventional energy technologies have been proposed. In addition, old ideas have been resurrected. Some of those receiving the most publicity have been tidal power, wave power, burning farm wastes, harvesting and burning ocean plants, and beaming solar energy to earth from satellites.

These and other systems have not been considered here. Plans for some of these systems are nebulous. Where they are more definite, estimates of the materials and labor required are often nonexistent. As soon as clearer estimates of the requirements of other systems become available, it would be useful to apply the present methodology to them.

L. Risk per Unit Energy, Power, or Worker

The results of this book use the unit of risk per unit energy, taken to be 1 megawatt-year. This unit was adopted because many previous studies of risk have used it. However, other units can be employed. For example, one can discuss the risk per unit power (or capacity) to generate energy. Results would be different from those using risk per unit energy. As noted in the appendices, some systems have a relatively low load factor (without storage or back-up), or ratio of energy produced in a given time to that which could be produced in that time. In other words, systems with low load factors have relatively high ratios of rated power to energy produced.

If all systems were evaluated on the basis of risk per unit power, those systems with low load factors would have a lower risk per unit power than risk per unit energy. Exactly how much the rankings of systems would change is not clear, since there are elements of risk which are not a direct function of load factor.

Risk per unit power is not evaluated here for at least two reasons. First, as noted above, the convention does not seem to be widely adopted by risk analysts. Second, to some extent the power rating of some facilities, such as windmills, is arbitrary. If the power rating is changed, it does not necessarily mean that more (or less) energy will be produced from a given facility. It seems likely that most consumers are concerned with the energy produced by a system, not its "nameplate" power rating. However, for those readers interested in the risk per unit power or capacity, with or without storage or back-up, the data in the appendices may be used to calculate it.

A second possible alternative measure of risk is the risk per worker, or unit time of labor. In this calculation, the total occupational risk is divided by the number of workers engaged in all phases of building and operating this system. Pochin in Britain has discussed this calculation, as has Provost in a 1978 article in *Quebec Science.*

This ratio has not been evaluated here for the following reasons. First, risk analysts have not generally adopted this convention. Second, focusing on occupational risk neglects public risk entirely. Many people interested in energy risk are more concerned about public than occupational risk. Third, considering only the risk per worker can lead to anomalies in results. Suppose a new and imaginary type of coal were discovered, exactly the same as regular coal except that it produced twice as much heat (and eventually electricity) per unit volume. Since mining this hypothetical coal is assumed to produce the same risk as ordinary coal, nothing has changed: the risk per worker remains the same. However, since the energy each worker produces has doubled, the risk per unit energy has dropped by half. Choosing the second definition seems to make more physical sense, although the question cannot be proved by logic.

For readers interested in the risk per worker, the data in the appendices can generally be used to calculate this quantity, although there are some gaps in the data. For example, the time to build hydro dams is not specified.

In summary, other denominators for the risk formula can be used. The risk per unit energy is now the most widely used formula. The data in the appendices can be used to evaluate other possible formulas.

M. Differential Risk

Another way of evaluating risk is to use a differential approach. The present report uses an absolute calculation, that is, the actual value of risk associated with unit energy is estimated and presented in graphs and tables. This implies that the risk is averaged over a wide range of facilities, both old and new. For example, some coal mines may have more risk per unit energy than others. These coal mines will tend to be older than average, although not always. One might contend that only the risk of coal mines, nuclear reactors and all energy facilities to be opened in the future (as opposed to those already open) should be considered. This would then be the differential approach. Since people tend to be most concerned about new or proposed energy facilities, there is some justification for this reasoning.

On the other hand, the energy produced both today and tomorrow will use a mix of both old and new facilities—coal mines, hydro dams, oil wells, and so forth. Even if the facilities themselves were all new, such as will likely be the case for some nonconventional systems, the plants where their materials are fabricated, such as steel and copper factories, will be a mix of old and new. Since this report is concerned with present-day conditions, it uses the absolute instead of the differential approach.

A further difficulty with the differential approach is the lack of data. Generally speaking, national data on occupational risk lump together both old and new facilities. While it might be of interest to calculate differential occupational risk, or the risk that would be produced in new factories, mines and plants, without data it cannot be done.

Another aspect of differential risk lies in the emissions from fossil-fueled electricity plants. As noted in Appendices A and B, there exist standards for these emissions in a number of countries. These standards may apply primarily to new fossil-fueled plants, so the public risk attributable to these plants can be called differential risk. On the other hand, it is also noted in these appendices that some existing plants exceed the emissions standards. Since this report evaluates present-day conditions, the emissions risk considered here takes into account plants that both exceed and do not exceed these standards.

In summary, differential risk is a useful concept. It is generally not used here because of lack of data and concentration on present-day rather than future risk. However, this concept could be employed in a more comprehensive study.

3. Results and Conclusions

The tables in the appendices contain much information. In order to simplify the results, they are presented in the following set of graphs and tables.

A. Summary of Main Assumptions

A number of assumptions had to be made to produce equitable treatment for all technologies. Most have already been discussed, but those which may require further emphasis are now listed.

First, much of the data apply to conditions in the United States, although in most cases Canadian conditions probably are similar. For example, requirements for producing 1 kilogram of steel or maintaining a solar thermal receiver are probably about the same throughout most of the Western world. One area of difference is that most Canadian coal is surface mined, in contrast to about half being surface mined and half underground in the Unied States. Whenever possible, data were adjusted for Canadian conditions. For example, account was taken, in considering solar systems, of the relatively smaller amount of solar insolation falling on much of Canada as compared to the U.S. Southwest.

Second, no attempt has been made to ensure that enough land, trees, steel, or other resources are available to support large use of a particular technology. For example, widespread use of methanol (wood alcohol) may require cutting down most of the forest in Canada. The amount of glass required for solar central receiver stations in the United States could be as much as 260% of present annual glass production (69). While these data are needed in planning overall energy strategies, this report considers only risk per unit energy.

Third, environmental and ecological factors or disturbances other than those which directly cause risk to humans are not generally discussed. Many of the environmental hazards of nonconventional technologies, such as air and water pollution, are mentioned in Reference 73. Those for conventional systems have been detailed in many sources. The major environmental effect relating to health considered here is the oxides of sulfur in air pollution. If and when more knowledge is developed on other air pollutants, their effects should be included in risk studies.

Fourth, most of the technologies have relatively small distances between where energy is produced and where it is consumed. As a result, risk incurred in producing, fabricating, and installing transmission lines has generally been neglected. Hydroelectric transmission lines can stretch up to hundreds of miles as in the cases of the Nelson River in Manitoba and the James Bay project in Quebec. However, Appendix L shows that transmission line risk per unit energy and length is fairly low for any land-based lines. Ocean thermal transmission lines will probably be both long and involve special systems so that these aspects may increase the overall risk from that technology by a significant factor.

Fifth, no attempt has been made to specify what the back-up systems for nonconventional technologies will be in the future. These produce a significant part of overall risk for some of these technologies. As noted in Appendix N, a weighted back-up has been assumed, reflecting recent Canadian electrical energy sources. The risk attributable to a weighted back-up would be higher in the United States, since coal forms a larger fraction of electricity production there.

Sixth, it should be stated again that conditions for each system as they existed at the time of writing have been assumed. This means that presently existing technology, rather than the best available technology, is implied. Advancing technology and improved safety procedures may decrease the risk for some systems, but it is by no means certain that the relative size of the risk will change.

Seventh, no attempt has been made to produce "scenarios" for future energy use in Canada. The resource mix used by Canadians in the future will depend on such factors as resource availability, cost, and other factors in addition to the risk estimates presented in this report. Also, some forms of energy production can cause long-term effects, as opposed to the generally short-term effects considered here. These effects include increased atmospheric concentrations of carbon dioxide, causing temperature changes, pollution due to radioactive isotopes like tritium and krypton-85, and health effects due to a wide range of industrial wastes. Their risk is not evaluated here for three reasons: (a) estimates differ widely on the size of the effects; (b) risk to human health is often not clear; and (c) any effect may be centuries or more in the future.

Eighth, the nuclear data refer only to light-water reactors. The public risk of these reactors has been discussed in the Reactor Safety Study (44) and associated literature. Reference 44 is the main source of data concerning the public risk of these reactors. It is based on only two operating reactors, and the assumption is made that the public risk calculated there is applicable to other light-water reactors.

Ninth, all systems considered are assumed to be supplying baseload energy. This is an important assumption since there are other modes such as intermediate or peaking which exist. For energy systems whose inputs are variable due to weather or other causes, storage, back-up, or both may be required to produce the reliability that baseload energy demands. It might be contended that requiring storage or back-up unfairly penalizes some systems in terms of risk. Since all technologies are being evaluated as baseload systems, all are treated on the same basis. At least one solar proponent (226) feels that baseload systems are desirable: "(solar systems) are the most economical when they are allowed to operate as much as possible (providing baseload power) because they are capital intensive (requiring a large initial investment) but have a relatively low operating cost." It is not clear that assuming intermediate, peaking, or "fuel saving" modes would always produce lower overall risk. While storage and back-up requirements would be less (or zero) in non-baseload modes, the load factor would also be less. Generally speaking, the lower the load factor, the higher the risk per unit energy, since less energy is produced per unit rated capacity, assuming all other conditions remain constant. Further research is needed to resolve this point.

On the other hand, some readers may be interested in what the results would be if no back-up (or low-risk back-up) were assumed. Results of both back-up and no back-up assumptions are shown in the concluding graphs. As well, the final table in the appendix for the three systems assumed to require back-up shows the back-up risk separately, so that readers can do the subtraction if desired. The question is discussed in Appendix N.

Tenth, units in this report are risk per unit energy. One might also calculate risk per unit capacity to produce energy (or power), or use other units.

Eleventh, air pollution effects are based on only one group of pollutants—sulfur oxides. The effects are based on a model dealing with populations in the U.S. Northeast and may not be applicable everywhere. See Section 2I.

Finally, no attempt has been made to perform a "parametric" study of the relationship of material use to risk. For example, substitution of plastics for steel in the construction of storage tanks for solar space heating might reduce public risk from emissions of steel making. On the other hand, occupational risk from the chemical industry might increase. Whether the overall risk would rise or fall would not be known until detailed analysis was undertaken.

B. Summary of Material Acquisition and Construction

The amount of materials used per unit energy output is a significant factor in computing the risk of each technology. In addition, construction times also play a key part. Although these quantities are noted in the appendices, it is of interest to summarize them here.

While the weights of material are important, the incidences of death, disease, or accident are not necessarily proportional to them. In Table E-2, for example, the risk in man-days lost from producing 39 metric tons of steel is 4 times higher than that incurred from mining 63 metric tons of iron ore. Nonetheless, the portrayal of the weights and times will give a feel for the quantities involved. Results are shown in Table 4. The four conventional technologies all have low material use and construction times. The material use and construction times of nonconventional systems vary widely. Labor and material requirements for any storage (except for solar space heating) and back-up are not included.

The high occupational risk of nonconventional systems is due primarily to the large amount of materials per unit energy output required in their construction. In turn, production of these materials requires comparatively long construction times, further increasing the risk.

While the construction time of an energy source is related to the amount of materials used, the relationship is not always proportional. Figure 6 shows this clearly. Both material and construction requirements are generally higher for nonconventional systems. Again, labor and material requirements for any storage (except for solar space heating) and back-up are not included.

As mentioned in Section 1B, some nonconventional technologies considered in this book will probably be of limited applicability to Canada in the forseeable future. Because these technologies—solar thermal, solar photovoltaic, and ocean thermal—have received considerable publicity in the last few years, the technologies not relevant to Canada are also presented in Figure 6. To indicate their difference, they are separated from the others with a dotted line.

C. Proportions of Risk

The final table in each appendix shows the source of each of the components of risk. Conventional technologies generally have their risk categorized as gathering and handling of fuels, transportation, and electricity production. Nonconventional technologies have six analogous categories. Figure 7 shows the proportion of total risk in each category for each technology. For simplicity, gathering and handling of fuels in conventional systems were equated with material acquisition and construction for nonconventional systems. The two concepts are similar, since they both refer to activities that must be performed before the first energy is generated. In addition, electricity production for conventional systems was equated with operation and maintenance for nonconventional systems, since in many ways the sources of risk are similar.

The maximum risk, as measured in man-days lost, was used as the base of the calculations shown in Figure 7. In effect, these maxima are normalized. As an example, Table A-2 indicates a maximum total of 1900 man-days lost per unit energy for coal-fired electricity. Of this, 1750 man-days (92%) are due to electricity production (primarily air pollution), 44 man-days (2.3%) are due to transportation, and 105 man-days (5.5%) are due to gathering and handling the necessary fuels. The data refer to the total risk, which includes both public and occupational sources. Graphs similar to Figure 7 may be drawn for the minimum of the risk range, number of deaths, and so on.

As shown in Figures 9–14, the man-days lost for each energy system are different. This should be kept in mind as Figure 7 is evaluated.

Figure 7 shows that the proportions of risk vary strongly with energy systems. Natural gas, for example, incurs most of its risk in gathering and handling fuels. This system is followed closely in its proportion of risk from this source by nuclear and ocean thermal. Economists would say that

Table 4. Summary of Material Acquisition and Construction Data

Energy system	*Materials (metric tons per megawatt-year net output over system lifetime)*						*Construction Time (man-hours per megawatt-year over system lifetime)*
	Steel	*Concrete*	*Aluminum*	*Glass*	*Copper*	*Total*	
Coal (g)	4.3(a)	6.8				11(n)	505
Oil (g)	3.2(a)	3.1				6.3	415
Natural gas (g)	1.5	2.4				3.9	302
Nuclear (g)	2.3(a)	12.7				15	633
Solar space heating	95		4.0	15	2.7	120(d)	12,000
Wind	23	5		3.4(j)	0.6	36(e)	2400(i)
Methanol (g)	89 (a),(h),(i)	23(h)				112	7300(i)
Solar thermal (b),(o)	65	290	3.7	11		370	10,000
Solar Photovoltaic (o)	30	10	91	4.8		140(c)	1350
Ocean thermal	12	80	2.8		0.5	96(f)	6800(i)
Hydroelectricity	0.7	92				630(k)	(1)

(a) Includes all metals.
(b) Not including 0.2 metric tons of silver.
(c) Including 4.8 metric tons of silicon semiconductors.
(d) Including 3.7 metric tons of fiberglass insulation.
(e) Including 3.5 metric tons of fiberglass insulation.
(f) Including 1.1 metric tons of ammonia.
(g) Does not include weight of fuel.
(h) Obtained by multiplying oil data by suitable factor. See Appendix J.
(i) This is an average of two values.
(j) Fiberglass and plastics.
(k) Including rock and earth.
(l) Not available.
(m) Most labor is fabrication, not construction.
(n) Excluding lime for scrubbers.
(o) Multiplied by appropriate factor for climatic conditions. See Section vii of Appendices E or F.

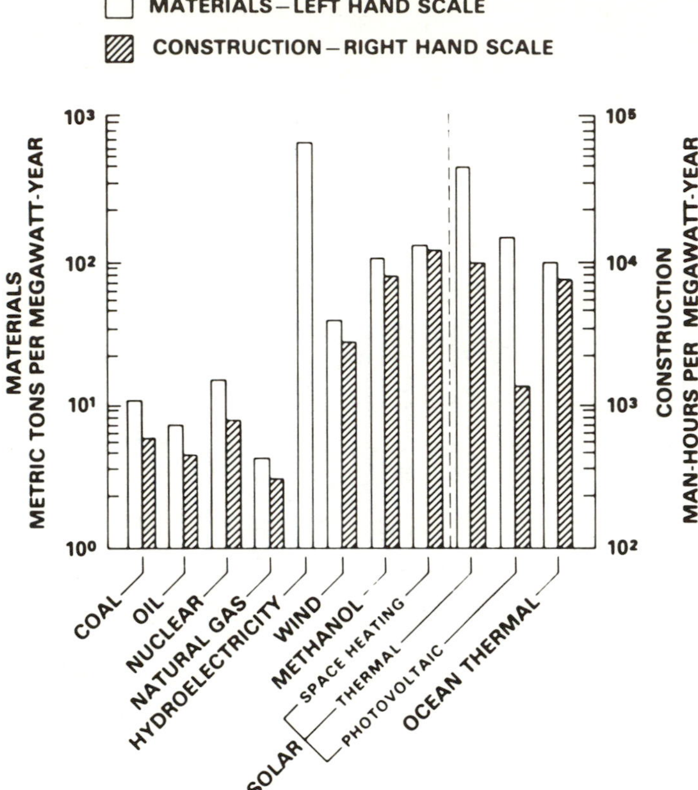

Figure 6. *Summary of material acquisition and construction requirements. Both material acquisition and construction time requirements are generally greater for nonconventional systems as compared to conventional systems. Natural gas apparently had the lowest requirements of both types. Hydroelectricity had the highest material requirements, mostly rock and earth.*

this shows that these systems are capital-intensive, although there is not usually a direct connection between capital and risk to human health.

As mentioned above, coal and oil have very little of their risk in gathering and handling fuels. Most of their risk is incurred in electricity production with air pollution as a consequence.

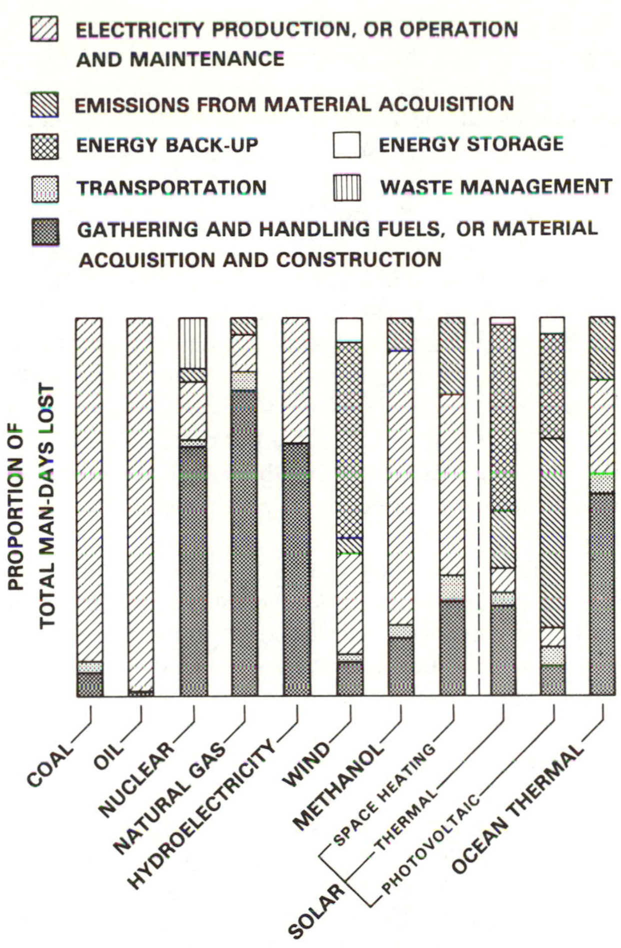

Figure 7. *Proportions of risk by source. The sources of risk vary considerably from one energy system to the next. Coal and oil have most of their risk due to electricity production (air pollution), whereas natural gas, nuclear, and ocean thermal have most of their risk from fuel or material acquisition. Wind, solar thermal, and solar photovoltaic have a large risk proportion from energy back-up. The total risk of each system has been normalized in order to show the differences clearly. See text for details of calculations.*

Wind, solar thermal, and solar photovoltaic have much of their risk produced by the back-up they require. However, this is to be expected. Caputo (373) finds that 78% of the deaths per unit energy in the solar thermal electric system he considers are attributable to back-up. The corresponding value for the solar photovoltaic system considered in Reference 373 is even larger at 90%. These values are higher than the proportions calculated in this book. Ocean thermal has the highest proportion in material acquisition of all the nonconventional systems.

Only nuclear power has calculated risk due to waste management. As noted in Appendix A, coal also has waste management risk, due to slag and fly ash, but this is not included in the computations due to lack of quantitative data. For coal, oil, nuclear, and natural gas, risk attributable to nonfuel material acquisition and construction has been assigned to the category of gathering and handling fuels. For solar space heating, risk attributable to energy for pumps has been assigned to operation and maintenance. Risk from hydroelectricity dam failure has been assigned to electricity production.

Each type of technology shown in Figure 7 is built and operated in a different way, so it is logical that each should have its own proportion of risk. The pair of technologies with the most similar risk structure is coal and oil (due to air pollution via electricity production). Most of the other systems are quite dissimilar.

D. Nuclear Risk

Because of the widespread interest in nuclear power, it is useful to consider its risk in more detail than those of some other systems. The data are shown in Figure 8. The total maximum number of man-days lost per megawatt-year is about 14. Of these, about 75% are due to occupational hazards of accident and disease. Only about 25% are incurred by the public, primarily through disease.

As noted in Appendix D, maximum values of nuclear risk were taken from various sources. This maximization procedure was not followed for other energy systems. These sources of data included a well-known critic of nuclear power. However, no attempt was made to consider all possible estimates.

The diagonal part of the bar on the extreme right of Figure 8 should be commented upon, since it represents the estimate of risk produced by catastrophic accident in light-water reactors. While not zero, it is low.

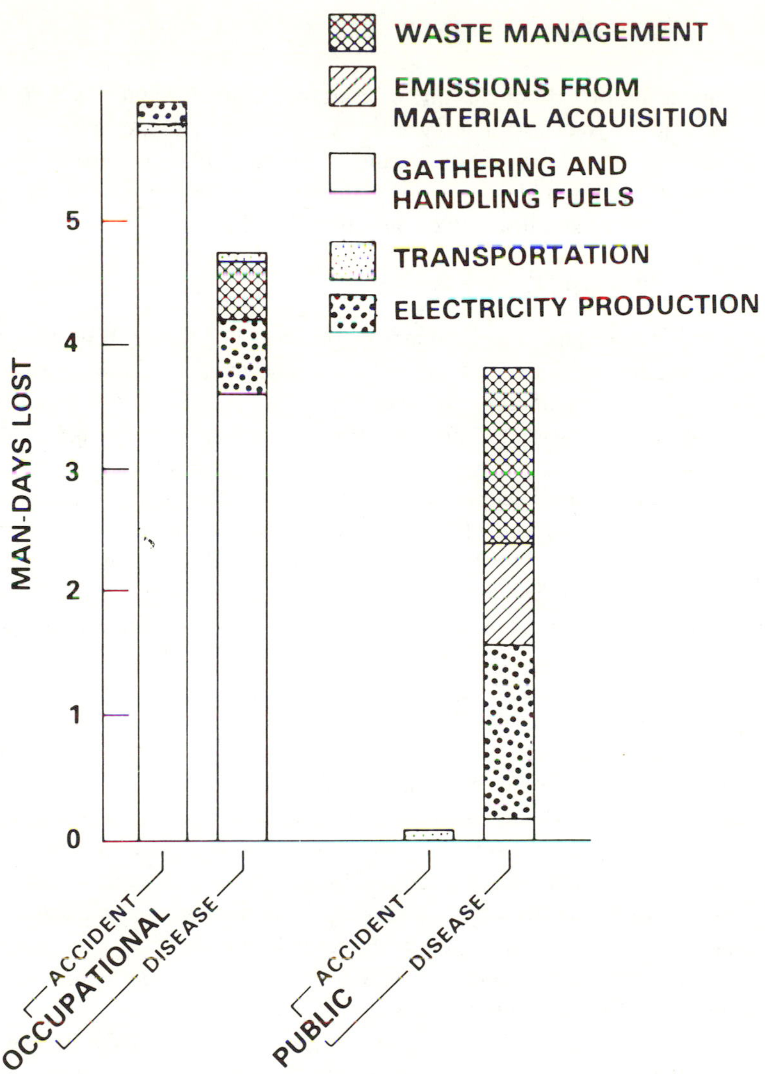

Figure 8. *Nuclear risk in man-days lost per megawatt-year net electrical output. The maximum of each component is shown here. Most of the risk is occupational, as compared to public, and is due to gathering and handling of fuels. A smaller proportion is due to disease caused by possible accidents at light-water reactors. See Appendix D for calculations.*

E. Conclusions

The most important conclusion to be drawn from this study is that the risk from nonconventional energy sources can be as high as, or even higher than, that of some conventional sources. In particular, it apparently is higher than that of natural gas or light-water nuclear power, although definitely substantially lower than that of coal and oil.

a. Results

Let us now consider the summarizing figures. The data can be divided into (a) occupational risk, borne by those who construct, fabricate, and maintain the energy sources, and (b) danger to members of the public. The total risk of a system is then the sum of the occupational and public risk.

Figure 9 shows the occupational deaths (times 1000) per unit energy averaged over the lifetime of each system, which for most systems is assumed to be 30 years. The "lifetime" concept is used to average the initial construction risk, which can be quite large in comparison to operation and maintenance risk. One should be aware that the logarithmic scale used masks large differences.

In Figure 9 and the following five graphs, the top of each column represents the maximum risk, and the dotted line the minimum. For some technologies, the spread can be large. As in Figure 6, those to the right of the dotted line are less likely to be used in Canada.

Figure 9 shows that the maximum number of occupational deaths results from methanol, followed by coal. Solar space heating and windpower follow. The lowest is natural gas, followed by nuclear. The bulk of the deaths for most systems is attributable to material acquisition and construction.

Figure 10 shows that the total occupational man-days lost per unit energy is similar to the numbers in the previous graph. Deaths are incorporated in the total at 6000 man-days lost per fatality.

The results of Figures 9 and 10 are similar to the conclusions reached by Brooks and Hollander (450). In *Annual Review of Energy,* they note that in terms of present-day systems, i.e., those considered in this report,

> The construction of most solar energy systems would require considerably more energy-intensive materials such as steel and cement, than a nuclear plant of equivalent capacity. The effluents

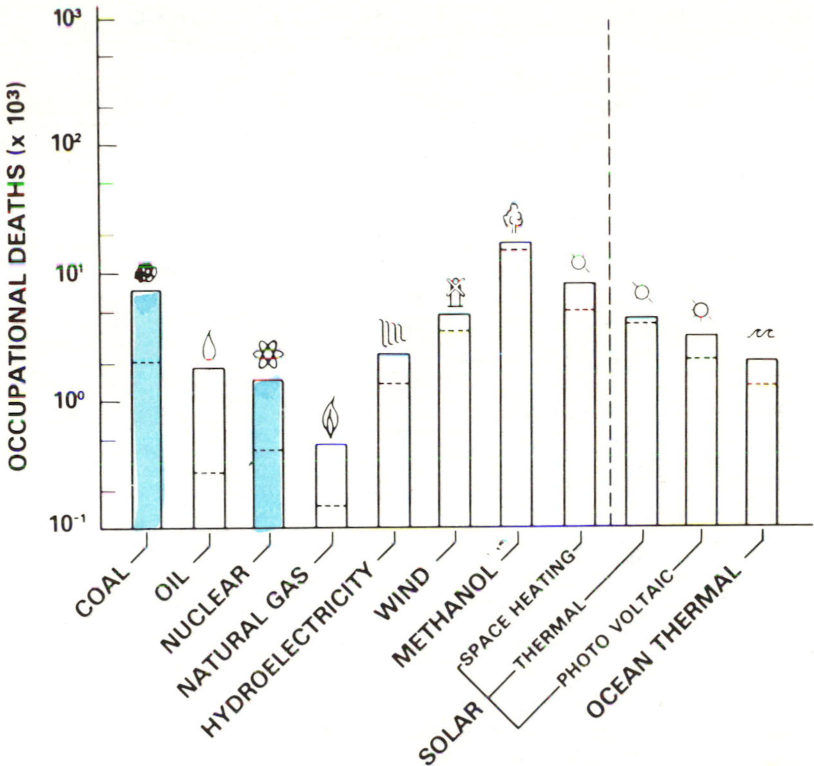

Figure 9. *Occupational deaths, times 1000, per megawatt-year, as a function of the energy system. The values refer to 1 megawatt, net output, over the life of the system. For example, coal would have a maximum of 7/1000 = 0.007 deaths per megawatt output per year over the 30-year system life. The tops of the bars indicate the upper end of the range of values; the dotted lines within the bars, the lower. Where no dotted line is shown, the upper and lower ends of the range are similar. Those bars to the right of the vertical dotted lines indicate values for technologies less applicable to Canada. This scheme of notation will be followed in Figures 10–14.*

involved in extracting and processing these materials, and the associated occupational hazards, would greatly exceed the corresponding effects for a nuclear plant in routine operation.

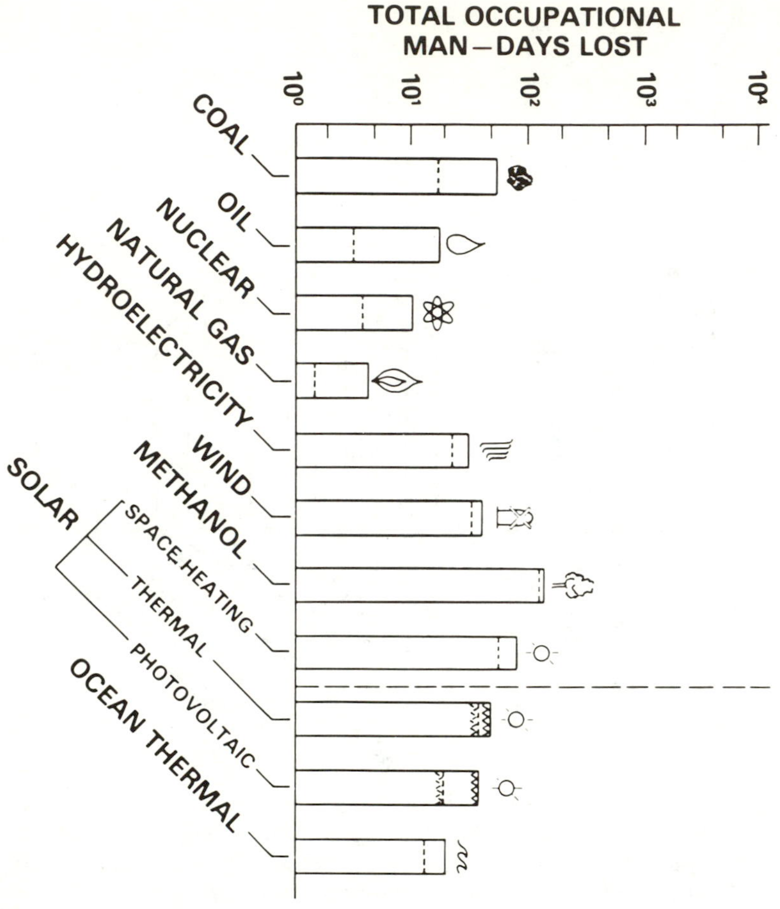

Figure 10. *Occupational man-days lost per megawatt-year net output over lifetime of system. (See explanation in caption to Figure 9.) As in the previous graph, these values refer to the risk incurred in particular activities related to gathering and handling fuels, acquiring material and equipment, and operation and maintenance of power plants. Jagged lines refer to the maximum and minimum values if low risk or no back-up is assumed. For calculation purposes, each death is counted as 6000 man-days lost. Methanol has the greatest values.*

The authors go on to note that it is possible in the future that solar systems will require less materials, but they do not evaluate this situation, and this is not done here either. Conclusions similar to those of Brooks and Hollander have been reported by Hoy (452). While the absolute values of risk as computed by Hoy differ from those here, the relative ranking of the systems he considered, both conventional and nonconventional, is approximately similar. In another study, Ramsay (491, 492) derives a risk per unit energy for parts of the solar space heating system higher than shown here. His results are discussed below.

Figures 11 and 12, which show risk to the public, are considerably different from Figures 9 and 10. Two of the conventional technologies, coal and oil, lead the list for both deaths and man-days lost, due to emissions produced by burning fuel. Some nonconventional technologies have moderate values of public risk, primarily due to the energy back-up required to provide baseload reliability. As noted in Appendix N, the back-up assumed is a weighted system, made up of present-day proportions of electrical energy capacity. Part of this back-up is coal and oil, with their attendant public risk.

Another component of public risk is due to the steel used in building systems. Coal is required for making most steel, and is the source of most of the oxides of sulfur produced industrially. The emissions traceable to those technologies, such as nonconventional systems, which do not produce any air pollution directly, can be called "pre-emissions." Those of coal and oil burned to make electricity directly can be called "post-emissions."

Natural gas-fired electricity has the lowest public risk, followed by nuclear and ocean thermal. For nuclear power, both the risk of waste management and possible reactor catastrophes were included. Nuclear waste risk is long-term, extending many centuries into the future. As mentioned in Section 3A, long-term risk from any technology considered here (except nuclear) is generally not evaluated. However, a case can be made that the long-term risk from nuclear waste management will be small with adequate planning (227, 228). Further study is needed to test this.

Figures 13 and 14 present total risk, comprising occupational and public categories. Because public deaths are higher than occupational deaths for most of the systems, the former dominates. The number of man-days lost from coal and oil is considerably higher than that of a "group of seven"

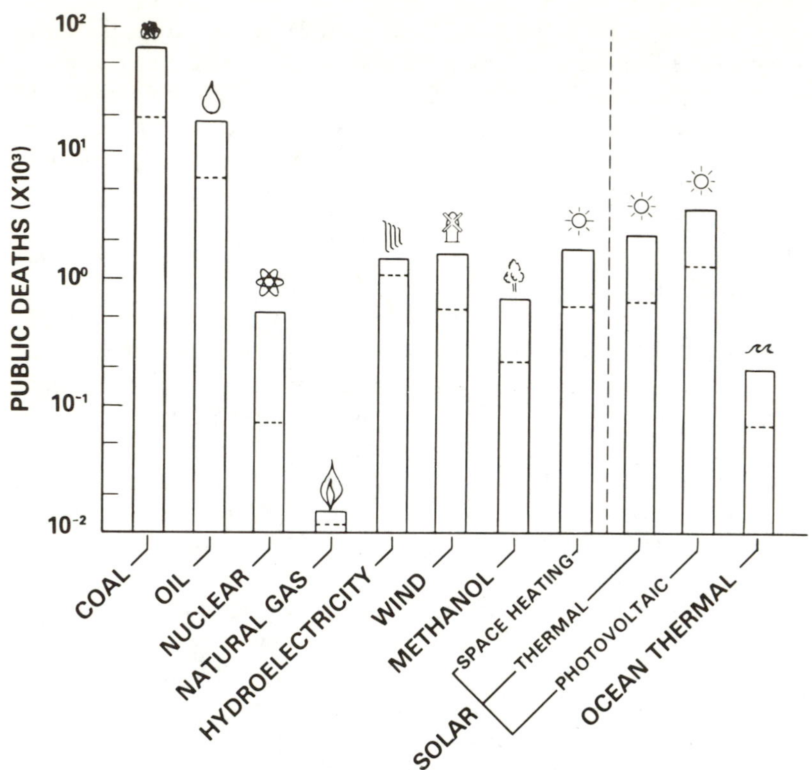

Figure 11. *Public deaths, times 1000, per megawatt-year, as a function of the energy system. (See explanation in caption to Figure 9.) Much of this risk is produced by emissions created after fuel is gathered (for the case of conventional technologies), as part of the back-up required for baseload reliability for some systems, or in the course of producing the materials for the systems.*

consisting of six nonconventional systems and hydroelectricity. There is not much reason to discuss the ranking within this group, since different models will probably produce somewhat different absolute values of risk. However, it appears that ocean thermal and hydroelectricity have generally lower risk than the other five in this group. Finally, natural gas used to produce electricity and nuclear power appear to have the smallest number of deaths and man-days lost.

Total risk is put into another perspective in Figure 15. Each slice represents the proportion of total man-days lost (maximum) for equal energy

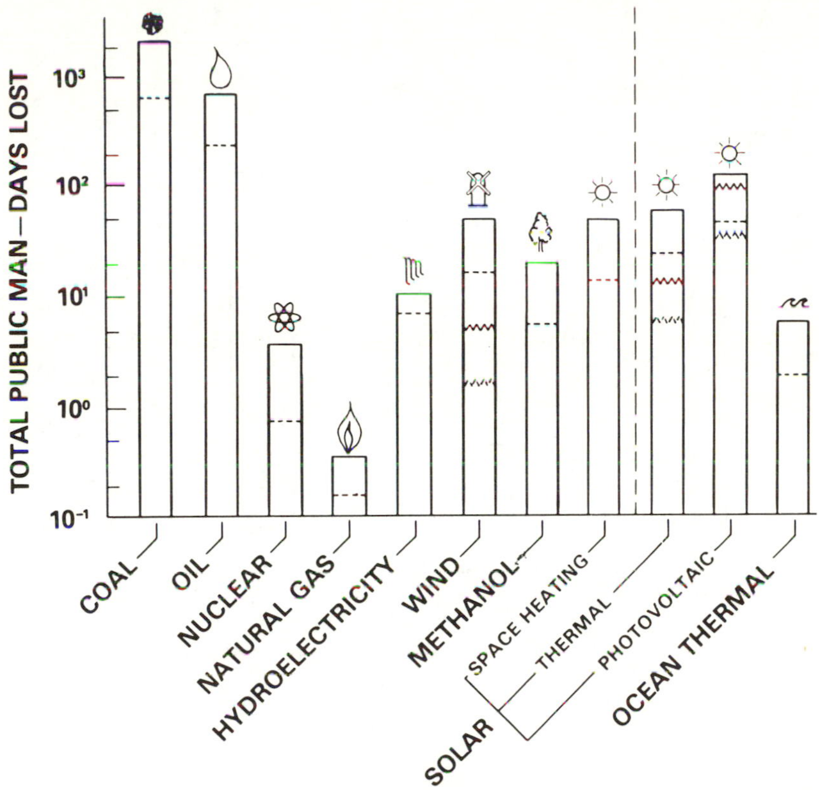

Figure 12. *Public man-days lost per megawatt-year net output over lifetime of the systems. (See explanation in caption to Figures 9 and 10.) As in the previous graph, ocean thermal has the lowest maximum of the nonconventional technologies. This is due to the lack of a requirement for energy back-up and storage, with their accompanying air pollutants, and the moderate requirement for materials. Jagged lines imply low risk or no back-up assumed.*

production. This transforms the logarithmic scale of Figure 14 into a linear one. Coal and oil produce by far the largest proportions of all the risk. Only natural gas, ocean thermal, hydroelectricity, and nuclear contribute small proportions of the total risk.

Figure 15 is only a recharting of the maximum values of Figure 14. For example, suppose systems A, B, and C produce 100, 40, and 60 man-days lost, respectively, for the same energy production, yielding a total of 200.

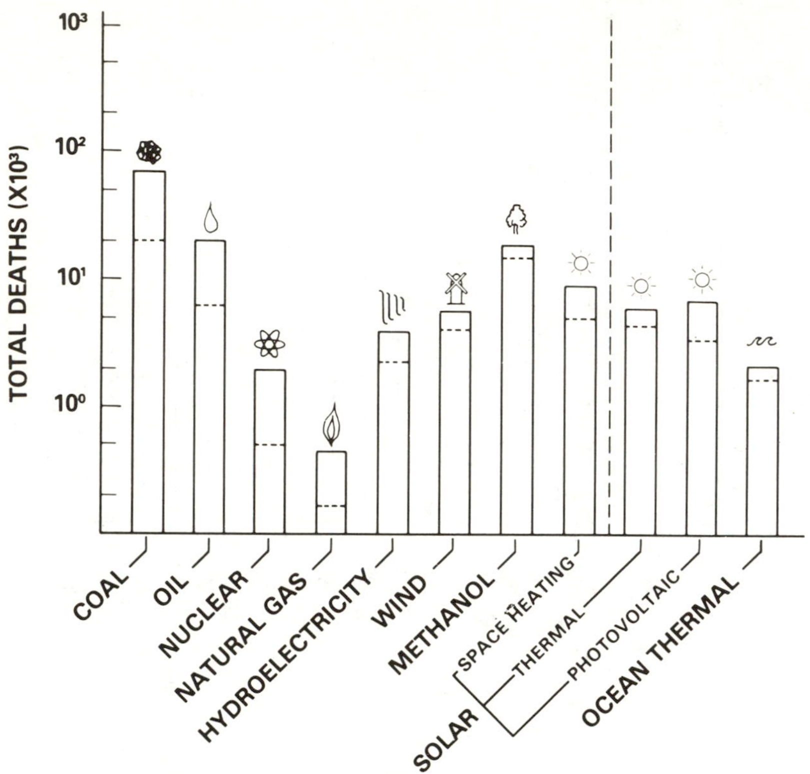

Figure 13. *Total deaths, times 1000, per megawatt-year as a function of the energy system. (See explanation in caption to Figure 9.) For this graph, the public and occupational deaths are combined. Natural gas-fired electricity has the lowest value, followed by nuclear.*

Their percentage of the total number of man-days lost is then 100/200 = 50%; 40/200 = 20%; and 60/200 = 30%, respectively. Similar calculations can be performed for the minimum risk value for each technology, the number of deaths, accidents, or illnesses, etc. These calculations are not shown here for reasons of space. While it is likely that the rankings of some systems would change, the overall conclusions probably would not be drastically altered.

Figure 16 shows the same results as Figure 15, except that the three technologies less applicable to Canada have been eliminated. Only hydroelectricity, nuclear, and natural gas have low proportions of risk.

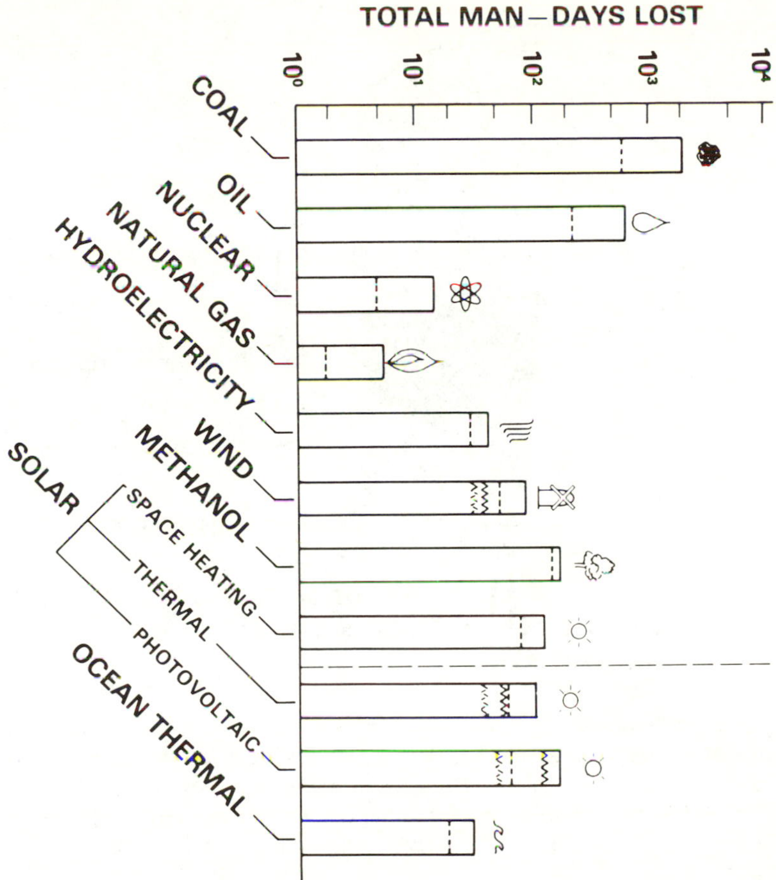

Figure 14. *Total man-days lost per megawatt-year net output over lifetime of the system. (See explanation in captions to Figures 9 and 10.) This graph is similar to Figure 13. However, there are some differences in ranking. Natural gas has the lowest values, followed closely by nuclear. Jagged lines imply low risk or no back-up assumed.*

b. Simplified Calculation for Solar Space Heating Risk

Because some of the above results may be contrary to the intuition of some readers, a brief summary for one energy system–solar space heating–is shown in Table 5. The methodology is due to Ramsay (491) but is basically a condensation of that used in this report. Only one aspect of

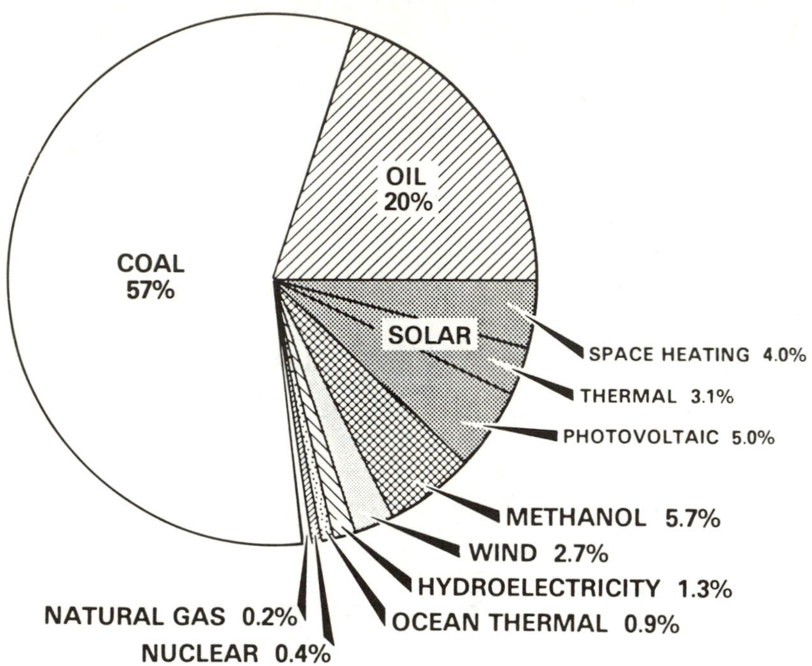

Figure 15. *Proportion of total risk for eleven conventional and nonconventional energy sources. Each source is assumed to generate the same unit energy, and data are obtained from the maximum values of man-days lost in Figure 14. The sizes of the smallest slices are exaggerated, for graphical reasons. Coal and oil have the largest proportions, with natural gas the smallest.*

risk is considered in this table, that from producing the final materials. As noted in Appendix G, there is also risk from producing raw materials, construction, operation and maintenance, energy for pumps, etc. However, this table should give some idea of the calculation of one aspect of risk.

By way of contrast, Smith, Weyant, and Holdren (8) estimate (1.4–3.1) $\times 10^{-4}$ accidental deaths for the entire nuclear fuel cycle, including harvesting, upgrading and transporting fuels, conversion into electricity, and management of final waste (reprocessing). Risk values of Table 5 are then (6.0/3.1)–(6.0/1.4) = 1.9–4.3 times this estimate for the total nuclear fuel cycle.

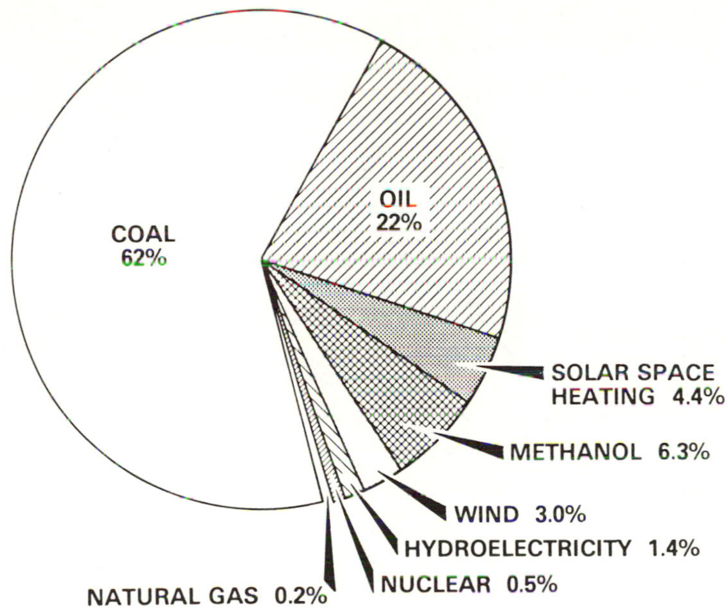

Figure 16. *Proportion of total risk for eight energy sources appropriate to Canadian conditions. The data of Figure 15, excluding solar photovoltaic, solar thermal, and ocean thermal, are used to calculate the proportions. As in Figure 15, the sizes of the smallest slices are exaggerated for graphical clarity.*

In terms of accidents, Smith *et al.* estimate $(7.1–17) \times 10^{-3}$ injuries for the total nuclear fuel cycle. Risk values of Table 5 are then (0.54/0.017)–(0.54/0.0071) = 32–76 times this estimate for the nuclear cycle.

In terms of total accidental man-days lost, the data of Smith *et al.* correspond to $(1.4 \times 10^{-4} \times 6000 + 7.1 \times 10^{-3} \times 18)–(3.1 \times 10^{-4} \times 6000 + 17 \times 10^{-3} \times 18) = 0.97–2.2$, using the assumptions of footnote (m) of Table 5. The risk value of this table is then (13.3/2.2)–(13.3/0.97) = 6.0–14 times this estimate of the nuclear fuel cycle.

If one assumes, as Smith *et al.* do, that each accident corresponds to 50 man-days lost, the total man-days lost using Table 5 is $6000 \times 6.0 \times 10^{-4} + 50 \times 0.54 = 31$, and that of Smith *et al.* is 1.2–2.7. The ratios of the Table 5 values to those of Smith *et al.* are then (31/2.7)–(31/1.2) = 11–26.

Table 5. Simplified Calculation of Material Acquisition Risk for Solar Space Heating (a), (p)

	Aluminum	*Steel*	*Glass*
(1) Weight of solar panels, in metric tons (b) (l)	0.25	0.50	0.36
(2) U.S. production in millions of metric tons (c)	5.0	120	1.36
(3) Solar weight/U.S. production [= (1)/(2)]	$0.25/(5.0 \times 10^{6})$ $= 5.0 \times 10^{-8}$	$0.50/(120 \times 10^{6})$ $= 4.2 \times 10^{-9}$	$0.36/(1.36 \times 10^{6})$ $= 2.7 \times 10^{-7}$
(4) Sum of aluminum and steel proportions	$5.0 \times 10^{-8} + 4.2 \times 10^{-9}$ $= 5.4 \times 10^{-8}$ (d)		
(5) U.S. deaths, by industry	201(e)		15(f)
(6) U.S. accidents, by industry	1.9×10^{5} (g)		1.3×10^{4} (h)
(7) Deaths due to panels production [= (4) × (5) or (3) × (5)]	$5.4 \times 10^{-8} \times 201$ $= 1.1 \times 10^{-5}$		$2.7 \times 10^{-7} \times 15$ $= 4.1 \times 10^{-6}$
(8) Accidents due to panels production [= (3) × (6)]	$5.4 \times 10^{-8} \times 1.9 \times 10^{5}$ $= 1.0 \times 10^{-2}$		$2.7 \times 10^{-7} \times 1.3 \times 10^{4}$ $= 3.5 \times 10^{-3}$
(9) Deaths/MW-yr for panels production [= (7)/lifetime energy] (i)	$1.1 \times 10^{-5}/0.025$ $= 4.4 \times 10^{-4}$		$4.1 \times 10^{-6}/0.025$ $= 1.6 \times 10^{-4}$
(10) Accidents/MW-yr for panels production [= (8)/lifetime energy] (i)	$1.0 \times 10^{-2}/0.025$ $= 0.40$		$3.5 \times 10^{-3}/0.025$ $= 0.14$

(11) Total deaths/MW-yr (j)	$(4.4 + 1.6) \times 10^{-4} = 6.0 \times 10^{-4}$ (n)
(12) Total accidents/MW-yr (k)	$0.40 + 0.14 = 0.54$ (o)
(13) Total occupational man-days lost/MW-yr (m)	$6000 \times 6.0 \times 10^{-4} + 18 \times 0.54 = 13.3$

(a) After Ramsay (491). Modified as noted.
(b) The area of collectors is 37 m^2, for a weight per unit area of 1110/37 = 30 kg/m^2, somewhat less than the value of 34 assumed in Appendix G. They produce 1.27 kW-yr annually, or $1.27 \times 8760/37 = 301$ kWh/m^2, about the same as is assumed here.
(c) From Table M-1. Values differ somewhat from those of Ramsay, but not enough to affect results significantly.
(d) Added together because occupational risk data are combined this way.
(e) 1978 death rate (per 10,000 employees) for blast furnace and basic steel was 1.4 (264); iron and steel foundries, 0.59; primary nonferrous metals, 1.3; nonferrous rolling and drawing, 1.3; fabricated metal products, 0.48. Employment in 1977 for these industries was (in thousands) 553, 231, 65, 199, and 1576, respectively (523). Total number of deaths was then $0.1\ (1.4 \times 553 + 0.59 \times 231 + 1.3 \times 65 + 1.3 \times 199 + 0.48 \times 1576) = 201$. Ramsay used a value of 235. This implies that the metals used are fabricated to about the same degree as the average for all metal use.
(f) 1978 death rate (per 10,000 employees) for stone, clay, and glass products is 0.75 (264). Employment in 1977 for flat glass, glass, and glassware and products of purchased glass was 199,000 (523). Number of deaths was then $0.75 \times 19.9 = 15$. Ramsay used a value of 70.
(g) 1977 accident rates (lost workday cases) per 100 employees for primary metal industries and fabricated metal products were 6.6 and 7.0, respectively (524). 1977 employment in millions in these industries was 1.179 and 1.577, respectively (523). Total accidents are then $6.6 \times 11{,}790 + 7.0 \times 15{,}770 = 1.9 \times 10^5$. Ramsay had a value of 1.0×10^5.
(h) 1977 accident rates per 100 employees for the three industries noted in the second sentence of footnote (f) were 4.7, 6.8, and 6.7, respectively (524). Employment, in thousands, was 21, 135, and 43, respectively (523), so the total number of accidents is $10(4.7 \times 21 + 6.8 \times 135 + 6.7 \times 43) = 1.3 \times 10^4$. Ramsay used a value of 5×10^3.
(i) Assumes a 20-year lifetime. If each collector produces 1.27 kW-yr annually [see footnote (b)], the energy produced in a lifetime is $1.27 \times 10^{-3} \times 20 = 0.025$ MW-yr.
(j) Equals sum of deaths for aluminum, steel, and glass.
(k) Equals sum of accidents for aluminum, steel, and glass.
(l) Excludes materials such as copper and fiberglass.
(m) Allowing 6000 man-days lost per death and 18 per average accident (264).
(n) Ramsay uses a value of 1.1×10^{-3}. (o) Ramsay uses a value of 0.30. (p) No storage assumed.

Smith *et al.* also calculate occupational risk from disease in the nuclear cycle. Since this is not considered in Table 5, no adequate integration of risk from accident and disease can be made. Smith *et al.* find (1.1–2.0) $\times 10^{-4}$ disease deaths/MW-yr for nuclear, and (2.2–3.9) $\times 10^{-4}$ disabilities. Allowing 16 man-days lost per disability (525), this is $(1.1 \times 10^{-4} \times 6000 + 2.2 \times 10^{-4} \times 16)$–$(2.0 \times 10^{-4} \times 6000 + 3.9 \times 10^{-4} \times 16) = 0.66$–$1.2$ man-days lost/MW-yr. The addition of these values to those of accidental risk would not change conclusions substantially.

In summary, the accidental risk attributable to producing only some of the final materials of a representative solar collector is 6–26 times that of an estimate for the entire nuclear fuel cycle.

The field of risk accounting is only beginning now. It is fully realized that the risk of energy production forms just one part of the selection criteria used to judge which system is most appropriate for a particular situation. While comparative risk studies should not be used to promote or oppose a given technology, without this knowledge a fully informed judgement cannot be made.

Appendices

APPENDIX A. COAL

i. Steam Cycle Plant

There are many ways of generating electricity using coal. Smith *et al.* (135) list 21, both present and potential. To keep the calculations within bounds, only one of those using present-day technology was chosen for consideration: the conventional coal-fired steam cycle plant (9). In this system, combustion of coal in a boiler produces steam at 540°C (1100°F). The steam is partially expanded through a turbine, driving an electric generator. The thermal efficiency of the plant is about 37%, although this can vary slightly. This assumption is probably high for systems with stackgas scrubbers. The U.S. efficiency in 1973 was 33% (421). British efficiency for all fossil-fueled stations was 31.5% in 1977–78 (626).

Table A-1 shows the labor and material requirements per megawatt-year net electrical output. Other assumptions and references are listed in the footnotes. The values used will depend to some extent on the type of coal used and the precise way it is processed and transported. To keep Table A-1 as close as possible to the organization of the original, a somewhat different scheme from that of Figure 1 is used.

The results indicate that, as expected, operations in the gathering and handling of fuels constitute about 80% of all labor. That is, most of the effort is expended in coal mining, one of the more dangerous industrial occupations (15).

For simplicity, the materials required were divided into metals and concrete, even though a wide range of other materials are used in practice. Most of the risk will come from the production and fabrication of these two categories of materials.

Results of risk calculations are shown in Table A-2 (22). Occupational risk is derived mainly from construction and coal mining, and public

Table A-1. Coal-Fired Electrical Production System Materials and Labor (per Megawatt-Year Net Electrical Output) (9)

Materials and Labor	*Gathering and Handling Fuel (a)*	*Transportation (g)*	*Electricity Production*	*Total (f)*
Labor (man-hours)				
Construction (b)				
Engineering	7.2	2.6	40.5	50
Administration	7.0	2.7	35.0	45
Unskilled labor	2.9	5.4	52.2	61
Skilled labor	32.5	15.9	301	349
Total	49.6	26.5	429	505
Operation and maintenance	1669 (j),(c),(k), (n)	92.1 (i),(d),(m)	407 (e)	2170 (h),(l)
Materials (metric tons)				
Construction—metals (b)				
Structural	0.49	0.029	0.74	1.26
Pipes	0.16	–	0.09	0.26
Major equipment	0.66	0.87	0.88	2.41
Other equipment	–	0.058	0.36	0.42
Total	1.31 (p)	0.95	2.08	4.34
Construction—concrete (b)	(o)	–	6.82	6.82
Operation and maintenance	–	–	247 (lime)	247

(a) Coal is assumed to be from Northern Appalachia. Mine life is assumed to be 20 years.
(b) From Ref. 10.
(c) From Refs. 11 and 12, assuming 1.03 million metric ton/yr per mine.
(d) From Ref. 13.
(e) From Ref. 14.
(f) Management of final wastes is included in corresponding power plant quantities.
(g) Assuming that all coal is shipped by rail. Train lifetime is assumed to be 30 years.
(h) Reference 240 has a value of 1400–12,400 man-hours per megawatt-year.
(i) Reference 240 has a value of 85–430.

risk is derived primarily from air pollution. For some of the entries, other data were available and were substituted for those of Reference 22. These are indicated in the footnotes. Generally speaking, as many data points were adjusted down as were adjusted up.

As noted in footnote (p) of Table A-2, there seems to be some divergence in the published literature on the public and occupational risk values of transportation. This should be kept in mind in any subsequent calculations based on these values.

Footnote (o) to Table A-2 shows that risk attributable to construction of the generating plant and acquisition of noncoal materials is included. In addition, risk attributable to the air pollutants created by these materials is also included, as mentioned in footnote (s).

For 700 megawatt-years, there are between 0.5 and 1.6 occupational deaths per year in gathering and handling fuels, 1.2 to 3.5 deaths in transportation, and less than 0.1 deaths in electricity production. These values are between 1 and 4 times that of comparable values in the United Kingdom (194), due presumably to greater emphasis on occupational safety in that country. In terms of impact on the public, there will be between 1 and 10 deaths from gathering and handling fuels, 0.6 to 1.3 deaths in transportation, and between 11 and 33 deaths from emissions due to electricity production. The range for this last category is large, with a factor of about 3 between the lowest and highest estimates. The highest value probably applies to areas of high population density.

As noted in the main text, risk attributable to coal mine wastes is not considered here. However, there have been accidents in Wales related to coal wastes which caused loss of life. In the United States, the coal refuse dam failure near Saunders, West Virginia, apparently was related to the Buffalo Creek accident in February 1972, which had a toll of 118 people (247).

(j) Reference 240 has a value of 850–11,400.

(k) Reference 481 has a value of 425–1340.

(l) Reference 485 indicates a value of 5200 for France.

(m) Reference 485 indicates a value of 242 for France.

(n) Reference 485 indicates a value of 4400 for France.

(o) Ferguson (632) estimates a weight of concrete 2×10^{-3} times that of coal produced. If 3500 metric tons are required for one MW-yr, this is $2 \times 10^{-3} \times 3500 = 7$ metric tons of concrete.

(p) Ferguson (632) estimates a weight of steel 5×10^{-3} times that of coal produced. If 3500 metric tons are required for 1 MW-yr, this is $5 \times 10^{-3} \times 3500 = 18$ metric tons of steel.

Table A-2. Coal-Fired Electrical Production System Risk (per Megawatt-Year Net Electrical Output) (22), (u)

Risks	*Gathering and Handling Fuel (j), (r)*	*Transportation (p), (v), (x)*	*Electricity Production*	*Total*
Occupational (q),(aa),(at)				
Accidental				
Death (h)	$(0.7–1.5) \times 10^{-3}$ (a),(ab),(an),(ap), (as),(aw),(bd)	$(1.6–5.0) \times 10^{-3}$ (n),(ac),(aq),(az)	$(1.3–9.0) \times 10^{-5}$ (ad),(bb)	$(2.5–6.7) \times 10^{-3}$ (ae),(ba)
Injury	0.04–0.07 (e),(i),(af),(an)	$(1.3–4.8) \times 10^{-2}$ (n),(ag)	$(1.6–8.5) \times 10^{-3}$ (ah)	0.056–0.083 (ai),(be)
Man-days lost	8–16 (aj)	10–32 (n),(ak)	0.16–0.97 (al)	20–51 (o), (am)
Disease				
Death (h),(z)	$(0–7.5) \times 10^{-4}$ (b),(ax),(bf)	–	–	$(0–7.5) \times 10^{-4}$
Disability (k)	$(4.2–8.4) \times 10^{-3}$ (f),(ay)	–	–	$(4.2–8.4) \times 10^{-3}$
Man-days lost	0.2–5.0 (ao)	–	–	0.2–5.0 (ao)
Public				
Accidental				
Death (h)	–	$(0.8–1.9) \times 10^{-3}$ (c)	–	$(0.8–1.9) \times 10^{-3}$

Injury	–	1.6×10^{-3} (bg)	–	1.6×10^{-3}
Man-days lost	–	5–11.6	–	5–11.6
Disease (t),(av)				
Death (h)	$(1.4–14) \times 10^{-3}$ (y)	–	0.016–0.047 (w),(d),(g),(m), (ar),(bc),(bh)	0.017–0.061
Disability (l)	–	–	94–280 (g),(m)	94–280 (s)
Man-days lost	8.4–84	–	580–1750 (g),(m)	590–1830 (s)

(a) In the original table, only underground coal mining was considered. To take account of the fact that strip mining is also used to gather coal, the data of Comar and Sagan (23) were used to calculate accidental death rates for gathering and handling fuels. Death rates from strip mining are considerably lower. Reference 23 showed rates of a 1000-MW (e) station operation for one year. The actual energy produced per year divided by this rated energy capacity is defined as the load factor, assumed to be 0.7 for all systems unless otherwise specified. The total energy produced is then 700 megawatt-years. The risk data of Ref. 23 are then divided by this value. The range in the data is the minimum and maximum from Refs. 14 and 25–28.

(b) Assuming 12,500 Btu/lb of coal and coal-to-electricity efficiency of 0.37. Data from Ref. 365 and 366. Best value is 1.1×10^{-4}. Data relate to post-1970 populations and assume that half of coal is mined underground (367), suface mining produces no disease and dust levels will decrease as much as regulations specify. As noted in the main text of this appendix, there is often a large difference between regulations and what actually happens, so these values can be taken as lower bounds.

(c) From Refs. 14, 25, 26, and 28.

(d) Reference 26 has 0.020–0.166 (for 1973).

(e) Allowance was made for varying types of coal, as noted in footnote (a). From Refs. 14 and 25–27.

(f) From Refs. 368 and 369. Same assumptions as footnote (b), except that data were collected during 1970–1971.

(g) Oxides of sulfur only (30). Other pollutants with an effect are not included for simplicity.

(h) 6000 man-days lost per death, from Ref. 31.

(i) 93 man-days lost per injury, from Ref. 31.

(j) Only 2/3 of coal is processed, from Refs. 26, 27, and 31.

Table A-2 continued

(k) WASH-1224 (499) assumes 1000 days lost per case of pneumoconiosis. However, only 50 days is assumed here.

(l) Between 1 and 5 days lost per disability.

(m) The range in values reflects the variation in plant location. See Ref. 131.

(n) From Ref. 22. These values are higher than those of another comprehensive evaluation (23), presumably because the former source includes risk attributable to building transportation systems in addition to operating them.

(o) Using Tables M-2 and A-1, the number of man-days lost due to accidents in construction of facilities is about $130 \times 505/100 \times 2000 = 0.33$; deaths are $0.042 \times 505/100 \times 2000 = 1.1 \times 10^{-4}$. Table A-1 states that 4.3 metric tons of metal (assumed to be steel) are used, implying 7.1, 6.5, and 4.3 metric tons of iron ore, coal, and fabricated metal products, respectively. Using, for example, Table E-2, the number of man-days lost to accidents is then $(7.1/63) \times 0.018 + (6.5/56) \times 0.067 + (4.3/8.3) \times 0.055 = 0.38$; deaths are $(7.1/63) \times 8 \times 10^{-6} + (6.5/56) \times 4 \times 10^{-5} + (4.3/8.3) \times 5.6 \times 10^{-5} = 3.5 \times 10^{-5}$. The total number of man-days lost due to accidents is then $0.33 + 0.38 = 0.7$; deaths are $1.1 + 0.4 = 1.5 \times 10^{-4}$. Allowing 6000 man-days lost per death, this is $1.5 \times 10^{-4} \times 6000 + 0.7 = 1.6$ man-days lost. By way of comparison, Ferguson (633) estimates $(8\text{–}100) \times 10^{-6}$ deaths per MW-yr for coal mine construction for British conditions.

(p) Values for occupational transportation risk were taken from Smith *et al.* (22), and those for public risk derived from Comar and Sagan (23). Report WASH-1224 (232) has approximately the same number of man-days lost per unit energy for public risk (about 5), but substantially lower occupational risk (about 1). Reference 23 has values of approximately 0.5–7 man-days lost for occupational risk. If the values of Ref. 232 were to be adopted, the total occupational risk of coal would be about 9–20 man-days lost per unit energy. This would be a decrease of about 50% of the total occupational risk shown in Table A-2. However, the total risk of coal-fired electricity would drop by only about 1% if the lower value for occupational transportation risk were adopted. Since the coal transportation risk is used as a model for the transportation risk of some nonconventional systems, a change in values will alter the risk of some of these latter systems. However, using the occupational transportation risk of Ref. 232 does not change the total risk of any nonconventional system by more than 10%. Further study is needed on this question.

(q) Reference 239 lists total occupational deaths as 3.1×10^{-3}; total occupational injuries as 0.18. This reference notes that the range for "other studies" has values of $(0.72\text{–}4.3) \times 10^{-3}$ and 0.035–0.10, respectively.

(r) Reference 27 indicates a total number of occupational man-days lost as 25 for surface mining and 52 for underground mining.

(s) Includes 0.04–0.12 disabilities and 0.24–0.8 man-days lost for emissions due to metals (assumed to be steel) in generating plant construction. See Table A-1 and Section 2I.

(t) Reference 23 indicates that values for electricity production may be a factor of 2 too low or 10 too high. Data apply to the U.S. Northeast, and may not be applicable to all parts of Canada. As far as is known, no comparable statistical study has been done nationally in Canada.

(u) Generally assuming a load factor of 0.7. The net capacity factor (a related concept) for baseloaded coal plants in the United States in 1978 was 0.55, which implies that some occupational risk in this table may be underestimated (298).
(v) Reference 370 estimates 5.3×10^{-5} occupational deaths and 3.9×10^{-3} occupational injuries per unit energy (assuming a load factor 0.75). This would produce 0.68 man-days lost per unit energy. This reference estimates 1.5×10^{-3} public deaths and 3.2×10^{-3} public injuries per unit energy, similar to values in Table A-2. Using unit trains would produce about one-quarter the above values. On the other hand, truck transport would produce a total of 2.4×10^{-3} deaths and 0.04 injuries per unit energy (371).
(w) Using Ref. 372, one can estimate $(3.6\text{–}19) \times 10^{-3}$ deaths per unit energy for present-day conditions (1.9% sulfur content). Populations within 80 km of the plant were assumed to be 0.62–3.3 million. A load factor of 0.75 is assumed. Health effects beyond the 80-km radius may be substantial, but are not included here.
(x) It can be deduced (416, 419) that transporting 3500 metric tons of coal by rail produces $(3.7\text{–}4.7) \times 10^{-4}$ occupational and $(4.6\text{–}5.6) \times 10^{-3}$ public deaths, based on 1970–73 statistics. While the proportions of the two categories are different from those in the above table, the total number of deaths is comparable.
(y) From Ref. 26.
(z) Schurr (435) estimates occupational disease deaths from coal as $(0\text{–}53) \times 10^{-4}$.
(aa) Schurr notes total occupational deaths from coal as $(4\text{–}67) \times 10^{-4}$ (435).
(ab) Reference 485 has a value of 1.8×10^{-3} for France.
(ac) Reference 485 has a value of 9.1×10^{-5} for France.
(ad) Reference 485 has a value of 2.3×10^{-5} for France.
(ae) Reference 485 has a value of 2.0×10^{-3} for France.
(af) Reference 485 has a value of 1.4 for France.
(ag) Reference 485 has a value of 1.7×10^{-3} for France.
(ah) Reference 485 has a value of 3.8×10^{-3} for France.
(ai) Reference 485 has a value of 1.4 for France.
(aj) Reference 485 has a value of 45.7 for France.
(ak) Reference 485 has a value of 1.5 for France.
(al) Reference 485 has a value of 0.2 for France.
(am) Reference 485 has a value of 48 for France.
(an) Reference 485 has a total value of deaths and disabilities of $1.3\text{–}11.4 \times 10^{-3}$ for France.
(ao) Reference 485 has a value of 0.5–3.8 for France.
(ap) Reference 536 has $(0.4\text{–}2.2) \times 10^{-3}$.
(aq) Reference 636 has 3.1×10^{-3}.
(ar) Reference 545 estimates 0.004–0.075 deaths from emissions for coal gasification, for a level of sulfur oxides emissions much lower than assumed here.

Table A-2 continued

(as) Reference 545 estimates (1–2.4) $\times 10^{-3}$ deaths.

(at) Reference 545 estimates (1.6–3.7) $\times 10^{-4}$ deaths for non-construction and non-fuel gathering activities.

(au) Oudiz (552) estimates 0.011–0.031 for France. The upper value is based on relatively high-sulfur coal. As noted in the main text of this appendix, this is a fairly widespread practice.

(av) See text for derivation of the values for electricity production.

(aw) Ferguson (634) estimates (0.6–2.3) $\times 10^{-3}$ for Britain during the 1970s. He notes that the death rate in mining coal for non-power uses is probably higher.

(ax) Ferguson (635) estimates (0–3) $\times 10^{-3}$ for present British conditions for lung disease. In addition he estimates (0–3) $\times 10^{-4}$ deaths from radon gas (636).

(ay) Ferguson (637) estimates about 1750 disabilities in the production of about 120 $\times 10^{6}$ metric tons. If 3500 metric tons are required for 1 MW-yr, this is 0.05 disabilities per MW-yr in Britain.

(az) Ferguson estimates (0–3) $\times 10^{-4}$ for British conditions (641).

(ba) Ferguson (646) estimates (1–2) $\times 10^{-4}$ deaths for plant construction, similar to that deduced in footnote (o).

(bb) Ferguson (646) estimates (1–3) $\times 10^{-4}$ for British conditions.

(bc) Ferguson (647) estimates 4 $\times 10^{-8}$ for British conditions due to oxides of sulfur and particulates. He uses a value of (0–2) $\times 10^{-4}$ for radioactive emissions (648).

(bd) A German study (665) uses a value of 3.7 $\times 10^{-3}$.

(be) A German study (665) uses a value of 0.14, if 50 man-days are allotted to each accident.

(bf) Black (663) uses a value of 1 $\times 10^{-4}$. A German study (665) uses 2.8 $\times 10^{-3}$.

(bg) Black (663) uses the range (1.2–1.5) $\times 10^{-2}$.

(bh) Black (663) uses a range of (3.2–22) $\times 10^{-3}$.

ii. Emissions

In Canada, oxides of sulfur, particulate matter, oxides of nitrogen, hydrocarbons, and carbon monoxide emissions were, in thousands of metric tons, 582, 193, 217, 3.6, and 12.0, respectively (220). These values apply to 1974, the last year for which data are available. A total of 30,400 megawatts of power capacity was available in that year. Fossil-fuel generating capacity was 31.4% of the total (275). The load factor for fossil fuels was about 0.73 (276). This implies that 0.73 × 0.314 × 30,400 = 6970 megawatt-years were produced from fossil fuels in 1974 in Canada. Of this, about 80% may have been derived from coal and oil (as opposed to natural gas), or 0.8 × 6970 = 5580 megawatt-years.

Then 582,000/5580 = 104 metric tons of sulfur oxides were emitted for each megawatt-year of electricity. Values, in the same units, for particulate matter, nitrogen oxides, hydrocarbons, and carbon monoxide were 35, 39, 0.65, and 2.2, respectively. Smith *et al.* (263) assumed 11-38, 1.5-3.8, 13-25, 0.5, and 0.3-1.7, respectively, in the same units. It can be seen that the emissions assumed by Smith *et al.* are 104/(11-38) = 2.7-9.5, 35/(1.5-3.8) = 9.2-23, 39/(13-25) = 1.6-3.0, 0.65/0.5 = 1.3, and 2.2/(0.3-1.7) = 1.3-7.3 times lower than recent Canadian conditions for sulfur oxides, particulate matter, nitrogen oxides, hydrocarbons, and carbon monoxide, respectively. In particular, for sulfur dioxide, the air pollutant considered in this book, WASH-1224 (185) notes that "the most promising (control) systems may achieve 80% efficiency in sulfur dioxide removal but the long-term reliability of these systems has not been demonstrated."

A Canadian Press story (429) suggested that Ontario coal-fired generating stations produced 500,000 tons of sulfur oxides in recent years. These stations produced about 2800 megawatt-years in 1977 (428). They then generated 500/2800 × 1.1 = 162 metric tons of sulfur oxides per megawatt-year.

Sulfur oxides emissions from fossil-fueled plants in the United States totaled 16.8 million metric tons in 1975 (273). The article referred to noted that "emission standards for the coal-fired plants already in existence today are lax to the point that for many of them SO_2 emissions are entirely uncontrolled." A total of 144,000 megawatt-years of electricity were produced by coal and oil in the United States in 1976 (274). The emissions then are about 120 metric tons of sulfur oxides per megawatt-year. A recent Norwegian study (262) noted that "emissions from modern

European plants put into operation during the last 5-6 years" are about 100 metric tons of sulfur oxides per megawatt-year output. A Finnish study uses a value of 60 metric tons (265).

It is of interest to note that another estimate (307) of sulfur dioxide emissions in the United States in 1974 from stationary fuel combustion was 24.5 million metric tons. This is about 50% higher than that noted above.

In 1972, before large-scale attempts to control oxides of sulfur, the U.S. Federal Power Commission (337) found that sulfur oxides weight per 1000 megawatt-hours generated from coal and totaled about 12.6 tons. This is then $12.6 \times 8.76/1.1 = 100$ metric tons per megawatt-year. To convert this value to weight per unit coal-generated electricity, one would have to know the proportion of coal to oil-fired electricity, and the sulfur content of each.

A value of 100 metric tons of sulfur oxides per unit energy will be assumed here, somewhat lower than recent Canadian and U.S. experience. The Smith *et al.* values (22) for 38 metric tons of sulfur oxides a year are $(6\text{–}18) \times 10^{-3}$, 36–108, and 225–675 deaths, illnesses, and man-days lost, respectively. These values may be a factor of 2 too low or 10 too high, and these extra factors are used in the final values of Smith *et al.* The quantities above are then multiplied by $100/38 = 2.6$ to take account of present-day levels of emissions, yielding 0.016-0.047, 94-280, and 580-1750 deaths, illnesses, and man-days lost, respectively. The variation of a factor of 3 is that used in Table 5 of Comar and Sagan (23).

In terms of national effects in Canada, using this model would produce $(590\text{–}1830) \times 4900 = (2.9\text{–}9) \times 10^6$ man-days lost to the public from disease due to air pollution, assuming about 4900 megawatt-years of coal-fired electricity were produced in Canada in 1977 (428). Included is risk due to gathering and handling fuels. Assuming a Canadian population of about 22,000,000, this corresponds to about $(2.9\text{-}9)/22 = 0.13\text{-}0.4$ man-days lost annually per person from this source.

APPENDIX B. OIL

While it is unlikely that oil will be used as extensively in the next few decades as it has been in the recent past, its risk can be calculated. The model plant considered here is similar to the coal-fired one, but there is virtually no ash produced. However, the oil fuel cycle is quite different from that of coal.

Oil may be transported to a refinery either by pipeline or tanker. At the refinery, crude oil is transformed into a number of petroleum products. The residual fuel oil fraction is then usually transported to the power plant by pipeline. The materials used to construct pipelines are included in the calculations.

The thermal efficiency of an oil-fired plant is assumed to be 37–39% (32). Average efficiency in 1973 was 31% (417, 421). The materials and labor required per megawatt-year net output given in Table B-1 show that requirements for oil are considerably lower than that for coal. Operation and maintenance man-hours for oil are about one-third that of coal. Metal requirements for construction are about three-quarters; concrete, about one-half.

These lower requirements are reflected in the lower incidence of risk, as shown in Table B-2. There is apparently little or no risk from occupational diseases or public accidents. However, Smith *et al.* (37) note that there could be tanker explosions killing between 100 and 1000 people (34, 40–42). In addition, there is no calculated public disease risk from gathering and handling fuels.

A recent Norwegian study (262) noted that "emissions from modern European plants put into operation during the last 5-6 years" are about 100 metric tons of sulfur oxides per megawatt-year output. A recent Finnish study also uses this value (265). As noted in Appendix A, recent U.S. experience has been about 120 metric tons per unit energy for coal and oil, and Canadian experience slightly lower. However, this value applies to coal and oil, and it is not clear how to separate out the two types in terms of effluents. A recent study (297) indicates a value of 37 metric tons per unit energy in the United States, and this will be assumed here.

In 1972, before large-scale attempts to control oxides of sulfur, the Federal Power Commission in the U.S. (336) found that the ratio of sulfur weight for oil compared to coal, per unit heat in the fuels, was 0.28. If the efficiencies of burning coal and oil are similar, this ratio should not change substantially per unit electricity generated. The ratio is similar to the ratio of 37/100 = 0.37 assumed here.

Occupational accidental risk in oil-fired electricity is much less than that of coal. Public disease risk is also less.

Table B-1. Residual Fuel Oil-Fired Electrical Production System Materials and Labor (per Megawatt-Year Net Electrical Output) (33)

Materials and Labor	*Gathering and Handling Fuel (a)*	*Transportation (b)*	*Electricity Production (c)*	*Total*
Labor (man-hours)				
Construction (d)				
Engineering	15.0	0.43	28	43
Administration	10.4	0.20	28	38
Unskilled labor	23.4	1.35	28	53
Skilled labor	113	1.41	167	281
Total	161	3.39	250	415
Operation and maintenance (e)	449 (g)	52	226	727 (f)
Materials (metric tons)				
Construction–metals (d)				
Structural	0.58	0.019	0.58	1.18
Pipes	0.64	0.18	0.093	0.92
Major equipment	0.30	0.007	0.56	0.86
Other equipment	0.05	0.006	0.21	0.27
Total	1.57 (h)	0.22	1.44	3.23
Construction–concrete (d)	0.78	0.009	2.42	3.11

(a) Gathering of fuels consists of offshore oil extraction and transport by oil pipeline. Computations involving oil sands technology would be different. The length of pipeline will affect both labor and materials. Its length is unspecified in Ref. 33. The offshore facilities were assumed to have a 20-year lifetime. The refinery was assumed to have a 30-year lifetime. For purposes of calculations, the refinery was treated as if its entire output was residual fuel oil.
(b) Pipeline assumed to have a 30-year lifetime.
(c) Power plant lifetime assumed to be 30 years.
(d) From Ref. 10. Reference 33 lists labor and materials requirements for importing oil from the Middle East.
(e) From Refs. 31, 34, and 35. Operation and maintenance labor requirements are about 200 man-hours for supertanker transport, from Ref. 36.
(f) Reference 241 has a value of 790 man-hours per megawatt-year, close to the present value.
(g) For France, 192 (485).
(h) Ferguson (638) estimates 2.1 for North Sea oil.

Table B-2. Oil Fired Electrical Production System Risk (per Megawatt-Year Net Electrical Output) (37)

Risks	*Gathering and Handling Fuel (g)*	*Transportation*	*Electricity Production*	*Total (i)*
Occupational (j)				
Accidental				
Death (e)	$(1.4–17) \times 10^{-4}$ (a),(m),(z),(aa),(ab)	$(4–14) \times 10^{-5}$ (b),(n),(ac)	$(1.3–5.0) \times 10^{-5}$ (o),(ae)	$(3.0–20) \times 10^{-4}$ (p),(ad)
Injury (d)	$(1.5–12) \times 10^{-2}$ (a),(q)	$(1.6–13) \times 10^{-3}$ (b),(r)	$(0.9–2) \times 10^{-3}$ (a),(s)	$(2.9–15) \times 10^{-2}$ (t)
Man-days lost	1.6–16 (u)	0.32–1.5 (v)	0.13–0.4 (w)	3.3–19 (x)
Disease				
Death	–	(ag)	(y)	–
Disability	–	–	(y)	–
Man-days lost	–	–	–	–
Public				
Accidental				
Death	–	–	–	–
Injury	–	–	–	–
Man-days lost	–	–		–

Table B-2 continued

Risks	*Gathering and Handling Fuel (g)*	*Transportation*	*Electricity Production*	*Total (i)*
Disease (h) (l)				
Death (e)	–	–	0.006–0.017 (c),(af),(ah)	0.006–0.017
Disability (f)	–	–	35–105	35–105 (k)
Man-days lost	–	–	215–650	215–650 (k)

(a) From Refs. 12, 14, and 25–27.
(b) From Refs. 12, 14, 25, and 26.
(c) Reference 26 has higher values.
(d) Implies loss of 55 man-days per injury. This is an average, since in Ref. 37 the implied loss varies from 31 to 80 days.
(e) Loss of 6000 man-days per death.
(f) Average disability between 1 and 5 days lost.
(g) Death, injury, and man-days lost rates for foreign production and tanker import are about 25 times those in the original table; from Refs. 12, 14, and 31.
(h) Oxides of sulfur only. See also Ref. 39.
(i) Includes, for occupational risk of accident, 0.9×10^{-4} deaths, 2.9×10^{-3} injuries, and 0.8 man-days lost. This results from the labor of 415 man-hours in building an oil-fired system (assumed to be miscellaneous contracting). By using Tables M-2 and B-1, the number of man-days lost due to accidents is about $130 \times 415/100 \times 2000 = 0.27$; deaths are $0.042 \times 415/100 \times 2000 = 8.7 \times 10^{-5}$. Table B-1 states that 3.2 metric tons of metal (assumed to be steel) are used, implying 5.3, 4.7, and 3.2 metric tons of iron ore, coal, and fabricated metal products, respectively. Using, for example, Table E-2, the number of man-days lost to accidents is then $(5.3/63) \times 0.018 + (4.7/56) \times 0.067 + (3.2/8.3) \times 0.055 = 0.28$; deaths are $(5.3/63)\ 8 \times 10^{-6} + (4.7/56) \times 4 \times 10^{-5} + (3.2/8.3) \times 5.6 \times 10^{-5} = 2.6 \times 10^{-5}$. The total number of man-days lost due to accidents is $0.27 + 0.28 = 0.6$,

and the number of deaths is $8.7 + 2.6 = 11 \times 10^{-5}$. Allowing 6000 man-days lost per death, this is $1.1 \times 10^{-4} \times 6000 + 0.55 = 1.2$ man-days lost.

(j) Reference 239 lists total occupational deaths as 0.0039; total occupational injuries as 0.014. This reference notes that the range for "other studies" has values of $(1.9–17) \times 10^{-4}$ and 0.016–0.125, respectively.

(k) Includes 0.06–0.18 disabilities and 0.4–1.2 man-days lost for emissions due to metals (assumed to be steel) in generating plant construction. See Table B-1 and Section 2I.

(l) For electricity production, see footnote (t), Table A-2.

(m) For France, 1.2×10^{-4} (485).

(n) For France, 1.7×10^{-4} (485).

(o) For France, 2.3×10^{-5} (485).

(p) For France, 3.0×10^{-4} (485).

(q) For France, 0.015 (485).

(r) For France, 2.7×10^{-3} (485).

(s) For France, 5.6×10^{-3} (485).

(t) For France, 0.023 (485).

(u) For France, 2.0 (485).

(v) For France, 1.2 (485).

(w) For France, 0.26 (485).

(x) For France, 3.4 (485).

(y) For France, total deaths and disabilities are 3.0×10^{-5} (485).

(z) Reference 194 has the value 3×10^{-3}.

(aa) Ferguson (639) estimates $(1.2–6.7) \times 10^{-4}$ for offshore North Sea oil.

(ab) Ferguson (642) estimates $(0–1) \times 10^{-4}$ for refinery operation in Britain.

(ac) Ferguson (642) states that transport deaths are negligible in Britain.

(ad) Ferguson (649) estimates $(0.9–2) \times 10^{-4}$, similar to that assumed in footnote (i).

(ae) Ferguson (650) estimates $(0.07–20) \times 10^{-5}$ for Britain.

(af) Ferguson (650) estimates the same values as for coal, a total of $(0–2) \times 10^{-4}$.

(ag) A German study (665) has a value of 2.6×10^{-5}.

(ah) Black (663) has a value of $(1.1–7.5) \times 10^{-3}$.

APPENDIX C. NATURAL GAS

Because Smith *et al.* (7) do not have a risk model for natural gas in the same way that they do for coal, oil, and nuclear power, an indirect method was used in risk computation. It is expected that this set of estimates will not differ greatly from that of a more direct method. As mentioned in the Introduction, risk is calculated for using natural gas to generate electricity, not for space heating use in the home.

To determine the materials and labor required per megawatt-year net output, the gasification (130) and residual fuel oil (33) data from Smith *et al.* were amalgamated. Comar and Sagan (23) indicate that natural gas has low overall risk compared to other conventional sources. In consequence, the lower of the two possible values for each category of material or labor was generally chosen. Results are shown in Table C-1.

Most of the labor goes into operation and maintenance, constituting about 75% of the total number of man-hours. About 1.5 metric tons of metals are used per unit energy, and about 2.4 metric tons of concrete. Almost all the concrete is used in electricity production.

Producing electricity from natural gas also incurs risk. The information of Comar and Sagan (23) was mainly used in compiling the first half of Table C-2. All of the risk mentioned in Reference 23 was occupational, as opposed to public.

The probable widespread use of liquefied natural gas shipped to North America by tanker will create the possibility of public risk. It is difficult to estimate its potential magnitude without further experience, but the maximum number of deaths is probably no less than that for oil tanker explosions. The latter events have been estimated to have maximum values of between 100 and 1000 deaths. Ninety people died in catastrophic accidents from 1943–1972 (495). (See Appendix B).

While the use of natural gas to produce electricity may generate air pollutants, sulfur dioxide is not generally among them. (This statement is, strictly speaking, true only for so-called "sweet" gas. "Sour" gas has considerable sulfur content, which must be removed before use.) Since it is assumed only sulfur dioxide generates public health risk, then the risk to the public from natural gas air pollution is negligible. If a stronger relationship is found between pollutants other than sulfur dioxide and health, the calculations will have to be altered.

As noted in footnote (g), Table C-2, the estimated risk for constructing the natural gas system is about 0.8 man-days lost per unit energy. This

Table C-1. Gas-Fired Electrical Production System Labor (per Megawatt-Year Net Electrical Output) (a)

Materials and Labor	*Gathering and Handling Fuel*	*Transportation*	*Electricity Production*	*Total*
Labor (man-hours)				
Construction				
Engineering	5.6 (b)	0.43	27.8 (b)	33.8
Administration	7.0	0.20	27.0	32.2
Unskilled labor	2.9	1.35	27.9 (b)	32.3
Skilled labor	33	1.41	167	201
Total	49	3.39 (e)	250	302
Operation and maintenance	449 (b)	52.3 (f)	226 (b)	727 (c)
Materials (metric tons)				
Construction–metals				
Structural	0.49	0.02	0.03	0.54
Pipes	0.16	0.18	0.09 (b)	0.43
Major equipment	0.24 (b)	0.007	0.14	0.39
Other equipment	0	0.006	0.12	0.13
Total	0.89	0.21 (d)	0.38	1.49
Construction–concrete	0	0.009	2.42 (b)	2.43

(a) Because data on natural gas were not available in the same forms as that for coal, oil, and nuclear power, modifications were made to data on gasification from coal. Generally speaking, the lower values from either residual fuel oil or gasification were used. The above values may be considered as upper limits for natural gas (130).
(b) Residual fuel oil.
(c) Reference 238 has a value of 810–1050 man-hours per megawatt-year.
(d) Reference 667 has a value of 2.2 for pipelines.
(e) Reference 667 has a value of 2.1.
(f) Reference 667 has a value of 125.

Table C-2. Gas-Fired Electrical Production System Risk (per Megawatt-Year Net Electrical Output) (e) (i)

Risks	*Gathering and Handling Fuel*	*Transportation*	*Electricity Production*	*Total (i)*
Occupational (h)				
Accidental				
Death	$(3.9–31) \times 10^{-5}$ (a),(k)	$(2.9–3.4) \times 10^{-5}$ (b),(l)	$(1.4–5.3) \times 10^{-5}$ (a),(m)	$(16–48) \times 10^{-5}$ (g),(c),(n)
Injury	0.004–0.031 (a)	0.0017–0.0019 (d)	$(0.9–2.1) \times 10^{-3}$ (a)	0.013–0.041 (g)
Man-days lost	0.43–3.4 (v)	0.18–0.22	0.13–0.42	1.5–4.8 (g)
Disease				
Death	–	–	–	–
Disability	–	–	–	–
Man-days lost	–	–	–	–
Public				
Accidental (f)				
Death	(o)	(q)	9×10^{-6} (s)	9×10^{-6}
Injury	(p)	(r)	4.7×10^{-5} (t)	4.7×10^{-5}
Man-days lost	–	–	0.06	0.06

Disease				
Death	–	(u)	–	$(2.4–7) \times 10^{-6}$ (j)
Disability	–	–	–	0.013–0.04 (j)
Man-days lost	–	–	–	0.086–0.27 (j)

(a) References 12, 14, and 25–27.
(b) References 14 and 25–27.
(c) Reference 536 has a value of 9.4×10^{-5}, excluding construction of generating plant.
(d) References 12, 14, 25, and 26.
(e) Assumed load factor of 0.7.
(f) The 1970–1975 period saw 13 deaths and 69 injuries to the public due to gas pipeline accidents in the United States (203). This is for transmission and gathering only, not distribution into homes. The number of deaths and injuries from distribution were 11 and 22 times higher, respectively. Assuming an average consumption of about 22×10^{15} Btu during this period (204) and an efficiency of gas turbines of about one-third, an annual total of about 2.4×10^{5} megawatt-years could have been produced if all the gas had been transformed into electricity. The number of deaths and injuries due to transmission are then 2.16 and 11.5 per year, respectively, yielding 9×10^{-6} deaths and 4.7×10^{-5} injuries per megawatt-year. It is assumed that about 50 man-days lost are attributable to each injury. If risk due to distribution had been included in addition to that of transmission, the number of man-days lost would have been about 0.70, still comparatively small.
(g) Includes risk from acquisition of nongas materials and construction of electrical generating plant. In terms of labor, if it is assumed that construction can be approximated by "miscellaneous contracting" (see, for example, Table M-2 and C-1), the number of man-days lost due to accidents is $130 \times 302/100 \times 2000 = 0.20$; deaths are $0.042 \times 302/100 \times 2000 = 6.4 \times 10^{-5}$. The number of man-days lost due to accident is about 0.20; deaths are 6.4×10^{-5}. Table C-1 states that 1.5 metric tons of metal (assumed to be steel) are used, so the weights of iron ore, coal, and fabricated metal products are 2.5, 2.2, and 1.5 metric tons per unit energy, respectively. The total number of man-days lost due to accidents for all material acquisition (see, for example, Table E-2) is then $(2.5/63) \times 0.018 + (2.2/56) \times 0.067 + (1.5/8.3) \times 0.55 = 0.11$; deaths are $(2.5/63) \times 8 \times 10^{-6} + (2.2/56) \times 4 \times 10^{-5} + (1.5/8.3) \times 5.6 \times 10^{-5} = 1.2 \times 10^{-5}$. The total number of man-days lost due to accidents is $0.20 + 0.11 = 0.31$; deaths are $1.2 + 6.4 = 8 \times 10^{-5}$. The number of man-days lost, allowing 6000 per death, is $8 \times 10^{-5} \times 6000 + 0.31 = 0.8$.
death, is $8 \times 10^{-5} \times 6000 + 0.31 = 0.8$.

Table C-2 continued

(h) Reference 239 lists total occupational deaths as 2.5×10^{-4} and total occupational injuries as 8.9×10^{-3}. This reference notes that the range for "other studies" has values of $(0.76–6.0) \times 10^{-4}$ and $(0.53–3.2) \times 10^{-2}$, respectively.

(i) As noted in the text, liquefied natural gas is not considered here. However, Ref. 239 lists total occupational deaths of this energy form as 3.6×10^{-4}, and total occupational injuries as 0.013. These are approximately mid-range of the values listed in this table.

(j) From emissions due to metals (assumed to be steel) in generating plant construction. See Table C-1 and Section 2I.

(k) Reference 536 has a value of 2.7×10^{-5}.

(l) Reference 536 has a value of 2.7×10^{-5}.

(m) Reference 536 has a value of 4×10^{-5}.

(n) A German study (665) has a value of 2.0×10^{-3}.

(o) Black (663) has a value of 1.5×10^{-4}.

(p) Black (663) has a value of 0.035, if 50 man-days are allotted to each injury.

(q) Black (663) has a value of 2×10^{-4}.

(r) Black (663) has a value of 0.02, if 50 man-days are allotted to each injury.

(s) Black (663) has a value of 3.5×10^{-4}.

(t) Black (663) has a value of 0.056, if 50 man-days are allotted to each injury.

(u) Black (663) has a value of $(2.5–17) \times 10^{-6}$.

(v) For offshore exploration in the United States, Ref. 666 suggests a value of 0.59.

crude estimate can be checked in another way. In WASH-1224 (299), the capital costs for natural gas plants were estimated to be about $42 million a year. This is about 0.82 that of coal, 0.89 that of oil, and 0.55 that of light-water nuclear.

It can be shown, using the methodology of Table A-2 [footnote (o)], Table B-2 [footnote (i)], and Table D-2 [footnote (o)], that the total man-days lost in construction of the electricity plants in Tables A-1, B-1, C-1, and D-1 (i.e., column 3 of each) is 1.1, 0.67, 0.54, and 1.4, respectively. If it is assumed, as a rough approximation, that man-days lost in capital construction are proportional to capital costs, the man-days lost in gas-fired electricity plants would be $0.82 \times 1.1 = 0.9$ (coal analogy), $0.89 \times 0.67 = 0.6$ (oil analogy), or $0.55 \times 1.4 = 0.8$ (nuclear analogy), i.e., 0.6–0.9. This is fairly close to the value of 0.54 deduced here for gas-fired plants.

A study by Black (663) uses a value of 0.21 and 4.9×10^{-5} accidental man-days lost and deaths, respectively, for construction. The total is then $0.21 + 4.9 \times 10^{-5} \times 6000 = 0.50$ man-days lost/MW-yr.

The highest occupational risk of Table C-2 comes from gathering and handling fuel. Overall, the number of man-days lost is low.

APPENDIX D. NUCLEAR

There are many ways in which nuclear power can be generated. Smith *et al.* (7) count at least 16 and consider four in greater detail: the light-water reactor with and without plutonium recycle, the liquid metal fast breeder reactor, and the high-temperature gas-cooled reactor.

For the foreseeable future, light-water reactors with or without plutonium recycle will be used throughout much of the world. In Canada, heavy-water-moderated reactors using natural uranium (the CANDU system) are the preferred method. Since the type of analysis in terms of materials and construction used in this book is not readily available for the CANDU system, only the light-water reactor is considered here. The amount of materials and labor involved in construction is probably about the same for both systems. Since enrichment of uranium is not required for the CANDU system, the data for gathering and handling fuels can be viewed as an upper bound. The CANDU system will, however, have additional risk due to heavy-water production. An energy accounting of heavy-water plants has been done by Winstanley and Emmett (129), but a risk accounting is not now available.

An area in which there may be significant differences between CANDU and light-water systems lies in risk from reactor accidents affecting the public. The latter system has been studied in detail (44), but the former has not had its risk evaluated in the same quantitative way. For purposes of this study, references relating to the Reactor Safety Study (44) have been used in estimating risk to the public from reactor accidents.

The light-water reactor system to which the present calculations apply has been described elsewhere (45). These reactors rely mainly on the isotope uranium-235 for their energy production. As used in a reactor after refining, enrichment, and fabrication, the uranium fuel weight is about 1/10,000 times that of coal for an equivalent amount of output energy. Bombarded by relatively slow moving neutrons, the uranium fissions (or splits), producing the thermal energy which converts water to steam, in turn driving electrical turbines.

Results of materials and labor calculations are shown in Table D-1. Nuclear construction man-hours are comparable to those of coal and oil. The operation and maintenance values in gathering fuel are considerably lower than coal, since a uranium miner mines about ten times more energy per hour than a coal miner, even though the weight of the coal is much greater (127). The man-hours required for both operation and construction are about the same for oil and nuclear, and about half that of coal.

Nuclear construction requires fewer metals than that for coal or oil, primarily because less is used for gathering and handling fuels. However, this is counteracted by the large amount of concrete used. This material forms a large part of the main reactor building.

Calculated risk for nuclear electricity generation from light-water reactors is shown in Table D-2. The data are obtained in part from a source not well known as a friend of nuclear power (123). An extra column has been added to take account of waste management. Although it is not known exactly what form or forms nuclear waste management will take in the future either in Canada or elsewhere, estimates were made of risk incurred by this aspect. Risk incurred in waste management constitutes only a small part of total risk. Even if waste management risk is miscalculated, this source probably will not sharply change the overall results.

The lifetime for a reactor has been assumed to be 30 years. However, one recent estimate (189) indicates a possible lifetime of about 80 years. This would lower the occupational risk in the third column of Table D-2 by a factor of over 2. As noted in footnote (o) to Table D-2, risk attributable to building nuclear facilities, as well as operating them, is included.

Although the data are not shown in Table D-2, Smith *et al.* (47) presented results on the calculated genetic risk for each of the four stages considered. In man-rem-equivalent per megawatt-year of electricity production, they are more than 0.035 for gathering and handling fuels, between 0.0025 and 0.042 for transportation, between 0.4 and 1.2 for electricity production, and between 0.04 and 0.24 for waste management. In addition, Smith *et al.* note that there is a total of 0.21 kg of fissile material per megawatt-year of electrical production, on average, in both storage and transit.

The unit "man-rem" is a measure of the number of people exposed to radiation times the average number of rems they receive. In turn, a rem is a measure of a dose of ionizing radiation which takes into account biological effects.

Excluded from consideration here are extremely long-term effects such as radon dose from mine or mill tailings. As noted in Section 3A, these effects are not evaluated for any energy system because their quantitative values are not known to any degree of certainty.

In Table D-2, the number of man-days lost per death is taken as 6000 in calculating the total number of man-days lost. This procedure is followed for other energy systems. However, Reissland and Harries (430) estimate that no deaths due to occupational radiation-induced cancer occur within five years of exposure, and that there is a linear distribution of these cancers in time between 5 and 50 years after exposure. It is not clear from this estimate how many man-days lost should be assigned to each radiation-induced death, but it is clear that it could be substantially less than 6000. For example, suppose one assumes (a) an average worker age of 41 years (see Appendix O), (b) the pessimistic estimate that all cancers will occur within $65 - (41 + 5) = 19$ years from onset, and (c) that the beginning of cancer removes a worker from the workplace immediately. Then the fraction of radiation-induced cancers can be visualized as a triangle rising from 0 at age $41 + 5 = 46$ to 100% (or 250 man-days lost per year) at age 65. The area under that triangle would be $0.5 \times (65 - 46) \times 250 = 2375$ man-days lost per death. While there is some justification for this reasoning, it is not employed in this report.

Where changes were made to the data of Table D-2 because more complete information was available, these were generally in the direction of increasing the risk. This "maximization" of risk was used only in the case of nuclear power, and not for other energy systems. In effect, the higher values of either Comar and Sagan or Smith *et al.* were used, unless other-

Table D-1. Light-Water Nuclear Electrical Production System Materials and Labor (per Megawatt-Year Net Electrical Output) (46)

Materials and Labor	*Gathering and Handling Fuel (b)*	*Transportation (a)*	*Electricity Production (c)*	*Waste Management (d)*	*Total*
Labor (man-hours)					
Construction					
Engineering	2.7	–	65	1.0	69
Administration	1.2	–	96	0.87	98
Unskilled labor	3.6	–	86	1.5	91
Skilled labor	13.8	–	357	4.6	375
Total	21.3	–	604 (r),(t)	8.0 (q)	633
Operation and maintenance (f)	205	–	250 (h)	31	486 (g),(s)
Materials (metric tons)					
Construction–metals (e)					
Structural	0.039	–	1.25	0.0008	1.29
Pipes	0.014	–	0.12	0.001	0.14
Major equipment	0.265	–	0.25	0.006	0.52
Other equipment	0.077	–	0.23	0.0003	0.30
Total	0.396 (j)	–	1.85 (l)	0.0008 (m)	2.25 (o)
Construction–concrete (e)	0.333 (k)	–	12.4 (l)	0.007 (n)	12.7 (i),(p)

(a) Included in "Gathering and Handling Fuel."

(b) Includes mining and milling. A 20-year lifetime is assumed for these facilities. The data pertain to light-water reactors (LWR) which require enriched uranium, as opposed to CANDU reactors using natural uranium. The enrichment through uranium hexafluoride comprises about 3% of the total labor, 18% of the total metal used in construction, and 2.5% of the total concrete used in construction. As a result, the occupational risk calculated in Table D-2 would be reduced by about 5–10% if the CANDU system were similar in every respect to the LWR except for uranium enrichment. To this must be added a risk from producing heavy water. Since a large part of the discussion of nuclear risk centers on public as opposed to occupational risk, this slight variation in the latter risk will not significantly affect conclusions.

(c) Includes a natural-draft evaporative cooling tower and assumes a thermal efficiency of 32%, lower than that of coal-fired plants.

(d) Includes spent-fuel shipping. Smith *et al.* (46) indicate that it also includes reprocessing. Note (g) of Smith *et al.* shows this is a correct assumption.

(e) See Ref. 10, as well as Refs. 49–52. Surface mining was assumed. If underground mining is specified, the manpower requirements for gathering fuels is about twice that of surface mining. The metals requirement is about seven times, and the concrete is about three times. The underground construction manpower would then be about 8% of total manpower, the metals would be about 5% of total metals, and the concrete about 0.5% of total concrete. We can conclude that the type of mining does not change construction risk significantly.

(f) From Refs. 14 and 52.

(g) Reference 238 has a value of 810 man-hours per megawatt-year.

(h) While, as noted in the text, the Canadian CANDU reactor is not directly comparable to the U.S. light-water system, the Pickering CANDU reactor employed about 690 people (excluding administration) in recent years. Since this reactor produced, on average, about 1700 megawatt-years per year, this was the equivalent of about 690 × 2000/1700 = 810 man-hours per megawatt-year per year.

(i) Assuming a load factor of 0.70 and a lifetime of 30 years, Ref. 280 implies 2.4 metric tons of cement, 1.9 of steel, 0.1 of copper, and 0.4 of lead per megawatt-year. This applies to British reactors only.

(j) Reference 470 indicates that the weight of iron or steel, in metric tons per MW-yr, is for mining 0.63–0.80; for milling, 0.05–0.07; for conversion, 0.007–0.008; for diffusion construction, 0.07; for diffusion equipment, 0.14 (using factor of 0.0082 relating diffusion plant size to energy output from section on diffusion plant construction); and for fuel fabrication, 0.02. The total is then 0.92-1.11. From the same reference, the weight of copper, in the same units, is for diffusion equipment, 0.006, and fuel fabrication, 0.0002, for a total of 0.006. Other materials required, in the same units, are zinc alloy (fuel fabrication), 0.042, and aluminum (diffusion equipment), 0.008.

(k) Reference 470 indicates that the weight of concrete, in metric tons per MW-yr, is for mining, 0.09–0.11; for conversion, 0.015–0.19; for diffusion construction, 0.62; and for fuel fabrication, 0.082. The total is then 0.81–0.83.

(l) Reference 471 indicates that the weight of steel, in metric tons per MW-yr, is 0.90. Copper, in the same units, is 0.066; other metals, 0.083, and concrete, 8.6.

(m) Does not include decommissioning, treated in Sec. v of this appendix. It is noted there that steel requirements are, in metric tons per MW-yr, 0.003. In the same units, steel for reprocessing is 0.012–0.043 (472) and waste repository, 0.0007–0.001. The total is then 0.012-0.044. Copper requirement for the waste repository is $(4-10) \times 10^{-6}$.

Table D-1 continued

(n) Reference 472 indicates that concrete requirements, in metric tons per MW-yr, are 0.11–0.36 for reprocessing; 0.006–0.12 for the waste repository. The total is then 0.12–0.37.

(o) References 470–472 indicate that the total metal requirements are about 2.0–2.2 metric tons per MW-yr.

(p) References 470–472 indicate that total concrete requirements are about 9.5–9.8 metric tons per MW-yr.

(q) Reference 473 indicates that $(1.6–3.5) \times 10^7$ man-hours would be required for a spent fuel waste repository. Smaller manpower would be required for a reprocessing waste repository. Reference 474 indicates that this repository is needed for the production of $(4.0–9.1) \times 10^6$ MW-yr of energy. The man-hours per MW-yr are then 16/4.0 to 35/9.1 = 4. Manpower to build reprocessing facilities is not included in these references.

(r) Reference 476 indicates construction time of 400 man-hours.

(s) Fagnani (485) estimates a total requirement of about 1440 for France.

(t) Mueller (484) states that 12.4 construction man-hours were required in 1977 in the United States to build 1 kW of capacity. With a load factor of 0.7 and a lifetime of 30 years, this is $12{,}400/0.7 \times 30 \times 1 = 600$ man-hours/MW-yr.

Table D-2. Light Water Nuclear Electrical Production System Risk (per Megawatt-Year Net Electrical Output) (23, 47, 123) (w)

Risks	Gathering and Handling Fuel	Emissions (sulfur oxides)	Transportation	Electricity Production	Waste Management (b), (ag)	Total (ad)
Occupational (r),(s),(af),(dl)						
Accidental	(ai)		(v)		(m)	(am)
Death (i)	$(1.2–5.7) \times 10^{-4}$ (a),(ao),(bk), (by),(cg),(du), (ee)		$(2.7–12) \times 10^{-6}$ (ch)	$(13–17) \times 10^{-6}$ (ap),(bf),(eb)	5×10^{-6} (aq),(bi)	$(3.0–7.6) \times 10^{-4}$ (o),(p),(ar),(br), (ci),(ea),(ef)
Injury (h)	$(3.4–16) \times 10^{-3}$ (f),(bb),(bl)		$(6.0–20) \times 10^{-5}$ (g)	1.7×10^{-3} (bf)	2.7×10^{-4} (bi)	$(16–29) \times 10^{-3}$ (o),(p),(bc), (bq),(bt),(ed)
Man-days lost	0.9–4.2 (bm),(bz)		$(1.9–8.2) \times 10^{-2}$	0.16–0.19 (bf)	0.04 (bi)	2.5–6.0 (o),(p),(bs)
Disease (l)	(ai)		(v)		(m)	(am)
Death (i)	$(2.2–60) \times 10^{-5}$ (c),(aa),(aj),(as),(bn), (ca),(cj),(cn), (co),(cv), (dv),(dx)		$(0.3–40) \times 10^{-7}$ (z),(au),(do)	13×10^{-5} (d),(at),(bg), (ck),(cp),(cq), (cw),(dq),(ec)	7.4×10^{-5} (n),(av),(bj), (cl),(cr),(cs), (dp),(dx)	$(2.4–8.1) \times 10^{-4}$ (p),(aw),(bp), (cm),(ct),(cu), (eg)
Disability (k)	$(1.1–1.6) \times 10^{-5}$ (bn),(cb)		$(0.6–80) \times 10^{-7}$	$(6–22) \times 10^{-5}$ (bg)	1.5×10^{-4} (bj)	$(2.2–3.9) \times 10^{-4}$ (p),(ba),(bp)
Man-days lost	0.1–3.6 (bn),(bq)		$(0.2–25) \times 10^{-3}$	0.8 (bg)	0.46 (bj)	1.6–4.9 (p),(bp)

Table D-2 continued. Light Water Nuclear Electrical Production System Risk (per Megawatt-Year Net Electrical Output) (23, 47, 123) (w)

Risks	*Gathering and Handling Fuel*	*Emissions (sulfur oxides)*	*Transportation*	*Electricity Production*	*Waste Management (b), (ag)*	*Total (ad)*
Public (dk)						
Accidental	(ai)		(v)		(m)	(am)
Death (i)	–		1.2×10^{-5} (dz),(eh)	–	–	1.2×10^{-5} (ax)
Injury	–		1.1×10^{-4} (j)	–	–	1.1×10^{-4}
Man-days lost	–		0.08	–	–	0.08
Disease (ah)	(ak)			(q),(t),(u),(y), (ab),(ac),(ae)	(b),(ag),(al),(bh), (ej)	(ad),(an)
Death (i)	31×10^{-6} (at),(bo),(bx) (cx),(dw)	$(3.6–10) \times 10^{-6}$	13×10^{-7} (au),(cy) (di),(dj),(dr)	$(3–23) \times 10^{-5}$ (x),(ay),(bu), (ca),(cc),(da), (dm),(dn), (dt)	$(0.5–25) \times 10^{-5}$ (az),(bw),(db), (dc),(dg),(dh), (ds),(dy)	$(7.0–52) \times 10^{-5}$ (bv),(cd),(dd), (de),(df),(ei)
Disability (k)	(bo)	0.020–0.063	44×10^{-8}	$(7–260) \times 10^{-7}$	2.5×10^{-4}	0.020–0.063 (bd),(ce)
Man-days lost	0.19 (bo)	0.13–0.41 (be)	7.8×10^{-3}	0.18–1.4 (e)	0.03–1.5	0.5–3.5 (cf)

(a) From Refs. 14 and 25–27.
(b) Waste management was not considered as a separate category by Comar and Sagan (23).
(c) From Refs. 14, 26, 43, and 48.
(d) Reference 38 has a value of 3.4×10^{-5}.
(e) From Refs. 14 and 25–27.
(f) Reference 47 has a value of $(0.2\text{–}8) \times 10^{-2}$.
(g) Implied 50 man-days lost per injury, on average.
(h) From Refs. 14, 25, and 27.
(i) 6000 man-days lost per death.
(j) Implied 50 man-days lost per injury.
(k) Implied 100 man-days lost per disability, except in the case of sulfur oxides emissions.
(l) Risk of cancer is taken to be 2×10^{-4} cases per man-rem for whole body exposure. Except for mining and accidents, only whole body routine exposures are considered. Half of cancers are assumed fatal, except for thyroid nodules, taken as 1% fatal. From Refs. 47 and 53–55.
(m) Reprocessing only for occupational risk (56).
(n) Reprocessing only.
(o) Includes approximately 1.5×10^{-4} deaths, 0.011 injuries, and 1.4 man-days lost due to acquisition of non-nuclear materials and construction of electrical generating plant. By using Tables M-2 and D-1, the number of man-days lost due to accidents is about $130 \times 633/100 \times 2000 = 0.41$; deaths are $0.042 \times 633/100 \times 2000 = 1.4 \times 10^{-4}$. [It can be deduced from Fig. 4 of Ref. 545 that the number of deaths for construction are $(0.9\text{–}13) \times 10^{-4}$, if the reactor lifetime is 30 years.] Table D-1 states that 2.3 metric tons of metal (assumed to be steel) are used, implying 3.8, 3.4, and 2.3 metric tons of iron ore, coal, and fabricated metal products, respectively. Using, for example, Table E-2, the number of man-days lost to accidents is then $(3.8/63) \times 0.018 + (3.4/56) \times 0.067 + (2.3/8.3) \times 0.055 = 0.17$; deaths are $(3.8/63) \times 8 \times 10^{-6} + (3.4/56) \times 4 \times 10^{-5} + (2.3/8.3) \times 5.6 \times 10^{-5} = 1.8 \times 10^{-5}$. The total number of man-days lost due to accidents is $0.41 + 0.17 = 0.48$; deaths are $1.4 + 0.2 = 1.6 \times 10^{-4}$. Allowing 6000 man-days lost per death, this is $1.6 \times 10^{-4} \times 6000 + 0.58 = 1.6$ man-days lost.
(p) Smith *et al.* (47) find the total number of occupational deaths between $(2.5\text{–}5.1) \times 10^{-4}$, injuries and disabilities between $(7.3\text{–}17.3) \times 10^{-3}$, and man-days lost between 1.9–3.9. Comar and Sagan (23) find (assuming a load factor of 0.7) the number of occupational deaths between $(1.4\text{–}12.3) \times 10^{-4}$, and the number of injuries and disabilities between $(6\text{–}19) \times 10^{-3}$. The number of man-days lost is not given. However, if one assumes that the number of man-days lost per injury and disability is approximately that which can be deduced from Smith *et al.* (50 man-days), the total number of man-days lost from the Comar and Sagan data is then deduced to be 1.0–8.4. The upper value is considerably higher than that of Smith *et al.*, although the lower value is not. In order to ensure that risk of light-water nuclear power is not inadvertently underestimated, the 1.6 man-days lost attributable to non-nuclear acquisition and construction is added to both the minimum and the maximum of the already maximized values of Refs. 23 and 47.
(q) Public cancer deaths per reactor year for two Norwegian sites are 0.01 (236). Assuming a load factor of 0.75 for a 1000–MW site, this is 1.3×10^{-5} deaths per megawatt-year. This value may be compared with the range of $(3\text{–}23) \times 10^{-5}$ used here. Early fatalities are

Table D-2 continued

about 1/100 of the Norwegian total. The uncertainties in the data of the present table do not apply here, since the Norwegian calculations were done for specific sites (237).

(r) Reference 239 lists total occupational deaths as $(0.8–5.3) \times 10^{-4}$; total occupational injuries as $(2.7–17) \times 10^{-3}$.

(s) Reference 27 lists total occupational deaths as 2.0×10^{-4} and total injuries as 7.2×10^{-3}. The total number of man-days lost can be deduced as about 1.6.

(t) A recent Finnish study (265) estimates the average number of cancer deaths from a 1000-MW station as 0.003–0.004. Assuming a load factor 0.7, this is $(4.3–5.7) \times 10^{-6}$ deaths per megawatt-year. If one allows a factor of 3 for consequences and 5 for probabilities either multiplying or dividing, this is $(0.03–8.6) \times 10^{-5}$ deaths.

(u) A recent Soviet study (266) notes that the estimated risk of fatality as a result of accidents in the use of 100 reactors for the population within 40 km of a station does not exceed 3×10^{-9} cases per year per person, and for a country like the U.S.S.R. is of the order of 10^{-10} cases per year per person. If one assumes a population of 10^8 within 40 km of 100 reactors in that country (probably high), this is 10^{-2} fatalities per year. If each of the 100 reactors has an average power of 700 MW (probably low, since the Soviet RBMK reactors are rated at 1500 MW), this estimate is then $10^{-2}/100 \times 700 = 1.4 \times 10^{-7}$ deaths per megawatt-year.

(v) Miettenin *et al.* (165) assume fatalities for transportation to be $(1.4–4) \times 10^{-6}$.

(w) Generally assuming a load factor of 0.7. The net capacity factor (a related concept) for baseloaded nuclear plants in the United States in 1978 was 0.68, close to the assumed value (298).

(x) From Ref. 23.

(y) Reference 427 uses a value of 2×10^{-5} deaths per unit energy.

(z) Erdmann *et al.* (427) have a value of 3×10^{-9}.

(aa) Erdmann *et al.* (427) use a value of 2×10^{-8} for routine mining and milling.

(ab) The Union of Concerned Scientists in the United States (431) estimates an average of 2.4 public deaths per reactor-year. Assuming this applies to a 1000-MW reactor with a 0.75 load factor, this is $2.4/1000 \times 0.75 = 3.4 \times 10^{-3}$ deaths per megawatt-year. The difference between this estimate and the maximum used in this report is then $(34 - 2.3) \times 10^{-4} = 32 \times 10^{-4}$ deaths. If each death corresponds to 6000 man-days lost, this is an increase in the maximum risk of $6000 \times 32 \times 10^{-4} = 19.2$ man-days. If nuclear electricity production in the western world was about 40,000 megawatt-years in 1978 (230), this assumption corresponds to $3.4 \times 10^{-3} \times 40{,}000 = 136$ deaths caused in 1978. Using this assumption would increase the maximum number of man-days lost from this energy source by about 130%. However, its relative rank would remain the same.

(ac) Caputo (432) estimates the effect of "public large accidents" as 0.003–10.8 man-days lost per megawatt-year.

(ad) Caputo (432) estimates the total number of man-days lost for nuclear as 2–15.6 per megawatt-year. He estimates the number of deaths per thousand megawatt-years as 0.3–2.3.

(ae) A report to the American Physical Society (437) quotes a radiation dose of 0.1 person-rem/MW-yr. Assuming 10,000 man-rems correspond to one death of 6000 man-days lost, the value suggests 0.06 man-days lost/MW-yr. The authors note that "this number must be regarded as quite uncertain, and we quote the order of magnitude for reference purposes only."

(af) A report to the American Physical Society (437) estimates a collective occupational dose from fuel cycle operations of 1–2 man-rems/MW-yr. The GESMO study (438) estimated 0.95, and Pochin (439) 4.2. The value of Ref. 437 is close to actual conditions in U.S. light-water reactors from 1973 to 1975, where 1.1–1.4 man-rem/MW-yr was recorded (440). Making the same assumptions as in footnote (ae), the APS, GESMO, and Pochin values correspond to 0.6–1.2, 0.6, and 2.5 man-days lost/MW-yr.

(ag) A report to the American Physical Society (437) implies that the world population dose commitment from waste management is negligible (presumably less than 0.01 man-rem/MW-yr, or 0.006 man-days lost/MW-yr).

(ah) The total world population radiation dose commitment from fuel cycle facilities, as estimated by a report to the American Physical Society, is 0.6 man-rem/MW-yr (437). The GESMO study (438) and Pochin (439) estimated 1.05 and 1.5, respectively. Using the assumptions of footnote (ae), the APS, GESMO, and Pochin studies correspond to 0.36, 0.7, and 0.9 man-days lost/MW-yr, respectively.

(ai) A United Nations report puts the total number of deaths from mining as (2.3–3.4) $\times 10^{-4}$ per MW-yr (444).

(aj) A United Nations report (444) estimates that radiation-related disease deaths are (3–10) $\times 10^{-5}$ per MW-yr.

(ak) A United Nations report (445) infers public cancer mortality resulting from radon and its daughters released after completion of mining of uranium ores as 4×10^{-5} per MW-yr. Assuming 6000 man-days lost per death and only a small contribution from disabilities, this is about 0.2 man-days lost per MW-yr. The inferred public cancer mortality in the same report (446) from radon releases after milling is 4×10^{-5} per MW-yr (for a 100-year dose commitment).

(al) The calculated cancer mortality per MW-yr from radioactive waste management, according to a United Nations report, is 6×10^{-10} (447).

(am) A United Nations report (448) estimates the occupational cancer mortality for the entire fuel cycle as 2×10^{-4} per MW-yr.

(an) A United Nations report (449) estimates the public cancer mortality from the entire fuel cycle, including mining, milling, enrichment, waste management, power generation, etc., is 1.2×10^{-4} per MW-yr.

(ao) Gotchy (458) has a value of 2.5×10^{-4}.

(ap) Gotchy (458) has a value of 1.25×10^{-5}.

(aq) Gotchy (458) indicates that the effects from this aspect are generally believed to be small.

(ar) Gotchy (458) has a value of 2.75×10^{-4}.

(as) Gotchy (458) has a value of 10×10^{-5}.

(at) Gotchy (458) has a value of 7.6×10^{-5}.

(au) Gotchy (458) indicates a value close to zero.

(av) Gotchy (458) indicates a value of 3.8×10^{-6}.

(aw) Gotchy (458) indicates a value of 1.8×10^{-4}.

(ax) Gotchy (458) indicates a value of 6.3×10^{-5}.

(ay) Gotchy (458) indicates a value of 1.4×10^{-5}.

(az) Smith *et al.* (47) have a value of (8–19) $\times 10^{-7}$.

(ba) Gotchy (459) has a value of 10×10^{-4}.

(bb) Gotchy (459) has a value of 13×10^{-3}.

(bc) Gotchy (459) has a value of 1.5×10^{-2}.

(bd) Gotchy (459) has a value of 0.001.

Table D-2 continued

(be) These values are based on emissions in making the steel used for construction. Reference 475 indicates that there are direct emissions of sulfur oxides from the following processes (in kg/MW-yr): mining, 6.9–9.1; milling, 69–91; diffusion plant, 23; reactor, 214; waste repository, 0.1–0.9. The total is then about 325. Air effluents from the reactor phase, which form the bulk of the total, are principally from oil or diesel support systems. In Sec. 21, it is assumed that 5.8–18 man-days lost are associated with one metric ton of sulfur oxides in the presence of particulates. This then implies that 0.325 (5.8–18) = 1.9–5.9 man-days are lost from these sources.

(bf) Reference 476 indicates deaths as 1.3×10^{-5}, injuries as 2.4×10^{-3}, and man-days lost as 0.18.

(bg) Reference 477 indicates average radiation exposure at U.S. light-water reactors from 1969 to 1976 as 1.3 man-rem/MW-yr. Making the assumptions of footnote (l), this corresponds to 1.3×10^{-4} deaths and an equal number of disabilities. In turn, this produces about $1.3 \times 10^{-4} \times 6000 + 1.3 \times 10^{-4} \times 100 = 0.8$ man-days lost.

(bh) From Ref. 478. Reference 47 has values of $(8\text{–}19) \times 10^{-7}$ deaths, $(1.6\text{–}3.8) \times 10^{-6}$ disabilities, and $5\text{–}12 \times 10^{-3}$ man-days lost. Reprocessing only. Control of carbon-14 emissions to 1% of normal releases, coupled with proposed radiation protection control on krypton-85, iodine-129, and plutonium would reduce these values to about 9.0×10^{-6} deaths, 9.0×10^{-6} disabilities, and 0.05 man-days lost. Assumptions of footnote (l) are used in estimating these last values.

(bi) From Refs. 473 and 479. Values from Ref. 47 are 2×10^{-7} deaths, 1.2×10^{-4} injuries, and 0.005 man-days lost. Spent fuel repository and reprocessing included. It is deduced from Ref. 474 that the energy from the fuel placed in this repository is $(1.3\text{–}3.2) \times 10^{6}$ MW-yr. A reprocessing waste repository would have a total number of man-days lost about one-third as great (473).

(bj) Reference 479 indicates a value of 0.034 man-rem/MW-yr. Using the assumptions of footnote (l), this corresponds to 3.4×10^{-6} deaths, 3.4×10^{-6} disabilities, and 0.02 man-days lost for reprocessing.

(bk) Total deaths from conversion, diffusion plant, and fuel fabrication are 4.4×10^{-6} in Ref. 480.

(bl) Total injuries from conversion, diffusion plant and fuel fabrication are 4.9×10^{-4} in Ref. 480.

(bm) Total man-days lost from conversion, diffusion plant, and fuel fabrication are 0.07 in Ref. 480.

(bn) Total man-rems from conversion, diffusion plant, and fuel fabrication are 0.015 (480). Using assumptions of footnote (l), this corresponds to 1.5×10^{-6} deaths, 1.5×10^{-6} disabilities, and 0.01 man-days lost.

(bo) Using the assumptions of footnote (l) the number of deaths from mining and milling, conversion, enrichment, and fuel fabrication are 5.5×10^{-2}, $(2.2\text{–}4.0) \times 10^{-8}$, $(5.2\text{–}5.9) \times 10^{-8}$ and $(4\text{–}10) \times 10^{-8}$, respectively (482). The total is then 5.5×10^{-2}. The respective number of disabilities are equal. The total man-days lost would be 330. The overwhelming proportion of the total values is due to long-term effects from mill tailings piles. As noted in the text, extremely long-term effects are not considered for any energy source. Covering the piles would reduce the total number of man-days lost to about 3 (482).

(bp) Fagnani (485) estimates a total of 2.9×10^{-4} effects from radioactivity for French conditions. He also estimates 1.7 man-days lost/MW-yr for this reason. Fagnani estimates 7.5×10^{-4} other nonradiological illnesses. Allowing 338 man-days per average illness (including deaths) (485), this is 0.25 man-days lost. The total is then 2.0 man-days lost.

(bq) Fagnani (485) estimates 0.045 for French conditions.

(br) Fagnani (485) estimates 1.8×10^{-4} for France.

(bs) Fagnani (485) estimates 2.3 man-days lost for France.

(bt) Fagnani (485) estimates 0.045 for France.

(bu) Fagnani (485) in France uses a total of 0.107 immediate, short-term and long-term deaths for an energy of 3.48×10^4 MW-yr, under accidental conditions. Thyroid cancers were expected to be 0.026 for the same energy. If they are 1% fatal (47), deaths from this source are 0.0026. Total deaths due to accidents are then $0.107 + 0.0026 = 0.11$. Deaths per unit energy are then $0.11/3.48 \times 10^4 = 3.2 \times 10^{-6}$.

(bv) Fagnani (485) estimates 1.8×10^{-5} deaths for France under normal operating conditions. Using the value of footnote (bu) for accidental conditions, the total number of deaths is then $1.8 + 0.32 = 2.1 \times 10^{-5}$.

(bw) Fagnani (485) estimates 9.1×10^{-5} deaths for France from reprocessing, assuming a dose commitment of 500 years. If this assumption is not made, a total of 1.4×10^{-7} deaths is indicated.

(bx) Travis (486) estimates a total of 17,000 man-rems to the lungs from radon-222 for the United States in 1978 due to active milling sites. Using the assumptions of footnote (l), this corresponds to about 1.7 fatal cancers. To put this value on the same basis as the other numbers in this table, it must be divided by the total energy that would be produced as a result of the materials that were milled. On the other hand, Erdmann (487) estimates a 50-year total dose commitment for milling, including lung, bone, whole body, etc. of 1.7 man-rem/MW-yr. Again using the assumptions of footnote (l), this would produce 1.7×10^{-4} deaths/MW-yr.

(by) Reference 47 has a value of $(1.2–2.7) \times 10^{-4}$.

(bz) Reference 47 has a value of 0.9–2.3.

(ca) Reference 47 has a value of $(3–6) \times 10^{-6}$.

(cb) Reference 47 has a value of $(0.7–1.2) \times 10^{-5}$.

(cc) Reference 47 has a value of $(0.03–1.3) \times 10^{-5}$.

(cd) WASH 1224 has a value of 1.3×10^{-4} (186).

(ce) WASH 1224 has a value of 8×10^{-3}(186).

(cf) WASH 1224 has a value of 1.2 (186).

(cg) Reference 536 has 1.3×10^{-4} for mining, milling, conversion, enrichment, and fabrication.

(ch) Reference 536 has 2.7×10^{-6}.

(ci) Reference 536 has 1.4×10^{-4}, excluding construction of generating plant.

(cj) Using the assumptions of footnote (l), the number of deaths would be about 1×10^{-5} (439).

(ck) Using the assumptions of footnote (l), the number of deaths would be about 5×10^{-7} (439).

(cl) Using the assumptions of footnote (l), the number of deaths would be about 3×10^{-6} (439), excluding reprocessing plants.

(cm) Using the assumptions of footnote (l), the number of deaths would be about 4.2×10^{-4} (439).

(cn) Using the assumptions of footnote (l), the number of deaths would be about 4×10^{-5} (438).

(co) Using the assumptions of footnote (l), the number of deaths would be about 2×10^{-5} (437).

(cp) Using the assumptions of footnote (l), the number of deaths would be negligible (437).

(cq) Using the assumptions of footnote (l), the number of deaths would be about 5.6×10^{-5} (438).

(cr) Using the assumptions of footnote (l), the number of deaths would be about 2.7×10^{-6} (438), including both reprocessing and waste management.

Table D-2 continued

(cs) Using the assumptions of footnote (l), the number of deaths would be about (0.08–6) × 10^{-5} (437), including both reprocessing and waste management.
(ct) Using the assumptions of footnote (l), the number of deaths would be about 1 × 10^{-4} (438).
(cu) Using the assumptions of footnote (l), the number of deaths would be about (1–2) × 10^{-4} (437).
(cv) Using the assumptions of footnote (l), the number of deaths would be about 1 × 10^{-5} (439).
(cw) Using the assumptions of footnote (l), the number of deaths would be about 4 × 10^{-8} (438).
(cy) Using the assumptions of footnote (l), the number of deaths would be less than 1.4 × 10^{-5} (437).
(cy) Using the assumptions of footnote (l), the number of deaths would be close to zero (437, 438).
(cz) Using the assumptions of footnote (l), the number of deaths would be about 1.5 × 10^{-5} (439), including both reactor accidents and normal operation.
(da) Using the assumptions of footnote (l), the number of deaths would be about 7.7 × 10^{-6} (438), including both reactor accidents and normal operation.
(db) Using the assumptions of footnote (l), the number of deaths would be about 1.4 × 10^{-4} for reprocessing (439).
(dc) Using the assumptions of footnote (l), the number of deaths would be about 3.5 × 10^{-5} for reprocessing (438).
(dd) Using the assumptions of footnote (l), the number of deaths would be about 1.5 × 10^{-4} (439).
(de) Using the assumptions of footnote (l), the number of deaths would be about 1.1 × 10^{-4} (438).
(df) Using the assumptions of footnote (l), the number of deaths would be about 6 × 10^{-5} (437).
(dg) Herrmann (542) notes that the exposure was 1000 man-rems in a reprocessing plant producing the equivalent of about 1000 MW-yr of energy in 1975–78. Using the assumptions of footnote (l), this is about 1000/1900 × 10,000 = 5 × 10^{-5} deaths/MW-yr for reprocessing. He notes that previous reprocessing plants (Windscale, 1972–73; West Valley, 1968–70) had higher values.
(dh) See Sec. iv of this appendix.
(di) Smith *et al.* (47) have a value of 2.2 × 10^{-7}.
(dj) Ericsson (544) estimates 3.5 × 10^{-3} cancers per year for transportation in Sweden. The proportion assumed fatal is not specified. Assuming all are fatal, and noting there were about 2700 MW-yr of nuclear energy produced in Sweden in 1978, this is 3.5 × 10^{-3}/2700 = 1.3 × 10^{-6} deaths/MW-yr.
(dk) Reference 545 has a value of 1 × 10^{-4} for gathering and handling fuels, transportation and noncatastrophic electricity production.
(dl) Reference 545 has a total number of deaths, for all phases except construction, of (0.9–5) × 10^{-4}.
(dm) Oudiz (552) uses a value of 4.2 × 10^{-6} for France, for normal operations, assuming the protection measures which are in place.
(dn) Heuser and Kotthoff (574), in reporting the results of the German Risk Study, estimated 10 late fatalities per year from a total of 25 plants. Neither the capacities of these plants nor their load factors was specified. Early fatalities form a very small fraction of late fatalities for all accidents considered, so they are ignored here. If each plant has a capacity of 1300 MW (608) and a load factor of 0.7, the fatalities per MW-yr are 10/25 × 1300 × 0.7 = 4.4 × 10^{-4}. This value is not used here because it related to primarily German condi-

tions. On the other hand, a numerical integration of Ref. 575 indicates only 4.6 late fatalities for the German Risk Study. This would yield $4.4 \times 10^{-4} \times 4.6/10 = 2.0 \times 10^{-4}$ deaths/MW-yr.

(do) A United Nations report (448) implies 0.6×10^{-7} deaths/MW-yr.
(dp) A United Nations report (448) implies 7×10^{-7} deaths/MW-yr for irradiated fuel storage and waste management.
(dq) A United Nations report (448) implies 9×10^{-5} deaths/MW-yr.
(dr) A United Nations report (449) implies 1.8×10^{-8} deaths/MW-yr.
(ds) A United Nations report (449) implies 2.8×10^{-9} deaths/MW-yr for irradiated fuel storage and waste management.
(dt) A United Nations report (449) implies 1.3×10^{-5} deaths/MW-yr, presumably for normal operation.
(du) Ferguson (640) estimates $(0–5) \times 10^{-4}$.
(dv) Ferguson estimates (640) $(0–5.7) \times 10^{-5}$ for lung disease and radiation.
(dw) Ferguson estimates $(0–1) \times 10^{-4}$ from radiation (640) for the next 100 years.
(dx) Ferguson (631) estimates a total of $(1–5) \times 10^{-4}$ for British manufacture and reprocessing of fuel.
(dy) Ferguson (643) estimates $(0.5–2) \times 10^{-5}$ for reprocessing only. He takes (644) the value for disposal underground to be $(0–4) \times 10^{-5}$.
(dz) Ferguson (645) estimates $(0–4) \times 10^{-7}$ for Britain.
(ea) Ferguson (651) estimates $(1–2) \times 10^{-4}$ for plant construction, similar to that of footnote (o).
(eb) Ferguson (651) estimates $(1–30) \times 10^{-5}$ for British conditions.
(ec) Ferguson (651) estimates $(2–8) \times 10^{-5}$ for British conditions.
(ed) Black (633) has a value of 11×10^{-3}. A German study (665) has a value of 3.8×10^{-3}.
(ee) Black (663) has a value of 9.4×10^{-5}. A German study (665) has a value of 9.1×10^{-5}.
(ef) Black (663) has a value of 1.7×10^{-4}. A German study (665) has a value of 2.3×10^{-4}.
(eg) A German study has a value of 1.9×10^{-4} (665). Black (663) has a value of $(4.2–4.9) \times 10^{-4}$.
(eh) Black (663) has a value of $(3–6) \times 10^{-6}$.
(ei) Black (663) has a value of $(2.7–3.2) \times 10^{-4}$ for all phases except reactor accidents. If the values for electricity production in this table are assumed to correspond to accidental risk, the present values are $(0.4–2.9) \times 10^{-4}$.
(ej) Black (663) assumes health effects of storage of nuclear wastes are close to zero.

wise specified. One entry where this rule was not followed was public disease-related deaths from electricity production, where the higher maximum values of Smith *et al.* (47) apparently refer to a "worst case" situation. As noted below in Section ii, "worst cases" are not generally used in this report.

The values for total deaths and man-days lost shown in Table D-2 are higher than some other estimates. For example, Hill (188) has a value between 3.0 and 5.8 deaths per 10,000 megawatt-years (excluding reprocessing). This compares with the values of between 6.2 and 21 deaths shown in Table D-2 for the same energy. In the same vein, Hill has a value for nonfatal injuries producing 0.44 man-days lost per megawatt year. This compares with values between 0.8 and 1.5 man-days lost per megawatt-year, derivable from Table D-2.

Electricity is required in the nuclear fuel cycle. One estimate (310) indicates that it is 3–4%, mostly in the fuel-enrichment aspect. In principle, the risk attributable to producing the electricity should be added to the other sources of risk. It is not clear which energy system or systems are generally used to produce this electricity. However, in Appendix J it is assumed that the electricity requirements to produce methanol can, in principle, be supplied by burning the methanol itself. If one makes the same assumption with respect to nuclear power, it implies that the risk calculated in Table D-2 should be multiplied by $1/(0.96–0.97) = 1.03–1.04$. For simplicity, this will not be done, but this point should be kept in mind if detailed calculations are done.

i. Public Risk from Accidents

Because of the controversy surrounding the risk from catastrophic nuclear reactor accidents, some discussion of it is in order. The average death rate per light-water reactor-year predicted by the Rasmussen report is 0.02 (132). If the reactor produces 1000 megawatts with a load factor of 0.7, this corresponds to 3×10^{-5} deaths per megawatt-year. This corresponds to the lower figure in column 4, row 10 of Table D-2.

The value of 0.02 deaths is made up of three factors multiplied together: the probability of meltdown (5×10^{-5} per reactor-year), the probability of breach of containment, and the average number of cancer fatalities per event. The probability of meltdown indicated here is an average estimated value. If a statistical confidence level were required, then a range, rather than a single number, would be used.

As an upper limit for the first factor, there have been at least 2000 power reactor years in the Western world without a meltdown. This excludes prototype and experimental reactors, in which accidents have occurred. No fatalities among the public have been linked to these accidents. A rough upper limit for meltdown probability may then be no more than 1/2000 = 0.0005 per reactor year. This implies that, if the other factors are held constant, an upper limit to death is 10 [= 0.0005/ 5×10^{-5})] × 0.02 per reactor year. In turn, this corresponds to about 0.0003 per megawatt-year, only slightly above the maximum value estimated for deaths due to disease in the public from electricity production from Table D-2.

The risk due to catastrophe is estimated to be extremely small in either case. The maximum risk from this source, measured in man-days lost, would have to increase by a factor of 7 to equal the risk from occupational losses. One would generally expect the risk to the public to be much lower than to workers.

The results of the U.S. risk study have been discussed by a recent German evaluation (460), which states, "the results of the German Risk Study more or less confirm those of the U.S. Reactor Safety Study . . . no fundamental difference can be seen in the risks which have been identified." A numerical integration of the median value curve in the German Risk Study (575) yields a value of 4.6 late fatalities for 25 plants. The early fatalities form a very small fraction of the late ones. If each plant had a 1300 MW (608) capacity and a load factor of 0.7, this would correspond to $4.6/25 \times 1300 \times 0.7 = 2.0 \times 10^{-4}$ deaths/MW-yr. This value is close to the upper limit used in Table D-2.

A similar approach is taken by Brooks and Hollander, using different numbers. They write (451):

> It is possible to combine the relative probabilities of events of varying severity of consequence estimated in WASH-1400 with reactor operating experience . . . If we assume 200 reactor years without a meltdown, then this upper limit is about one (1) latent cancer death per reactor year. The true statistical expectation almost certainly lies between this value and the estimate of 0.025 latent deaths per reactor year given in WASH-1400.

The value of 200 reactor years differs from that quoted above, possibly because different countries and time periods are evaluated. If one assumes that 1 reactor-year equals about 700 megawatt-years, then the Brooks estimate corresponds to about $1/700 = 1.4 \times 10^{-3}$ deaths per megawatt-year. This upper limit is about a factor of 6 higher than the upper limit in Table D-2. If one assumes, as do Smith *et al.* (47), that disabilities due to large accidents make up 3% of the total man-days lost, one can disregard disabilities as opposed to deaths. Under the Brooks analysis, the upper limit would be raised by $1.4 \times 10^{-3} - 23 \times 10^{-5} = 1.2 \times 10^{-3}$ deaths, yielding an increase of about 7.2 man-days lost per unit energy. This would raise the upper limit of this energy form by about 50%, but its relative rank would remain almost the same.

As mentioned in the Introduction, the estimates of public risk due to potential accidents in light-water reactors and in other stages of the nuclear cycle are only that—estimates. At present, there is not enough experience with nuclear systems to confirm these values.

ii. "Worst Cases"

In connection with these estimates, no claim is made here that the highest possible values in the scientific literature have been chosen. For example, a Ford Foundation study (224) indicates that under the most pessimistic assumptions, estimates of the public risk from potential reactor accidents could be much higher than those stated in the Rasmussen report. On the other hand, it can be deduced from Cohen (225) that assuming the "worst claim by critics of the nuclear industry," the relative rank of light-water nuclear power in this report does not change (although of course the absolute value of risk does rise).

These worst claims are not included in Table D-2 for at least two reasons. First, and more important, there is apparently no empirical evidence for some of these claims in terms of nuclear power. Secondly, a worst case can easily be thought of for many, if not all, energy systems. In order to treat all systems equally, these cases would have to be applied to each technology. It may well be that using these worst cases will produce risk for many energy systems which, if not infinite, will be very large. This worst case assumption would indicate that most systems had high and relatively uniform levels of risk, which would be little proof of anything in particular. Morris (127) states that the comparison between "upper limits" (or worst cases) and best estimates is "not a fair one."

A few examples of worst cases can be suggested. In terms of hydroelectricity, one might consider public risk only for Italian dams in 1963, where and when the largest number of deaths attributable to hydro production occurred. In terms of ocean thermal electricity, one might consider the lifetime of the system only in terms of the two relatively short-lived attempts which have taken place (see Appendix I). This would increase the material acquisition risk per unit energy substantially, since it varies inversely with the lifetime. In terms of coal occupational risk, one might consider only that associated with a shift in which a mining disaster occurred. In terms of oil occupational risk, one might consider it in terms of a month in which a large oil refinery fire took place. In terms of windpower, one might consider material acquisition and construction risk in terms of those projects which had relatively short lifetimes (see Appendix H for a discussion). If, as in the case of one windmill mentioned there, its lifetime was 1 year, and the estimated lifetime use in the calculations of Appendix H was 20 years, the material acquisition and construction risk per unit energy would be increased by a factor of 20.

A further distinction can be added. What is considered in the previous paragraph is primarily the worst historical case. This is not, for some systems, the worst conceivable case. For example, it has been claimed that if some dams in California failed, the loss of life would be much greater than has occurred in past dam failures. Generally speaking, the worst conceivable case would have risk greater than the worst historical case.

Similar examples could be suggested for many, if not all, of the systems considered here. The nature of the examples given indicate that it would not be reasonable to use the worst case for only one system and not for others. The approach of Smith *et al.* (7) may be a reasonable way of handling the problem of contrasting worst case situations (which, it must be emphasized, can exist) with what could be termed "reasonable" or "most likely" averages of risk. They present the worst case risk for fossil-fuel and nuclear technologies separately from their estimates for most likely risk. In this way, the two concepts are shown in isolation, for distinct consideration. As might be expected, there is not a linear correlation between the worst case risk they estimate and their "most likely" values. However, this approach is a way of reducing the confusion between these two concepts which sometimes exists.

iii. Three Mile Island

In view of the accident at the Three Mile Island reactor in the United States which occurred in early 1979, it is of some interest to calculate how the consequences of that event relate to the theoretical values of public risk noted above. It should be emphasized that what follows is a crude evaluation, based on a number of major assumptions. As a result, it can only be viewed as semiquantitative and not to be used as proof of anything. Part of the problem in this evaluation is that reports of the accident's effects were still incomplete at the time of writing. Later evidence may change some of the calculations.

A value of interest is the number of man-rems incurred by the public. A preliminary report indicates that it was around 3300 (231). [A later report (541) suggests a value of 1600–3300.] There is some reason to believe that this value is pessimistic, since no reduction in the estimate was made to account for (a) the shielding by buildings when people were inside, and (b) the people who left the area of the accident. The number of fatal cancers to be expected is given as 0.7 (central estimate), equal to the number of nonfatal cancers. The dose-effect relationship is based on a 1972 report of the U.S. Committee on the Biological Effects of Ionizing Radiation (53). A more recent evaluation (229) gives roughly similar values. It is unlikely that using the newer evaluation would change the health effects to be expected drastically.

Smith *et al.* (8) assume 100 man-days lost per nonfatal cancer. This would imply a total of about $100 \times 0.7 = 70$ man-days lost for nonfatal cancers.

Any cancers produced as a result of the Three Mile Island accident will be delayed for 10–30 years. That is, the number of man-days lost will probably be less than the 6000 assumed per death in the rest of this report. If one makes the pessimistic assumption that any cancers are not delayed, then the number of man-days lost to death as a result of this accident are $6000 \times 0.7 = 4200$. If one adds the man-days lost due to nonfatal cancers, the total is around $4200 + 70 = 4270$.

This quantity may be put on a historical or actuarial basis, as is done in the case of hydroelectricity. To do this, one may make another assumption: that the accident at Three Mile Island was the most serious event to date affecting public health in the history of civilian nuclear reactors. This assumption is the subject of some controversy.

The total amount of energy produced to the end of 1978 in reactors in the Western World was about 170,000 megawatt-years, according to *Nucleonics Week* (230). Dividing the estimated number of man-days lost due to the Three Mile Island accident by the total amount of nuclear energy produced, one obtains $4270/170{,}000 = 2.5 \times 10^{-2}$ man-days lost per megawatt-year. These values are approximately 1.8–14% the corresponding values in the fourth column, last line of Table D-2. The values noted in Table D-2 include both normal and accidental risk to the public from reactor operation. It then could be concluded that the public risk values in Table D-2 probably do not substantially underestimate this aspect of risk.

Again, it must be emphasized that this calculation is only tentative and preliminary. It is likely to be revised in the future.

iv. Waste Management

As mentioned in Section 3E, it is not clear from Reference 47, which estimates waste management risk, whether or not long-term as well as short-term risk is included. A study from the Environmental Protection Agency in the United States (227) indicates that the long-term risk per unit energy probably will be small with appropriate planning, although the exact value is apparently not specified. Another approach may be taken. In what follows, the cautions applicable to the calculations dealing with the Three Mile Island accident also hold, i.e., the results are tentative, preliminary, and highly approximate. They cannot be used as "proof" of anything, but merely suggest a line of reasoning.

According to a Norwegian description of an assessment of public risk associated with high-level waste disposal (243), the assumptions used are "extremely pessimistic," i.e., in the direction of maximizing risk. These assumptions, restated here, are: (a) groundwater starts dissolving waste 100 years after final disposal; (b) it dissolves 0.3% of the wastes each year: (c) the groundwater moves 0.3 km/day and runs 16 km before reaching the surface; (d) there is a population of 100,000 within 80 km of where the waste reaches the surface; (e) the contaminated water enters a river, with a flow of 2.5×10^6 m^3 a day, flowing through the populated area. The river is used for both drinking and irrigation. Reprocessed waste from 167,000 ton fuel was assumed to be stored in the facility. If 30 metric tons of fuel are assumed to produce about 750 megawatt-years (245), then the high-level reprocessed wastes correspond to about $167{,}000 \times 750/30 = 4.2 \times 10^6$ megawatt-years. Based on the above assumptions, the calculated maximum annual doses from this weight of fuel are 0.4 millirem to individuals or 30 man-rems to the population (243). If there are $(0.7\text{–}3.5) \times 10^{-4}$ excess deaths per man-rem (229), this corresponds to

$[(0.7–3.5) \times 10^{-4}]$ $(30)/4.2 \times 10^6 = (5–25) \times 10^{-10}$ deaths per megawatt-year.

Since this is an annual rate, it must be multiplied by the time over which the risk extends. This is a question of great uncertainty, although it is known that the radiotoxicity of waste will decrease with time. Some estimates indicate that the hazard of high-level waste from reprocessed fuel relative to natural uranium drops to about 1 in 10,000 years (244, 246). If one makes the assumption that the waste radiotoxicity remains constant for $10^4–10^5$ years, there are $5 \times 10^{-10} \times 10^4–25 \times 10^{-10} \times 10^5 = (0.5–25) \times 10^{-5}$ deaths per megawatt-year. This is 0.7–12% of the total number of deaths shown in Table D-2, when minima are compared to minima and maxima to maxima. Addition of this quantity to those of Table D-2 would not change the relative ranking of this energy form.

With respect to the length of time that wastes constitute a hazard, a Canadian report (374) indicates that irradiated fuel, with 99.5% of plutonium removed, has a toxicity similar to that of uranium ore containing 0.2% uranium after about 500 years. The estimate of 10,000 to 100,000 years used above is much higher than this lower bound.

Based on the assumptions of WASH-1297 (243), this calculation is apparently an example of a worst case. It is not implied here that there are no worse cases. As mentioned above, a worst case cannot be equated with a representative or average case. The calculation is presented here solely to illustrate the orders of magnitude involved.

Erdmann *et al.* (427), in another study, suggest a value of 5×10^{-18} latent cancers due to radiation effects for post-closure of a waste repository, per megawatt-year. This is based on 30-year individual dose rates integrated over one million years and a population of one million. As an upper bound, the number of latent cancers can be taken to equal the number of deaths. Even if the man-days lost per death were taken as 6000, the man-days lost per unit energy would be extemely small. This value is then much smaller than that of the Norwegian study noted above. Erdmann *et al.* also note a pre-closure number of radiation-induced latent cancers of 2×10^{-13} per unit energy. The sum of the two sources is still small.

A recent comprehensive study of waste disposal in Sweden (500) found that the dose for the most unfavorable 500-year period was 0.007 man-rem/MW-yr. This dose was to the entire world population. A dose of 0.1 man-rem/MW-yr was expected for the first 10,000 years of waste disposal. These values apply to conditions described as "unfavorable": for example, the encapsulation of the waste containers is penetrated after 1000 years; all waste glass bodies are completely exposed to the ground-water after another 5000 years; the glass is leached at a rate which leads to complete dissolution in 30,000 years, etc. Based on the information given, this could be an example of a worst case. It is suggested that the most probable case produces a dose about 30–100 times smaller. Assuming that 1 man-rem corresponds to 10^{-4} fatal cancers and 1 death corresponds to 6000 man-days lost, the most unfavorable 500-year period corresponds to $0.007 \times 10^{-4} \times 6000 = 0.004$ man-days lost/MW-yr; the first 10,000 years corresponds to $0.1 \times 10^{-4} \times 6000 = 0.06$ man-days lost/MW-yr. The most probable case would be substantially lower.

Another example of what is apparently a worst case estimates that the maximum radiological exposure from repository operation is 8.2×10^{-4} man-rem dose commitment per metric ton of heavy-metal waste (300). Assuming 38 metric tons of heavy-metal spent fuel and 7.6 metric tons of heavy-metal reprocessed waste per 1000 megawatt-years (315), this is about 0.046 metric tons of heavy metals per megawatt-year. Then a total of $8.2 \times 10^{-4} \times 0.046 = 3.8 \times 10^{-5}$ man-rems of exposure are produced per megawatt-year. If one sets 1×10^{-4} deaths per man-rem and 6000 man-days lost per death, this is $3.8 \times 10^{-5} \times 10^{-4} \times 6000 = 2.3 \times 10^{-5}$ man-days lost per megawatt-year. As noted above, the effects could persist for a long time. However, even taking this factor into account, the overall risk appears to be fairly small.

v. Decommissioning

There have been comparatively few studies dealing with the risk associated with decommissioning or dismantling any energy system, although there have been some studies of the economic cost of decommissioning nuclear facilities. One of the few recent studies is by Science Applications (461). It is realized that the discussion which follows is incomplete, but it should give some idea of the magnitude of some risks associated with decommissioning.

The energy requirements for decommissioning are as follows (in MW-yr of electricity required per MW-yr output from the system): mining, $(1.2–1.5) \times 10^{-7}$ (462); milling, $(5.4–6.5) \times 10^{-8}$ (463); fuel fabrication, 1.6×10^{-9} (465); reactors, 1.3×10^{-3} (466); and reprocessing, 2.5×10^{-6} (467). In the case of reprocessing, the value was found by using the ratio of reprocessing plant size to final output, from Reference 468. The total electricity requirement is then 1.3×10^{-3} MW-yr. Using the weighted risk of Appendix N, this corresponds to 0.0013 (316–948) = 0.4–1.2 man-days lost per MW-yr for the Canadian case, and 0.8–2.5 man-days lost per MW-yr for the U.S. case.

In addition to electrical requirements, decommissioning also requires fossil fuels, with which risk is associated. However, in References 462–467 neither the type of fossil fuel nor the mode of use is given. As a result, the associated risk cannot be estimated from these sources. However, based on the estimated heat values of fuel in References 462–467, it is probably small.

Materials are also employed in decommissioning. In Reference 461, the only material identified is steel. In kg/MW-yr, the weights are: fuel fabrication, 1.4 (465); and reprocessing, 2/(467). The total weight of steel is then 3.4 kg/MW-yr. In order to translate this into risk, the assumptions of Table E-2 with respect to coal and iron ore can be used. Then 5.0 kg of coal and 5.5 kg of iron ore are required to produce 3.4 kg of steel. Assuming that the risk attributable to this steel, coal, and iron ore is proportional by weight to that of Table E-2, the risk in man-days lost is $(3.4/39000)(0.24 + 0.31 + 0.06) = 5 \times 10^{-5}$. If one assumes that all this steel is made into fabricated metal products, one can further assume that the risk of producing steel metal products in decommissioning is proportional, by weight, to that of these products in Table E-2. The risk is then $(3.4/8300)\ 0.90 = 3.7 \times 10^{-4}$ man-days lost. The total risk from material acquisition is then $(3.7 + 0.5) \times 10^{-4} = 4.2 \times 10^{-4}$ man-days lost/MW-yr.

Steel also produces risk from the fuels used in smelting. In Section 2I, it is assumed that 1 metric ton of steel has 0.12–0.36 man-days lost associated with this mode. If 3.4 kg of steel are used in decommissioning per MW-yr, then the risk is $(3.4/1000)(0.12\text{–}0.36) = (4\text{–}12) \times 10^{-4}$ man-days lost.

Construction risk of decommissioning is shown only for the reactor phase (469). It is likely to be small for other phases. For the reactor, assuming a lifetime of 30 years, deaths are 1.4×10^{-6} and injuries 2.3×10^{-4}, both per MW-yr. Assuming 6000 man-days lost per death and 50 per injury, this is $1.4 \times 10^{-6} \times 6000 + 2.3 \times 10^{-4} \times 50 = 0.02$ man-days lost/MW-yr.

In addition to the above, there is radiological risk. One decommissioning in India (537) produced about 3000 man-rems. Using the assumptions of footnote (l) of Table D-2, this corresponds to $3000 \times 6000/10{,}000 = 1800$ man-days lost. Since Reference 537 does not indicate the total energy that was produced from this facility, the data cannot be placed on a risk per unit energy basis. As well, it is not clear if Indian conditions would be the same as North American ones.

In summary, the risk attributable to fuel cycle phases other than electricity and fossil-fuel requirements is about 0.02 man-days lost/MW-yr, primarily in construction risk in decommissioning reactors. Including the electricity requirements, total risk rises to 0.4–1.2 (in the same units) for the Canadian mix of electrical sources and 0.8–2.5 for the U.S. mix.

As noted in the main text, decommissioning risk is not included for any system discussed in this work. In the case of nuclear power, this type of risk might add about 20% to the total from other sources. This risk is apparently primarily attributable to the pollution effects of electricity used in the process, rather than from construction accidents or related activities.

APPENDIX E. SOLAR THERMAL ELECTRIC

Appendices E through J deal with "nonconventional" energy sources. Less is known about them, in terms of risk, than more conventional sources. These new sources generally rely on renewable energy, such as that of the sun and wind. Most do not generate direct pollution, such as that emitted from a smoke stack, but they are not risk-free.

Six ways to generate electricity using solar power have been considered as candidates for risk evaluation (6). Of these, only two were analyzed in detail in Reference 6. The first is the central receiver or solar thermal electric, considered here. The second, photovoltaic, is discussed in Appendix F.

The central receiver, or "power tower" concept, is described fully elsewhere (57, 58). Briefly, a tall tower stands in the center of a large field of heliostats or mirrors. The mirrors are moved automatically as the sun crosses the sky, to reflect light to the top of the tower. This tower holds a fluid which is heated to steam, which then runs conventional electrical generating turbines. The system is shown in Figure E-1.

i. Material Acquisition and Construction

The materials required for construction of a 1000-MW plant with a load factor of 0.7 and a baseload reliability are shown in Table E-1. It should

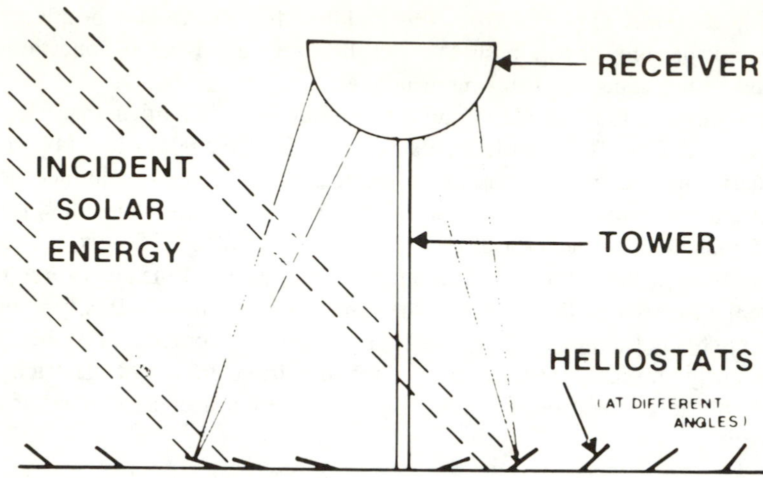

Figure E-1. *Solar central receiver. Incident solar energy strikes the mirrors around the central tower. The mirrors are automatically tilted so they focus on the tower. At its top is a cavity containing a fluid which is superheated and drives a turbine (not shown).*

not be assumed, however, that these values are the only ones that can be chosen. For example, Davidson (407) notes four designs for solar thermal systems. In kg/m^2 of collector, steel requirements range from 28 to 51; concrete from 124 to 274; and glass from 0 to 16. Material requirements thus vary widely.

The next step is to find the quantities of basic materials required to make the materials of Table E-1. Because silver plays a minor part in these risk calculations, it will be disregarded. The following assumptions are made: (a) 1.5 tons of coal are required for 1 ton of steel (16); (b) the iron ore grade is 60% (20); (c) 4 tons of bauxite are required for 1 ton of aluminum (20) (Reference 233 assumes 5.8 tons); and (d) sand is approximately the same density as concrete. Using these assumptions, the following materials, in 100,000 metric tons, are mined to yield the quantities of Table E-1, assuming that the "mechanical" category is primarily steel: $1.5 \times 8.2 = 12.3$ of coal; $8.2/0.6 = 13.7$ of iron ore; $4 \times 0.46 = 1.84$ of bauxite; and $36 + 1.3 = 37$ of sand. The amount of iron ore will depend on the ore grade. The values shown probably apply to high-grade hematite ores. Magnetic taconite, the principal ore presently used in the United

Table E-1. Materials Required for 1000-Megawatt Central Receiver (16), in 10^5 Metric Tons (a–g)

Receiver Parts	*Steel*	*Mech-anical*	*Alum-inum*	*Silver*	*Glass*	*Con-crete*
Heliostats	5.0	1.5	0.46	0.034	1.3	17
Tower and piping	1.1					17
Receiver and generator	0.3	0.3				2
	6.4	1.8	0.46	0.034	1.3	36

(a) The materials required will depend somewhat on the system design. For example, Ref. 81 also lists significant quantities of copper, plastic, and insulation in addition to the materials noted above.

(b) Original data were scaled to 1000-MW average power. Since other systems are considered as having a load factor of 0.7, original data were multiplied by this number.

(c) These values are similar to those of Caputo (361) to within about 30%.

(d) Risk due to the actual photovoltaic materials themselves (silicon, cadmium, gallium, etc.) is not evaluated here due to lack of data. This risk is discussed non-qualitatively by Stang (551).

(e) Grimmer (617) notes that a 400-kW (apparently peak, rated thermally) system in Georgia is expected to produce 2.12×10^9 Btu of heat annually. If its electrical generating efficiency is 28% (617), its annual electrical output would be $2.12 \times 10^9 \times 0.28 \times 1055/3600 \times 8760 \times 10^6 = 0.02$ MW-yr. Its steel, aluminum, glass, and concrete requirements are 64, 0.5, 3.5, and 98 metric tons, respectively. To make these values correspond to those above, they must be multiplied by $1000 \times 0.7/0.2 = 35{,}000$. This yields, in 10^5 metric tons, 22, 0.18, 1.2 and 34 for steel, aluminum, glass, and concrete, respectively.

(f) Lawrence (618) estimates requirements for steel, aluminum, silver, glass, and concrete, in 10^2 metric tons per MW (rated), as 4.5–6.4; 0.18–0.45; 0.00005–0.005; 0.45–0.93; and 13.6–22.7, respectively. There are also requirements for copper and chromium. To put these values on the same basis as those in the table, the load factor must be known. Reference 619 suggests typical load factors of 0.2–0.3 without storage. If the values in the table assume a load factor of 0.7, the quantities of Ref. 618 should be multiplied by $0.7 \times 1000/(0.2\text{–}0.3) = 2330\text{–}3500$, yielding, in 10^5 metric tons, 10.5–22; 0.42–1.6; 0.00012–0.0018; 1.05–3.3; and 32–79 of steel, aluminum, silver, glass, and concrete, respectively.

(g) Baron (619) expects that steel, aluminum, glass, and concrete requirements would be, in 10^5 metric tons, 10.5, 0.75, 1.5, and 38, respectively. These are scaled up from a 100-MW system. Values do not include 136 of concrete (in the same units) for a dust-reducing mat. He implied a load factor of 0.3, in contrast to the value of 0.7 assumed in this table. To make the two quantities comparable, the values of Ref. 619 are multiplied by $0.7/0.3 = 2.33$, yielding, in 10^5 metric tons, 24, 1.7, 3.5, and 88 of steel, aluminum, glass, and concrete, respectively.

States, has a lower iron content. To take proper account of present conditions, probably about 3 tons of iron ore would be required for 1 ton of steel (233). This would tend to increase the risk of those systems requiring large amounts of steel.

Aluminum is a material that requires 0.0025 MW-yr of electricity per metric ton (493). In principle, the risk attributable to that electricity production should be added to the risk attributable to other aspects of solar thermal electricity. This could be done, for example, by (a) adding the appropriate fraction of risk from the "average grid" of Appendix N, or (b) determining the actual sources of electricity used to produce aluminum, probably mostly hydroelectricity, and calculating the associated risk. Some of this risk is included in Section ii below.

Table M-2 in Appendix M shows that both death rates and man-days lost can vary strongly from one industry to the next. Roofing and sheet-metal work ranks high in man-days lost in the industries considered. This fact has implications for a number of technologies that require climbing, such as solar space heating and windmills.

The occupational and health data are used to determine the risk due to material acquisition and construction. To put the data on the same footing as that presented previously, the risk is divided by the estimated lifetime of the central receiver, 30 years. [On the other hand, a Japanese report (441) assumes a lifetime of 15–30 years.] Results are shown in Table E-2. Inclusion of the category "Fabricated Metal Products" implies that the metals used are fabricated to about the same level, in terms of stamping, forging, etc., as the average for typical metal uses. This is clearly an approximation.

The construction time may be compared with another estimate derivable from Reference 488. A 10-MW system with an average load factor of 0.4 (intermediate mode) was assumed. A total of 300 man-years of construction is required (488). Assuming a lifetime of 30 years, $300 \times 2000/10 \times 0.4 \times 30 = 5000$ man-hours/MW-yr are needed.

This apparently experimental value is higher than estimates that can be based on some prior theoretical studies. The Office of Technology Assessment (576) estimates an installation cost of \$10–\$30/m^2 for heliostats. The cost of foundations and columns was \$11.27/m^2. If half of the latter value was for labor and half for materials, the labor cost would be (10–30) + 11.27/2 = \$16–\$36/m^2. Assuming (a) an optical efficiency of 0.75 (576), (b) a thermal-to-electricity efficiency of 0.30, (c) \$8 per man-hour for labor (this value is somewhat lower than those assumed elsewhere in this report, since overhead and profit are excluded), (d) an average solar insolation of 250 W/m^2 for U.S. Southwest conditions, and (e) a 30-year plant lifetime, this is $(16\text{-}36) \times 10^6/8 \times 250 \times 0.75 \times 0.30 \times 30 = 1200\text{–}2700$ man-hours/MW-yr. These values are lower than that used in Table E-2, possibly because the labor required to construct the solar tower is not included.

By way of evaluating the accident rate used here (as opposed to the number of man-hours per unit energy), Reference 364 lists 1.8 man-days lost per unit energy in accidents for construction of a 1000-MW plant. A total of 5.4 deaths are estimated. In order to relate deaths to man-days lost per unit energy, the load factor must be known. This is not given explicitly in this reference. However, it is stated that the total man-days lost per unit energy for construction and material acquisition is 5.8, and that for illness and accident is 3.0, in the same units. The total number of deaths for construction and material acquisition is 7.7. If one death corresponds to 6000 man-days lost and the system life is 30 years, the load factor is $7.7 \times 6000/(5.8 - 3.0) \times 1000 \times 30 = 0.55$. Using this value, the man-days lost per unit energy in construction due to deaths is $5.4 \times 6000/1000 \times 0.55 \times 30 = 2.0$. The total number of man-days lost is then $2.0 + 1.8 = 3.8$. The plant construction time is 1900 man-hours (67), so the man-hours lost per man-hour worked is $3.8 \times 8/1900 = 0.016$. Table E-2 notes that 13.1 man-days are lost for 6250 hours of construction, so the man-hours lost per man-hour worked is $13.1 \times 8/6250 = 0.017$. This suggests that the risk per unit time in the present work is about the same as in Reference 364.

There is some evidence that risk of material acquisition may be underestimated. Ramsay (491) notes that 0.50, 0.25, and 0.36 metric tons of steel, aluminum, and glass, respectively, correspond in the United States to proportions 5×10^{-9}, 6×10^{-8}, and 2×10^{-7} of national production. Using the values of Table E-2, proportions for 1 MW-yr of solar thermal electric energy are $(39/0.50)\ 5 \times 10^{-9} = 3.9 \times 10^{-7}$, $(2.2/0.25)\ 6 \times 10^{-8} = 5.3 \times 10^{-7}$, and $(6.3/0.36)\ 2 \times 10^{-7} = 3.5 \times 10^{-6}$ for steel, aluminum, and glass, respectively. The total proportion for steel and aluminum is $(3.9 + 5.3) \times 10^{-7} = 9.2 \times 10^{-7}$. The number of annual injuries for steel and aluminum, and glass are 1×10^5 and 5×10^3, respectively. Then the total number of injuries attributable to the weights of steel, aluminum, and glass in Table E-2 is $9.2 \times 10^{-7} \times 1 \times 10^5 + 3.5 \times 10^{-6} \times 5 \times 10^3 = 0.11$. The numbers of deaths for steel and aluminum, and glass, are 235 and 70, respectively. The total number of deaths attributable is $9.2 \times 10^{-7} \times 235 + 3.5 \times 10^{-6} \times 70 = 4.7 \times 10^{-4}$. Allowing 6000 man-days lost per death and 50 per injury, this is $4.7 \times 10^{-4} \times 6000 + 0.11 \times 50 = 8.3$ man-days lost/MW-yr.

Table E-2. Material Acquisition and Construction Risk for Solar Thermal Plants (per Megawatt-Year Net Electrical Output) (16), (g),(m)

	Materials (Metric Tons) (l)	*Man-Hours (per Metric Ton)*	*Man-Hours (d)*	*Man-Days Lost, Accident*	*Man-Days Lost, Illness*	*Deaths (X 10^3)*	*Man-Days Lost, Total*
Material and Equipment Acquisition							
Iron ore mining (e)	63	0.44	27	0.018	0.0003	0.008	0.06
Bauxite (e)	8.7	0.50	4.4	–	–	–	–
Sandstone (c)	180	0.50	90	0.025	0.0006	0.016	0.12
Hard coal mining (a)	56	1.20	67	0.067	0.0014	0.040	0.31
Cement (k) (c)	174	0.84	146	0.049	0.0001	0.006	0.08
Flat glass (c)	6.3	38	240	0.007	0.0001	0.007	0.05
Steel	39	10	390	0.155	0.0039	0.014	0.24
Aluminum	2.2	12.4	270	0.011	0.0003	0.001	0.02
Fabricated metal products	8.3	149	1240	0.553	0.0167	0.056	0.90
Total	538		2480	0.885(n)	0.0234	0.148(p)	1.78
Construction (b),(h)							
Plumbing (i)	5.0	132	660	0.232	0.0079	0.076	0.70
Electrical	8.7	56	490	0.131	0.0029	0.043	0.39
Roofing and sheet metal	25.3	162	4100	3.53	0.0509	1.15	10.5
Concrete	173	0.33	57	0.024	0.0007	0.010	0.08
Miscellaneous contracting	8.3	114	950	0.613	0.0083	0.196	1.80

Total	6250(j)	4.53 (o)	0.071	1.48 (q)	13.1

(a) If soft coal mining is specified instead of hard coal mining, results are modified somewhat. The man-days lost due to accidents decrease by 6%, that due to illness by 4%, and deaths by 22%. This will affect conclusions only slightly.
(b) The total weight of materials used in construction is not the same as in material acquisition, due to the intermediate fabrication stages.
(c) Sandstone weight equals sum of flat glass and cement.
(d) This column is the product of the previous two.
(e) Iron ore and bauxite mining are combined into the category of metal mining for risk assessment.
(f) Death is assumed to equal 6000 man-days lost.
(g) Original data were for a 1000-MW plant with a load factor of 0.7. Since a lifetime of 30 years was assumed, data were divided by 30 × 1000 × 0.7.
(h) See footnote (e), Table M-1.
(i) Reference 16 indicates that 1170 man-hours per metric ton are required for installation of plumbing materials. However, Ref. 219 suggests a lower value. The latter reference states a cost of \$935–\$1320 per metric ton for piping construction. Allowing \$10 per man-hour for plumbing, the prevailing wage rate at the time of writing (219), this implies 94–132 man-hours per metric ton. The exact type of construction will determine the ratio of time material weight, but it is unclear why the values are so different. The higher value of Ref. 16 is derived in turn from a Jet Propulsion Laboratory study (19) which is apparently unavailable, so it cannot be verified. In the interest of giving solar thermal electric the benefit of the doubt, the value of 132 man-hours per metric ton will be assumed, the upper limit of the lower data set. This is comparable to other "productivity coefficients."
(j) This differs from the value of Ref. 67, which does not indicate how its value was derived.
(k) Includes cement, sand, gravel, etc.
(l) Reference 78 estimates a weight of 300–600 kg of nonconcrete final materials per installed kW, and 300–1000 kg concrete. Allowing 20% efficiency and a lifetime of 30 years, this is (300–600)/0.2 × 30 = 50–100 metric tons/MW-yr of nonconcrete final materials and (300–1000)/0.2 × 30 = 50–165 metric tons of concrete/MW-yr.
(m) Grimmer (617) estimates the steel, aluminum, glass, and concrete requirements for a facility that could produce 0.02 MW(e) annually are 64, 0.5, 3.5, and 98 metric tons, respectively. If the lifetime of the system is 30 years, each of these values is divided by 0.02 × 30 = 0.6, yielding 107, 0.8, 5.8, and 163 metric tons/MW-yr of steel, aluminum, glass, and concrete, respectively.
(n) Black (663) uses a value of 0.29.
(o) Black (663) uses a value of 4.6.
(p) Black (663) uses a value of 0.055.
(q) Black (663) uses a value of 1.07.

ii. Emissions

As noted in Section 2I, there are emissions from the processes used to produce materials such as steel. Only the effect of sulfur oxides is considered here, although in principle other pollutants can be evaluated. The methodology is to relate the sulfur oxides emissions to health effects. In turn, these emissions are compared to those from materials used in this system.

Only emissions from Canadian plants are considered in the evaluation. Appropriate emission factors could be used for other countries, if desired.

Solar thermal electric uses steel and aluminum. The health effect factors from Section 2I, per unit weight of each material, are multiplied by the weights from Table E-2. Man-days lost are 39(0.058–0.18) + 2.2 (0.3–0.9) = 3–9. Values differ from those of Caputo (432), primarily because Canadian emission factors were used.

iii. Operation and Maintenance

The risk produced from operation and maintenance of a plant will depend on the type of labor used. Reference 59 indicates an annual operation and maintenance force of 1200 for a 1000-MW capacity system producing 435 MW on average. This is then 1200/435 = 2.8 man-years/MW-yr. Reference 349 indicates that about 1.3 man-hours per kilowatt of capacity is required. Allowing 2000 man-hours per man-year, this is 1.3 × 1000/2000 = 0.7 man-years per megawatt of capacity. Assuming that the load factor is 0.7 and that there are no other losses, this translates to 0.7/0.7 = 1 man-year per megawatt-year. On the other hand, a later table in the same document (21) indicates between 10 and 37 man-hours per year per 39 m^2 are required for preventive and corrective maintenance. Assuming an overall efficiency of 20% and a solar insolation in the U.S. Southwest of 250 W/m^2, this is $(10–37)10^6/39 \times 250 \times 0.2 \times 2000$ = 2.6–9.5 man-years per megawatt-year. Giving the benefit of the doubt to solar thermal electric, the value of 1 man-year per megawatt-year will be chosen. This is close to a value of 0.95 man-years mentioned elsewhere (67).

Caputo (346) notes an annual economic cost for operation of a 100-MW (rated) plant at a load factor of 0.7 of $3.16 million, excluding mirror cleaning. Allowing $10 an hour for labor, and assuming that 75% of costs are for labor as opposed to parts (345), the number of man-years per unit energy for labor other than mirror cleaning would be $0.75 \times 3.16 \times 10^6 / 10 \times 100 \times 0.7 \times 2000 = 1.7$. This implies that there are labor requirements in addition to mirror cleaning. Another study (62) for plants with load factors of 0.44–0.57 estimates a maintenance cost of about $20 per installed kW. Making the same assumptions as above, the man-years per unit energy would be $0.75 \times 20 \times 1000/10 \times (0.44–0.57) = 1.3–1.7$. Schwing (443) states that 20 operating personnel will be required for a proposed system in California of 10-MW net electrical output. If the system operates in an intermediate mode (see Section iv) with a load factor of 0.4, then $20/0.4 \times 10 = 5$ man-years per megawatt-year will be required.

Herrera (349) indicates that supervisory personnel probably constitute 16% of one man-year, so the operating personnel can be assumed to be about 1–0.16 = 0.84 man-years per unit energy. Because most of the labor is consumed in cleaning mirrors, this labor, for purposes of risk calculation, can be assigned to the roofing and sheet-metal category (see Table 2). The result is $0.84 \times 172/100 = 1.45$ man-days lost per unit energy due to accident, 0.02 due to illness, and 0.00047 deaths. Allowing the usual 6000 man-days lost per death, the total is $1.45 + 0.02 + 6000 \times 0.00047 = 4.3$ man-days lost per unit energy.

In an independent study, Black (663) found 3.45 man-days lost per unit energy due to accidents, and 9.6×10^{-4} deaths. This would produce $3.45 + 9.6 \times 10^{-4} \times 6000 = 9.2$ man-days lost per unit energy, in contrast to the 4.3 used here.

By way of comparison of time lost per unit time worked, Smith *et al.* assume 407 man-hours of operation and maintenance per megawatt-year for coal with desulfurization (22). They estimate this produces 0.16–0.97 man-days lost, or $2000 \times 0.16/407–2000 \times 0.97/407 = 0.78–4.8$ man-days lost per man-year. This averages to $(0.78 + 4.8)/2 = 2.8$ man-days lost per man-year. Values for fluidized-bed coal-fired electricity are 407 man-hours, 0.61–1.16 man-days lost, 3.0–5.8 man-days lost per megawatt-year, and an average of 4.4 man-days lost per man-year (130); for coal with low-Btu gasification, 407 man-hours, 0.98 man-days lost, 4.8 man-days lost per man-year (426); fuel oil, 226 man-hours, 0.20–0.68 man-days lost, 1.8–6.0 man-days lost per man-year, and an average of 3.9 man-days lost per man-year (37, 33). The average of these four energy systems is then $(2.8 + 4.4 + 4.8 + 3.9)/4 = 4.0$ man-days lost per man-year of operation and maintenance time.

iv. Energy Back-up

If a baseload system is assumed, it is likely that energy back-up from conventional systems will be required. The exact amount of energy needed is still a matter of debate. Many studies show results in terms of back-up power (or capacity) which would be needed, rather than energy.

There appears to be some variability in the quantity of energy back-up specified, probably due to factors such as plant design, location, amount of storage, etc. Reckard (436) assumes that back-up energy for solar electric (probably thermal electric) forms 30% of total energy. Iannucci *et al.* (442) specifies 16% back-up energy for a particular system; it is not clear how representative this is. In a later comprehensive analysis based in part on Iannucci's work (570), the proportion of back-up ranged from 25% for Miami, Florida to 43% for Madison, Wisconsin. Reference 67 (footnote 23) indicates a range of 0% to 20% back-up energy. Footnote 15 of the same reference can be taken to imply a value of about 22%. On the other hand, Table 6-6 of the same document uses 10% back-up energy, presumably midway between 0 and 20%. In a detailed study, Manvi (65) says that with an annual load factor of 0.70, the required back-up power is 32% of the solar plant rated power. That is, a 1000-MW(e) solar installation requires 320 MW(e) back-up capacity. However, the energy required is a smaller fraction because the back-up will not be in use continually. The back-up energy is about 19% of the rated energy (65).

As confirmation of this, Figure 4-8a of Caputo (348) shows that (a) back-up energy divided by rated energy plus (b) solar load factor divided by conventional load factor equals 1.0. When (b) = 1.0, no back-up energy is required. However, this reference assumes a conventional load factor of 0.864. If the solar load factor is 0.7, then (b) = 0.7/0.864 = 0.81. Then (a) plus 0.81 = 1.0, and (a) = 0.19.

In order to calculate the extra risk due to back-up energy, and to give systems requiring back-up the benefit of the doubt, only 10%, rather than the more probable 19%, of the risk incurred by the weighted system, as noted in Table N-1, is added. See Appendix N for discussion concerning no back-up (or low-risk back-up).

As mentioned in Section 3A, it is assumed that the mode of energy production is baseload. There are, of course, other modes which are possible. One model (58) meets "intermediate" load requirements, but has a load factor of about 0.4. One may make a crude approximation of what the risk of this system might be, if it is constructed in the same way as that assumed here. If this is the case, one might dispense with the back-up of Table E-4, and multiply all other components by 0.7/0.4 = 1.75, assuming that the present load factor is 0.7. The non-back-up risk of Table E-4, for the present system, is 43–57 man-days lost per unit energy. Multiplied by 1.75, this is 75–100 man-days lost. The present total risk, including back-up, is 60–104 man-days lost. It must be emphasized that this is only a rough calculation.

v. Energy Storage

The amount and types of energy storage needed for solar plants are not clear. The National Academy of Sciences (66) states in a section on solar thermal electric stations:

> Base-load plants operating for as long as 7500 hours per year will require storage for 15–18 hours to supply power at standard reliability criteria (a loss of load probability of 1 day in 10 years).

As another example, Loftness (318) indicates that a baseload system requires 12 hours storage, an intermediate load system 6 hours, and a peaking system 3 hours. Iannucci (442) assumes 90–250 hours of storage are required for a Madison, Wisconsin location and 89–148 for a Miami, Florida location, if no back-up is assumed. Fairly realistic weather conditions are postulated. If back-up is assumed, these times decrease to 12–75 hours for Madison, and 13–50 for Miami. A subsequent comprehensive analysis (568), in part based on Iannucci's work, suggested 200–400 hours of storage with no back-up, and 15–30 hours with back-up. However, in these latter calculations the proportion of back-up was greater than assumed here, and between 0.25 and 0.43 of the total energy produced by the system (see Section iv). This implies that if 10% back-up were assumed, the hours of storage required would be somewhere between 15 and 200 and 30 and 400. A value of 38 days (without back-up) is recommended for the United Kingdom (205). Daey Ouwens (78) assumes that storage is required, but does not specify a value.

Hall *et al.* (553) note:

> Pumped storage hydro is the only storage system which has had an appreciable application in the United States, and while some of these systems have storage as small as 6 hours, it is generally believed that 10 to 12 hours is a safer specification. The weighted

average storage capacity for existing pumped storage plants is 9.5 hours.

The degree of storage required has been variable in other studies. For example, McKoy (223) found that for a hypothetical solar system supplying the U.S. Atlantic States, required back-up capacity decreased only from 80% to 75% of solar plant penetration of the network as storage capacity increased from 3 to 9 hours. That is, there seems to be some insensitivity of back-up capacity to the storage time. Of course, the actual back-up energy used (as opposed to the capacity) may be less insensitive. As another example, Caputo (346, 347) specifies 9 hours storage for a load factor of 0.7 (corresponding to a load factor of 0.81 without maintenance). On the other hand, a storage time of 3 hours would produce a load factor of 0.3 (619).

Because of the many types of storage, there is no simple way of calculating risk due to these systems. Caputo (67) indicated that a system of 2.3 million metric tons of rock and 0.29 million metric tons of heat transfer oil would provide about 6 hours of storage capacity at 70% rated power for a 1000-MW(e) solar plant. However, it is unclear exactly how to convert this into risk data.

An indirect approach can be taken. Table 1(a) of Reference 16 indicates the risk due to an average production of 1000 MW(e), without storage. Reference 68 shows the same risk, but modified in two respects: it applies to a load factor of 0.54 and has 6 hours storage. If Table 1(a) of Reference 16 is multiplied by 0.54 and subtracted from the table of Reference 68, the risk incurred to produce 6 hours worth of storage is obtained. Results for this and 16.5 hours, the average recommended by Reference 66, are shown in the first eight lines of Table E-3.

It is still unclear whether to use the value of 16.5 hours suggested by the National Academy of Sciences study or 6 hours. In the interest of minimizing calculated risk from this system, the lower value was chosen, and the results are shown in the last four lines of Table E-3.

vi. Transportation

It is assumed, as mentioned in the Methodology, that the materials used in construction travel about the same distance as those used in the coal fuel cycle. Also, it is assumed that the risk they produce is proportional on a weight basis. However, see Appendix P for more discussion.

Table E-3. Risk from Storage for Solar Thermal Electric System (16, 66, 68)

Risks	*Material Acquisition*	*Construction*
6 Hours		
Man-days lost, accident	500–2800	$(1.7–6.7) \times 10^4$
Man-days lost, illness	0–200	690–1560
Deaths (b)	0.33–0.35	8.4 (d)
Total man-days lost	2500–5100	$(6.8–12) \times 10^4$
16.5 Hours (a)		
Man-days lost, accident	1380–7700	$(4.7–18) \times 10^4$
Man-days lost, illness	0–550	1900–4300
Deaths (b)	0.91–0.96	23
Total man-days lost	$(0.68–1.4) \times 10^4$	$(19–33) \times 10^4$
Per Megawatt-Year (c)		
Man-days lost, accident	0.03–0.13	0.81–3.2
Man-days lost, illness	0–0.01	0.03–0.07
Deaths (b)	$(1.6–1.7) \times 10^{-5}$	0.0004
Total man-days lost	0.12–0.24	3.2–5.7

(a) Storage time recommended in Ref. 66.
(b) Assuming 6000 man-days lost per death.
(c) Assuming 30-year lifetime, load factor of 0.7. Each of the first four lines is then divided by 30 × 1000 × 0.7 = 21,000.
(d) Misprint in original. Deaths assumed proportional to man-days lost due to accident.

What is the weight of material transported for solar thermal electric? Table E-2 shows that the nonsand and cement construction materials weigh 184 metric tons per unit energy. The approximate weight of all materials for the coal system is 3500 metric tons per unit energy, so the coal transportation system risk is multiplied by 184/3500 = 0.053. On the other hand, Myers and Vant-Hull (615) assume that cement and aggregate travel significant distances in their model and are assumed to travel by truck. As can be deduced from Appendix P, this mode of transport has a greater risk per ton-mile than rail. In addition, Myers and Vant-Hull assume that the distance traveled for glass and steel is 4–5 times that specified in Appendix P. As a result, the risk in the Myers–Vant-Hull model of transportation is probably higher than assumed here.

There is other evidence that risk estimated here is low. Ramsay (491) notes that 1.11 metric tons correspond to 8×10^{-8} of all trucking in the United States. Then 56 metric tons of finished materials (steel, glass, aluminum, etc.) correspond to $56 \times 8 \times 10^{-8}/1.11 = 4.1 \times 10^{-6}$. Total occupational truck and warehousing accidents and deaths were 8×10^4 and 380, respectively. Accidents and deaths attributable to the present weight are then $4.1 \times 10^{-6} \times 8 \times 10^4 = 0.33$ and $4.1 \times 10^{-6} \times 380 = 1.6 \times 10^{-3}$, respectively. Allowing 6000 man-days lost per death and 50 per injury, this is $0.33 \times 50 + 1.6 \times 10^{-3} \times 6000 = 26$ man-days lost/MW-yr. Risk due to transportation of raw materials would be extra.

vii. Other Considerations

The location of the proposed solar plant will govern how much solar energy is received. For example, a much bigger solar collector area is needed for a plant located in the Canadian North as compared to one in the Sahara Desert.

Caputo (71) indicates that the proposed plant would be located in the U.S. Southwest, around California or Arizona. What is the relationship of this to Canadian sites? Cockshutt (215) notes that about 90% of the Canadian population lives in areas where the annual total insolation is about 150 W/m^2. The corresponding value for the U.S. Southwest is about 250 W/m^2. This implies that the risk from some of the preceding parts of this appendix should be multiplied by $250/150 = 1.67$ to make them correspond to Canadian conditions.

All components of risk are multiplied by this factor with the exception of storage and back-up. These latter two are not related to the physical size of the system (which in turn depends on the insolation), whereas the other components are related, directly (as in the case of material acquisition) or indirectly (as in the case of transportation or maintenance). No attempt is made here to take account of such factors as snowfall, which would provide a complication in gathering energy.

viii. Summary

Results of the calculations are shown in Table E-4. The arrangement of risk tables for nonconventional sources is somewhat different from those of conventional sources, because the origin of risk is not always the same. This point is discussed in Section 2G. Because man-days lost (excluding deaths) is a common calculation for occupational accidents, it is included. However, this does not change the calculations in any way.

For a few entries, only total man-days lost were available. In order to keep Table E-4 as similar as possible to risk tables for conventional sources, the deaths and disabilities producing this quantity were inferred by an indirect process, explained in the footnotes.

It is of interest to note that Caputo (67) estimates a total of 14,400 man-hours are required per unit energy. This includes construction, operation and maintenance, and so forth. This compares with 2480 for material acquisition, 6250 for construction, and 2000 for operation and maintenance, for a total of 10,700 man-hours, assumed here. There are, in addition to this, labor requirements for storage and back-up which are not included in the latter total, although their risk is included in total risk. These omissions are probably small. For example, Caputo (67) estimates the labor requirements for the coal cycle he considers to be 2640 man-hours per unit energy. If 10% back-up were assumed, the labor attributable to back-up would be 0.10 × 2640 = 264 man-hours, for a total of 14,400 + 264 = 14,700. The general results indicate that labor requirements assumed here may be underestimated.

APPENDIX F. SOLAR PHOTOVOLTAIC

A photovoltaic power plant generates power by directly converting solar energy to electricity using photocells. Although the system to be considered is described elsewhere (72), a brief description will be given here. Individual photocells are arranged in modules, which are used in a switchable network to maintain a constant voltage output level. From each array in the collector field, there are transmission lines to the central power station. The storage is located there. A schematic diagram of the system is shown in Figure F-1.

Table E-4. Solar Thermal Electricity Risk (per Megawatt-Year Net Electrical Output) (b)

Risks	*Material Acquisition and Construction*	*Emissions (c)*	*Operation and Maintenance*	*Energy Back-up (g)*	*Energy Storage*	*Transportation (h)*	*Total*
Occupational (e)							
Accidental							
Death	2.70×10^{-3}	–	0.83×10^{-3}	$(1.4–3.1) \times 10^{-4}$	3.8×10^{-4}	$(1.4–4.5) \times 10^{-4}$	$(4.2–4.7) \times 10^{-3}$
Injury	0.18	–	0.048	0.022	0.02–0.07	$(1.2–4.3) \times 10^{-3}$	0.27–0.32
Man-days lost, excluding deaths (d)	9.0	–	2.4	1.1	1.0–3.5	0.06–0.22	13.6–16.2
Man-days lost, total	25.2	–	7.3	2.0–3.2	3.3–5.8	0.90–2.9	39–44
Disease							
Death	(a)	–	4.4×10^{-6} (f)	$(1.2–19) \times 10^{-6}$	$(4.4–11) \times 10^{-6}$ (f)	(h)	$(1.0–3.5) \times 10^{-5}$
Disability	0.0031	–	1.0×10^{-4} (f)	$(0.9–1.6) \times 10^{-4}$	$(11–27) \times 10^{-5}$ (f)	(h)	$(3.4–3.6) \times 10^{-3}$
Man-days lost	0.157	–	0.031	0.012–0.12	0.033–0.083	(h)	0.23–0.39

Public							
Accidental							
Death	–	–	–	$(8–13) \times 10^{-5}$	–	$(7–17) \times 10^{-5}$ (i)	$(1.5–3.0) \times 10^{-4}$
Injury	–	–	–	$(1.0–7.7) \times 10^{-4}$	–	1.4×10^{-4} (j)	$(2.4–9.1) \times 10^{-4}$
Man-days lost	–	–	–	0.48–0.81	–	0.45–1.0	0.93–1.8
Disease							
Deaths	–	$(1.4–4.0) \times 10^{-4}$	–	$(4.2–14) \times 10^{-4}$	–	–	$(5.6–18) \times 10^{-4}$
Disability	–	0.8–2.4	–	2.3–6.9	–	–	3.1–9.3
Man-days lost	–	5–15	–	15–43	–	–	20–58

(a) All occupational deaths in material acquisition and construction assumed to be accidental and not disease.
(b) All risk except storage and back-up multiplied by 1.67 to take account of Canadian insolation, as explained in Sec. vii.
(c) See footnote (t), Table A-2.
(d) This calculation is included because so many preliminary computations produce this result.
(e) Occupational risk differs from that of Caputo (364). Part of the difference is probably due to the somewhat different assumptions in the present report. Calculations leading to the values of Ref. 364 are not shown, so their origin is not clear.
(f) Since ratio of deaths to disabilities is not known, this ratio is taken from Ref. 22.
(g) See Appendix N for discussion assuming no back-up or low-risk back-up.
(h) See footnote (p), Table A-2.
(i) Black (663) has a value of $(3–8) \times 10^{-5}$.
(j) Black (663) has a value of 1.5×10^{-3}, if 50 man-days are allotted to each injury.

i. Material Acquisition and Construction

The methodology followed in this section is similar to that of Appendix E. One problem with calculating risk of photovoltaics lies in determining the ratio of peak to average power, since the data of Reference 16, used in Table F-1, concern a peak power of 1000 MW(e). Reference 75 states that the ratio of peak power to average overall power is about 4 to 1. The original materials data of Reference 16 should be multiplied by a factor of 4, since calculations in this report refer to average as opposed to peak power. Reference 545 notes that for a decentralized photovoltaic system the ratio is about 8. Caputo (79) indicates that the ratio is over 5. Williams (609) assumes a ratio of 5.

Material requirements are shown in Table F-1. It can be seen that both the quantities and types of materials are different from those of the solar thermal system listed in Table E-1. Reference 78 indicates that "well-en-

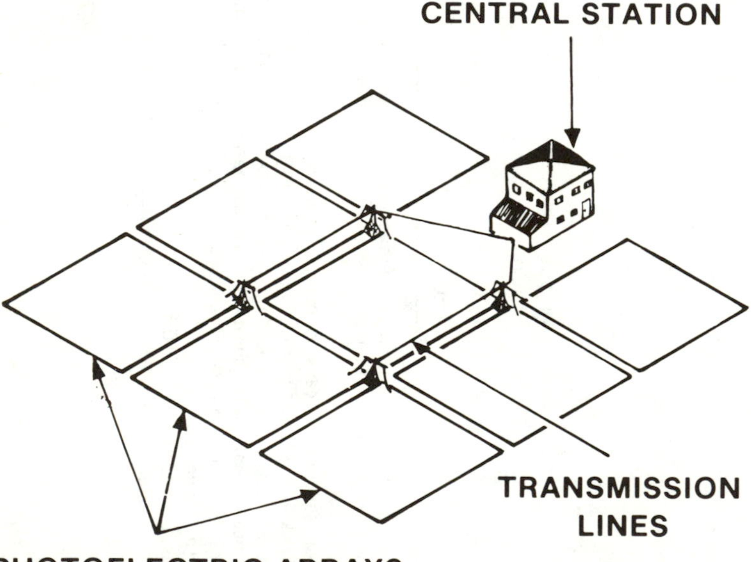

Figure F-1. *Solar photovoltaic system. Thousands of cells make up the system and convert solar energy directly to electricity. Bus bars as well as a central station are also shown.*

Table F-1. Materials Required for 1000 Megawatt (Average) Photovoltaic System, in 10^5 Metric Tons (16), (a),(c),(d)

Aluminum (b)	*Glass*	*Steel (c)*	*Cement*
16.3	0.87	5.4	1.76

(a) Original data were for 1000-MW peak power. Since peak power is about 4 times average power, original values were multiplied by this number.
(b) About one-fifth for frames and the rest for supports. While steel could also have been used, the original data specified aluminum.
(c) Another source (200) indicates a requirement for 42×10^5 metric tons of steel for the same average power, using all-steel structures.
(d) Values for aluminum, glass, and cement are about twice that of Caputo (361). This reference does not supply a justification for the values chosen.

capsulated photocells have a weight of 1100 kg/kW" installed, without tracking systems or foundations. Allowing a ratio of peak to average power of 4, this is $1100 \times 4 \times 10^3 = 44 \times 10^5$ metric tons/MW (average). This may be contrasted with the value of 24×10^5 in Table F-1. Tracking equipment and foundations would be extra.

There seems to be a difference in the cement requirements as noted in Table 8b and Figure 2b of Reference 16 (on which Table F-1 of this book is partially based), with the former indicating values 10 times that of the latter. The difference is not significant in terms of overall risk, since the material acquisition and construction risk of cement and concrete is apparently small. In addition, transportation risk is considered here only for nonsand and cement materials. In the interest of giving solar photovoltaic the benefit of the doubt, the lower value for cement will be assumed. Appropriate reduction will also take place in the category "nonmetal mining" as shown in Table F-2.

The photovoltaic cells are assumed to be silicon. Sources of risk incurred by producing this semiconductor material are detailed in Reference 74. Because some of this risk is not yet quantified, only a rough estimate can be made of direct occupational risk in Table F-2. Inclusion of the category "Fabricated metal products" implies that the metals used are fabricated to about the same level, in terms of stamping, forging, etc., as the average for typical metal uses. This is clearly an approximation.

The efficiency, or ratio of output to incoming solar radiation, plays a part in calculating material requirements. Reference 609 assumes 10% efficiency. By way of contrast, Watkins (616) found efficiencies of 7–10% for

Table F-2. Material Acquisition and Construction Risk for Photovoltaic Electricity (per Megawatt-Year Net Electrical Output) (16)

	Materials (metric tons)	*Man-Hours (per Metric Ton)*	*Man-Hours (c)*	*Man-Days Lost, Accident*	*Man-Days Lost, Illness*	*Deaths ($\times 10^3$)*	*Man-Days Lost, Total*
Material and Equipment Acquisition (b)(e)(n)							
Iron ore mining	30.7	0.44	13.3	0.007	0.0001	0.003	0.03
Bauxite mining	261	0.50	131	0.073	0.0013	0.031	0.26
Nonmetal mining	11.7	0.50	5.8	0.002	–	0.001	0.01
Steel	18.1(j)	10.0	181	0.073	0.0018	0.007	0.11
Hard coal mining (a)	26.7	1.2	32	0.031	0.0006	0.019	0.14
Fabricated metal products (f)	18.1		9,000–19,000	1.7–3.6	0.05–0.11	0.17–0.36	2.8–5.8
Aluminum	54.4(k)	12.4	675	0.222	0.0057	0.024	0.37
Semiconductors	2.9	8.6	25	0.002	0.0003	–	–
Flat glass	2.9(l)	38	110	0.003	–	0.004	0.03
Cement (g)	5.9(m)	0.84	5.0	0.002	–	–	–
Total	432		10,200–20,200 (o)	2.1–4.0 (p)	0.06–0.12	0.26–0.45 (i),(q)	3.8–6.8
Construction (d)							
Concrete	5.9	0.33	1.9	–	–	–	–
Electrical			670	0.18	0.004	0.060	0.54
Miscellaneous contracting			138	0.09	0.001	0.029	0.26

Total	810 (r)	0.27 (s)	0.005	0.09 (h),(t)	0.80

(a) As in Table E-1, the substitution of soft coal for hard coal would change the overall risk only slightly.
(b) As in Table E-1, both intermediate and final products are considered.
(c) Equals the product of the two previous columns.
(d) See footnote (e), Table M-1, for discussion.
(e) Values for final materials like aluminum and glass are similar to those of Caputo (67).
(f) See text for explanation. Risk values are 0.42 times that of fabricated metal products, per man-hour.
(g) Includes cement, sand, gravel, etc.
(h) It can be deduced from Fig. 4 of Ref. 545 that the number of deaths from construction are (0.36–0.52) $\times 10^{-3}$, if a lifetime of 30 years is assumed.
(i) It can be deduced from Fig. 4 of Ref. 545 that the number of deaths from direct and indirect manufacturing labor is (0.11–0.23) $\times 10^{-3}$, if a lifetime of 30 years is assumed.
(j) Moskowitz *et al.* (652) have a range of 29–59 for four systems, with an average of 40.
(k) Moskowitz *et al.* (652) have a range of 25–69 for four systems, with an average of 50.
(l) Moskowitz *et al.* (652) have a range of 0.005–16 for four systems, with an average of 4.
(m) Moskowitz *et al.* (652) have a range of 13–101 for four systems, with an average of 72.
(n) Moskowitz *et al.* (652) have a range of 3.3–5.4 of copper for four systems, with an average of 4.4. They have a range of 0.1–4.6 of plastic for these systems, with an average of 1.2.
(o) Moskowitz *et al.* (654) estimate 3900–5700 for four systems, with an average of 4800. However, this excludes fabrication.
(p) Moskowitz *et al.* (654) estimate 4.8–10 for four systems, with an average of 7.6. This excludes fabrication.
(q) Moskowitz *et al.* (654) estimate 0.33–0.45 for four systems, with an average of 0.40. This excludes fabrication.
(r) Moskowitz *et al.* (655) estimate 6400–22,200 for four systems, with an average of 17,000. This may include fabrication.
(s) Moskowitz *et al.* (655) estimate 2.2–12 for four systems, with an average of 7.7.
(t) Moskowitz *et al.* (655) estimate 0.42–4.5 for four systems, with an average of 2.5. This may include fabrication.

a testing facility in New Mexico. A system in Texas (589) had an efficiency of about 5% during 1978–1979. A recent conference (668) found average efficiencies of about 8%.

Pulfrey (415) estimates 22.5 kg of glass and 5.9 kg of aluminum are required for a module producing 247.5 watts (probably maximum). If the ratio of maximum output to average is about 4, the glass required for 1000-MW average output is $22.5 \times 4 \times 10^6/247 = 3.6 \times 10^5$ metric tons. The aluminum requirement is 0.95×10^5 metric tons. In addition to this, supports and machinery will be required, likely made of steel and aluminum. These latter requirements are unspecified in Reference 415.

Table F-1 indicates that considerable amounts of aluminum are apparently required. As noted in Appendix E, production of this metal takes considerable electricity. Risk attributable to this electricity is not included in the calculations of this appendix, although it should be in a comprehensive analysis.

Calculations of risk per unit energy are shown in Table F-2, again assuming a 30-year plant lifetime. This may be overoptimistic, since present arrays are guaranteed for only one year (206). Another source (77) indicates a lifetime of 5 years for CdS cells and 10 years for silicon and GaAs cells. Flaim *et al.* (560) state that government tests indicate 10–15 year lives for panels commercially available in 1978. They later (562) assumed a lifetime of 20 years for economic calculations. Williams (609) assumes a lifetime of 20 years for solar cells. It is clear that the lifetime assumed will be important in risk calculations.

Table F-2 uses the same procedure as Table E-2, except in fabricated metal products. It can be deduced from Reference 422 that the 10–20 man-hours of fabrication per square meter imply 9000–19,000 man-hours per megawatt-year. This value is used in Table F-2. There is some reason to believe that the risk per unit time in fabricating these facilities will be less than fabricating ordinary metal products. Heavy electrical equipment production has a total risk per unit time about 0.42 that of fabricated metal products for 1973–75 (419), but higher than that of light electrical equipment production. It is likely that fabrication of photovoltaic systems will combine elements of light electrical equipment for the electrical portion and fabricated metal products for frames and supports, so the risk assumed is that of heavy electrical equipment. It is assumed that the distribution of deaths, accidents, and disease within the "severity index" for heavy electrical equipment is the same as for fabricated metal products.

As an example of what fabrication time per unit energy might be in the future, Bickler (669) estimates the man-power needed to produce photovoltaic cells at a cost of \$2 per peak watt, much less than now. He states that 75 workers would produce a 140 watt (peak) panel every 3 minutes. If the ratio of peak to average power is 5 and the panels last for 30 years, this is $75 \times 5 \times 3 \times 10^6/140 \times 30 \times 60 = 4500$ man-hours/MW-yr.

Construction times are taken from Reference 67. The next step is to allocate the construction time among trades. Reference 16 indicates that only two trades will apparently be used, electrical and miscellaneous contracting, with 83% and 17% of the nonconcrete construction time, respectively. Results are shown in Table F-2. In summary, most labor is for fabrication, not construction.

Williams (609) finds that labor and overhead costs for building solar panels are 61¢/peak watt. If the ratio of peak to average is 5, labor is \$10 per hour, and the silicon cell lifetime is 20 years, this corresponds to $0.61 \times 5 \times 10^6/20 \times 10 = 15{,}300$ man-hours/MW-yr.

Flaim (561) estimates installation (taken to be equivalent to construction) costs of \$1–\$6 per peak watt, averaging about \$3, for present-day photovoltaic uses. Assuming that the ratio of peak to average power is about 4, that labor is \$10 an hour, and the lifetime is 30 years, this is a construction time of $(1\text{–}6) \times 4 \times 10^6/10 \times 30 = 13{,}300\text{–}80{,}000$ man-hours/MW-yr. Williams (609) estimates a cost of \$2.50 per average watt, with a lifetime of 20 years. Making the same assumption about the cost of labor, this is $2.5 \times 10^6/10 \times 20 = 12{,}500$ man-hours/MW-yr.

A report by the U.S. Office of Technology Assessment (576) estimates an installation cost of \$10–\$30/m^2 for air-cooled photovoltaics in a field. In addition, there was a \$0.90/m^2 cost for site preparation (assumed to be all labor) and \$5.02/m^2 for foundations. If it is assumed that half of the last value is for labor and half for materials, the total labor cost is about $(10\text{–}30) + 0.9 + 2.5 = \$13\text{–}\$33/m^2$. Assuming (a) a photovoltaic efficiency of 15% (576), (b) \$8 per man-hour for labor (this value is somewhat lower than those assumed elsewhere in this report, since overhead and profit are excluded), (c) an average solar insolation of 250 W/m^2 for U.S. Southwest conditions, and (d) a 30-year plant lifetime, this is $(13\text{–}33) \times 10^6/8 \times 250 \times 0.15 \times 30 = 2{,}400\text{–}3{,}700$ man-hours/MW-yr.

There is some evidence that risk of material acquisition may be underestimated (see Appendix E, Section i). Using the values of Ramsay (491), the steel, aluminum, and glass specified in Table F-2 correspond to $(18.1/0.50)\ 5 \times 10^{-9} = 1.8 \times 10^{-7}$, $(54.4/0.25)\ 6 \times 10^{-8} = 1.3 \times 10^{-5}$, and $(2.9/0.36) 2 \times 10^{-7} = 1.6 \times 10^{-6}$ of U.S. production, respectively. The total proportion for steel and aluminum is $(1.8 + 130) \times 10^{-7} = 1.3 \times 10^{-5}$. The total number of injuries attributable to the weight of steel, aluminum, and glass in Table F-2 is $1.3 \times 10^{-5} \times 1 \times 10^{5} + 1.6 \times 10^{-6} \times 5 \times 10^{3} = 1.3$. The total number of deaths is $1.3 \times 10^{-5} \times 235 + 1.6 \times 10^{-6} \times 70 = 3.2 \times 10^{-3}$. Allowing 6000 man-days lost per death and 50 per injury, this is $3.2 \times 10^{-3} \times 6000 + 1.3 \times 50 = 84$ man-days lost/MW-yr.

ii. Emissions

As in Appendix E, emissions are assumed to be proportional to the weight of steel and aluminum. From Section 2I, man-days lost per metric ton are 0.058–0.18 and 0.3–0.9, respectively. Using the values of Table F-2, this is 18 (0.058-0.18) + 54 (0.3-0.9) = 17-52 man-days lost. In addition, there are other environmental effects from silicon production, such as silicon dust and hydrochloric acid in the atmosphere (74). These are not considered here.

Moskowitz *et al.* (653) estimate the sulfur oxides emissions for four systems to range from 0.79 to 3.7 metric tons/MW-yr, with an average range of 1.1–2.9. From Section 2I, there are 6–18 man-days lost per metric ton, so there are (1.1-2.9) (6–18) = 6.6-52 man-days lost using these estimates.

iii Operation and Maintenance

There is no simple way of evaluating the manpower required for operation and maintenance. Reference 67 states that the manpower required for photovoltaics is the same as for solar thermal electric, and this approximation of 1 man-year/MW-yr will be used here. There is no clear indication whether the trades required for operation of this system would be different from that of solar thermal electric, so it will be assumed that they are similar for lack of evidence to the contrary. Both systems will require cleaning of surfaces. Solar photovoltaic will require electrical work, and solar thermal electric will probably require some mechanical work.

In the same document, Caputo (346) stated that the annual operation and maintenance cost of a 100-MW plant at a load factor of 0.7 was $1.36 million, excluding collector cleaning. Allowing $10 an hour for labor, and assuming that 75% of costs are for labor as opposed to parts (345), the number of man-years per unit energy for labor other than cleaning would be $0.75 \times 1.36 \times 10^{6}/10 \times 100 \times 0.7 \times 2000 = 0.73$. This implies that there are labor requirements, with associated risk, in addition to collector cleaning.

The data of Flaim (556) can be used to estimate operation and maintenance costs. For the three 100-MW (peak) systems he considers, the annual costs, in millions of 1975 dollars, are \$1.86, \$2.44, and \$2.10. The capacity factors (taken to be equivalent to load factors in this report) for the three systems are 0.28, 0.26, and 0.30, respectively. If labor costs in 1975 were \$8 an hour, and most costs were due to labor, the operation and maintenance costs for the three systems would be $1.86 \times 10^6/8 \times 2000 \times 100 \times 0.28 = 4.2$; $2.44 \times 10^6/8 \times 2000 \times 100 \times 0.26 = 5.9$; and $2.10 \times 10^6/8 \times 2000 \times 100 \times 0.30 = 4.4$ man-years/MW-yr.

In a recent study, Moskowitz *et al.* (655) estimate 0.87–6.3 man-years/MW-yr for operation of four systems, with an average of 4.5 man-years/MW-yr. This is much higher than used here.

iv. Energy Back-up

It is not clear whether the back-up requirements for solar photovoltaic will be different from that of solar thermal electric. The amount of back-up will depend on the distribution of solar insolation, the type of and the time lag in the storage system, and other factors. Reference 67 makes the assumption that the amount of back-up energy will be the same for solar photovoltaic and solar thermal electric, although no detailed justification is given. For lack of evidence to the contrary, the same assumption is made here, although Reckard (436) assumes a value of 30% in comparison to the value of 10% used here.

It is realized that other assumptions could be made. If photovoltaics were used in a non-baseload mode, the calculations would definitely change. The "weighted" system described in Appendix N is assumed as back-up. It can also be specified that no back-up is needed, or that low-risk back-up is used. The subject is discussed in Appendix N.

v. Energy Storage

The amount of storage required for photovoltaics is probably comparable to that for solar thermal electric. Lovins (577) uses a value of 10 hours, greater than the 6 hours assumed here. Caputo (346, 347) specifies 9 hours storage for a load factor of 0.7 (corresponding to a load factor without maintenance of 0.81). Reference 545 specifies 20 hours of storage for a decentralized photovoltaic system. Darlington *et al.* (521) assume 4–28 days storage. A British paper (205) assumes 38 days. Daey Ouwens (78) assumes storage is required, but does not specify a value.

What type of storage should be used? There are at least a score which have been proposed, and Reference 66 lists their features, advantages, and drawbacks.

Energy produced for photovoltaics will be in the form of electricity, whereas it is first in a superheated fluid in the case of solar thermal electricity. It seems logical to consider storage systems designed for electricity, rather than for heated fluids or gases. Although Herrera (69) specified a "redox battery" storage system for photovoltaics, this type of battery is only in the development stage. Without knowing the materials required and the time needed to acquire, fabricate, and construct such batteries and their links to the photovoltaic system, the risk involved cannot be calculated.

A likely candidate in energy storage for photovoltaics will be pumped storage. The risk attributable to this form will depend on factors such as the type of construction, type of terrain, and so on. For some areas with unsuitable terrain, such as parts of the U.S. Southwest, the risk per unit energy stored in pumped storage could be high. For those parts of Canada where hydroelectricity is common, the present hydro dams could, in principle, be used as pumped storage. This would incur comparatively little risk. However, it has proved difficult to determine a typical value for risk of pumped storage.

A recent method of energy storage is compressed air (89). While the risk associated with it is not yet calculable, its efficiency may be determined. Each kWh of output requires 0.8 kWh of electricity for air compression and a fuel-energy input of 5300 Btu for air reheating. Assuming that fuel can be burned with perfect efficiency and noting that 1055 W-s are in 1 Btu, the ratio of energy out to energy in is $1/(0.8 + (5300 \times 1.055/3600)) = 0.42$.

Some evidence suggests that storage systems more sophisticated than the rock and oil system may have higher risk. For example, in the steel storage system outlined by Turner (80), steel is heated either electrically or by liquids to store energy. Hall *et al.* (29) note that while this is a costly system, it is the most available storage at present. About 750 metric tons of steel are required to store the equivalent of 1 megawatt-electrical at 70% load factor for 6 hours. Over a 30-year lifetime, this is equivalent to 750/30 = 25 metric tons of steel per megawatt-year. The risk incurred in

Table F-3. Risk from Material and Equipment Acquisition for Steel Energy Storage System

Material and Equipment Acquisition	*Materials (Metric Tons)*	*Man-Hours (per Metric Ton)*	*Man-Hours*	*Man-Days Lost, Accidents*	*Deaths (X 10^3)*
Iron ore	42 (d)	0.44	18.5	0.07	0.001
Hard coal (a)	37 (c)	1.2	44	0.047	0.027
Steel	25	10	250	0.10	0.009
Fabricated metal products	25 (b)	149	3700	1.72	0.169
Total				1.87	0.206

(a) Results affected only slightly if soft coal is used.
(b) Fabricated metal products assumed to equal weight of steel. See Table F-2.
(c) 1.5 kg of coal per kg of steel.
(d) 1.66 kg of iron ore per kg of steel.

producing the steel alone, without consideration of construction, electrical, or liquid connections, or other additional equipment can then be computed. As shown in Table F-3, the days lost by deaths equal 6000 $\times$ 0.206 $\times$ 10^{-3} = 1.24, and the total number of man-days lost solely in material and equipment acquisition is 1.2 + 1.9 = 3.1.

Previous calculations have shown that risk from construction is greater than that for material acquisition. For example, Table E-2 indicates construction risk of about 7 times material acquisition. It can then be said that the total risk of the steel storage system should be of the order of 3.1 $\times$ 7 = 22 man-days lost per unit energy. This is higher than the total of 3.3–5.9 shown for the rock and oil system in Table E-3.

Another method of energy storage would be batteries. Loftness (303) notes the energy density of a lead–acid battery is about 22 W-h/kg. [Another report (483) assumes 26 W-h/kg.] if 1000 MW of power is produced for 6 hours and the energy is stored, the weight of batteries required is 6 $\times$ $10^6/22 \times 10^3$ = 2.7 $\times$ 10^5 metric tons. If batteries last about 5 years, and assuming a load factor of 0.7, a total of 0.7 $\times$ 1000 $\times$ 5 = 3500 megawatt-years of energy are produced over that period. This is 2.7 $\times$ 10^5 /3500 = 77 metric tons of battery per megawatt-year. Of this, about 92.5% is lead, 7% sulfuric acid, and 0.5% antimony (38). Reference 622 assumes 35 W-h/kg and a lifetime of 3.3 years, yielding 6 $\times$ $10^6/35 \times 3.3 \times 0.7 \times 1000$ = 74 metric tons/MW-yr. In addition, not all the energy that enters the battery can be extracted. McGeehin and Jensen (383) note that the ratio of energy out to energy in is about 0.6 to 0.75, as does Jensen (578). The required weight is then 77/(0.6–0.75) = 103–128 metric tons per megawatt-year.

Pulfrey (413) states that "it is necessary that the photovoltaic system include either an energy storage unit or a tie-in with a non-solar generating plant." He notes (414) that "there is no practical alternative" to lead–acid battery. He estimates the energy density of these batteries to be 11–12 Wh /kg, with a cycle life of 10 years. [References (545) and (562) also assume a battery lifetime of 10 years. On the other hand, Jensen (578) assumes 5–10 years.] This set of assumptions would require 6 $\times$ $10^6/(11\text{-}22)10^3 \times 10$ = 27–55 metric tons of batteries per megawatt-year for 6-hour storage.

The weight of batteries per unit energy is then of the order of the final materials in the rest of the solar photovoltaic system. It is logical to assume that the risk to gather the raw materials for these batteries, and to build and install them, would be large. In consequence, the system whose risk in outlined in Table E-3 is assumed to be a lower bound for storage systems. It is assumed that photovoltaics incur that risk.

vi. Transportation

For lack of more definite information, it is assumed that the transportation risk is proportional to the weight of nonsand and cement materials. The finished products used in construction will add to this risk. It is assumed that the risk in transporting the storage system is small, although this may also be an underestimate. See also Appendix P.

Table F-2 indicates that the weight of nonsand and cement materials is about 414 metric tons per unit energy. The corresponding value for coal is 3500 metric tons, so the coal transportation risk is multiplied by 414/3500 = 0.12.

There is some evidence that risk estimated here is low (see Appendix E, Section vi). Ramsay (491) notes that 1.11 metric tons corresponds to 6.4×10^{-3} trucking and warehousing accidents and 3.0×10^{-5} deaths. Using the present weight of finished products (steel, aluminum, glass, etc.) yields $(96/1.11)6.4 \times 10^{-3} = 1.1$ accidents and $(96/1.11)3.0 \times 10^{-5} = 2.6 \times 10^{-3}$ deaths. Allowing 6000 man-days lost per death and 50 per injury, this is $6000 \times 2.6 \times 10^{-3} + 1.1 \times 50 = 71$ man-days lost/MW-yr. Transportation of nonsand and cement raw materials by rail would produce another (330/3500) (15–47) = 1.4–4.4 man-days lost, for a total of 72–75.

vii. Other Considerations

Another assumption in these calculations is that the photovoltaic plant is in the U.S. Southwest. As described in Appendix E, all risks except that of back-up and storage are multiplied by 1.67 to make them correspond to insolation conditions throughout much of Canada.

viii. Summary

Results of the preceding calculations and assumptions are shown in Table F-4. Three of the columns–operation and maintenance, energy back-up, and energy storage–are the same as the corresponding ones of Table E-4.

Table F-4. Solar Photovoltaic Electricity Risk (per Megawatt-Year Net Electrical Output) (b),(g)

Risks	*Material Acquisition and Construction*	*Emissions (c)*	*Operation and Maintenance*	*Energy Back-up (i)*	*Energy Storage*	*Transportation (h)*	*Total*
Occupational							
Accidental							
Death	$(59–90) \times 10^{-5}$ (a),(e),(j)	–	8.3×10^{-4} (j)	$(1.4–3.1) \times 10^{-4}$	3.8×10^{-4}	$(3.2–10) \times 10^{-4}$	$(23–34) \times 10^{-4}$
Injury	0.08–0.15	–	0.048	0.022	0.02–0.07	$(2.6–9.6) \times 10^{-3}$	0.17–0.30
Man-days lost, excluding deaths (d)	4.0–7.4 (k)	–	2.4 (k)	1.1	1.0–3.5	0.13–0.48	8.6–15
Man-days lost, total	7.5–12.8	–	7.4	2.0–3.2	3.3–5.8	2.1–6.5	22–36
Disease							
Death	(a)	–	4×10^{-6} (f)	$(1.2–19) \times 10^{-6}$	$(4.4–11) \times 10^{-6}$	–	$(0.9–3.4) \times 10^{-5}$
Disability	0.002–0.004	–	1.0×10^{-4} (f)	$(9–16) \times 10^{-5}$	$(11–27) \times 10^{-5}$ (f)	–	$(2.2–4.6) \times 10^{-3}$
Man-days lost	0.09–0.19	–	0.029	0.012–0.12	0.032–0.080	–	0.16–0.42

Public							
Accidental							
Death	–	–	–	$(8–13) \times 10^{-5}$	–	$(16–38) \times 10^{-5}$	$(2.4–5.1) \times 10^{-4}$
Injury (c)	–	–	–	$(1.0–7.7) \times 10^{-4}$	–	3.2×10^{-4}	$(4.2–11) \times 10^{-4}$
Man-days lost, total	–	–	–	0.48–0.81	–	1.0–2.3	1.5–3.1
Disease							
Death	–	$(7.5–22) \times 10^{-4}$	–	$(4.2–14) \times 10^{-4}$	–	–	$(1.2–3.6) \times 10^{-3}$
Disability	–	4.5–14	–	2.3–6.9	–	–	6.8–21
Man-days lost	–	28–87	–	14.5–43	–	–	42–130

(a) All occupational deaths in material acquisition and construction assumed to be accidental and not disease.
(b) All risk except back-up and storage multiplied by 1.67 to take account of Canadian insolation conditions.
(c) See footnote (t), Table A-2.
(d) This calculation is included because so many preliminary calculations produce this result.
(e) It can be deduced from Fig. 4 of Ref. 545 that the value of deaths is $(7.9–12) \times 10^{-4}$, if a lifetime of 30 years is assumed and the climatic correction mentioned in Sec. vii is applied.
(f) Since ratio of deaths to disabilities is not known, the ratio is taken from Ref. 22.
(g) Values differ from those of Caputo (364). Part of the difference is probably due to the somewhat different assumptions in this report. Calculations leading to the values of Ref. 364 are not shown, so their justification is not clear.
(h) See footnote (p), Table A-2.
(i) See Appendix N for discussion assuming no back-up or low-risk back-up.
(j) Moskowitz *et al.* (656) estimate $(9.5–47) \times 10^{-4}$ for material acquisition, construction, and operation for four systems, with an average of 36×10^{-4}.
(k) Moskowitz *et al.* (656) estimate 3.6–26 for material acquisition, construction, and operation for four systems, with an average of 16.

APPENDIX G. SOLAR SPACE AND WATER HEATING

Solar energy can be used without converting it to electricity. Employing the sun's rays to warm water or other fluids, which in turn heat a building, has been a feature of a number of projects in North America and elsewhere. However, comparatively little discussion has been held on the risk involved. Part of this has been due to the wide variety of systems of solar space heating. In addition, it has been difficult to obtain accurate data on the materials required, construction times, and other information required for adequate risk analysis.

To simplify this problem, a well-known design was chosen. Estimates were then made of the thermal energy produced per unit weight of this system. All the previous systems considered produced electricity as their final product, while the present one produces thermal energy. It is assumed, to make data comparable, that the thermal energy corresponds to the electrical energy that would have been required to heat a building. There are, of course, losses in (a) transmitting and (b) distributing electricity. WASH-1224 (118) assumes an average transmission loss of 8%. Lapedes (304) notes that the ratio of U.S. electrical sales in 1975 to output was 0.904. The difference between output and sales can be taken to be transmission and distribution losses. This implies that in order to compare solar space heating to heating by electricity, the risk from solar heating should be divided by 1/0.904 = 1.11. This assumes that electrical heating has an efficiency of 1.0.

This procedure is followed in Table G-4. In order to make the reader's task as simple as possible, the rest of this appendix discusses the risk undivided by the factor of 1.11. The division is carried out only in Table G-4, and results are shown in applicable graphs.

As an example of Canadian conditions, Wardrop (351) notes that transmission losses for Vancouver of oil-fired electricity is 2.8%; coal-fired, 3.3%; and hydroelectricity, 11.5%. The last value is high apparently because hydro dams are fairly far from population centers. For Ontario, comparable values were nuclear, 2.8%; oil, 3.3%; coal, 4.1%; and hydroelectricity, 10.8% (352). Wardrop finds (354) that the average total transmission and distribution losses for Ontario are about 12%.

On the other hand, circulating losses in solar space heating can apparently be up to 10% (340). It is not clear from many of the references to be cited whether or not this is taken into account. If it is not, then the ratio of heat collected to area of collector would have to be reduced by up to 10%, and appropriate adjustments made in risk calculations.

Of course, buildings are usually heated by means other than solar space heating or electricity. Comparison could be made to heating oil, natural gas, or coal furnaces. This is not done here. While the risk associated with obtaining fuel can be found in Appendices A–C, there is comparatively little information on the air pollution from household furnaces. As noted in Appendices A and B, air pollution effects can form a substantial part of the total risk. Evaluation of the risk of building heating by coal, oil, or natural gas is left for the future.

Energy back-up and storage, as has been seen, contributed to the total risk incurred by solar thermal and photovoltaic production. Almost all designs have storage facilities, although the degree to which the storage evens out weather cycles to make the solar system comparable in reliability to conventional systems is rarely specified. Very few designs for solar space heating specify the amount of back-up which would be required. Requirements for back-up can be eliminated by making the on-site storage facilities large enough. This is what is done in a space heating system to be considered.

The previous systems all were assumed to have one plant with a 1000-MW rating. This is clearly not feasible for space heating. It has been assumed that the materials required for large-scale use of space heating are proportional to that required for small-scale installations, per unit energy output.

i. Material Acquisition and Construction

The space heating system considered is based on a 1.67-m^2 flat plate panel manufactured by the Honeywell Corporation. Its specifications in terms of material requirements are given in Reference 82.

a. Energy per Unit Area

The energy gathered per unit area is a function of many variables such as climate, size of storage, tilt and type of collector, etc. It would be impossible to consider and evaluate every possible factor, so a value of annual energy per unit area was chosen which is probably optimistic: 308 kWh/m^2 (214). This collector has single glazed low absorption glass with selective surface and has a large seasonal storage tank. If an average Canadian insolation (total) is about 150 W/m^2, this corresponds to an average efficiency of 23% (215). Lof (402) uses the "rule of thumb" of 100,000 Btu/ft^2, or 318 kWh/m^2 per year. This is also used by McNamara (455) for a Chicago location.

A system in Portland, Oregon was expected to collect 237 kWh/m^2 per year (319). Lenchek (342) found that an installation in Fort Collins, Colorado was expected to collect 26,200 kWh per year over an area of 71 m^2, or 367 kWh/m^2. This is larger than the value assumed here. However, the total solar insolation at this site is higher than at typical Canadian locations. Loftness (343) notes that the insolation in the Fort Collins area is about 1550 Btu/ft^2/day, or $1550 \times 1055/3600 \times 24 \times 0.093 = 204$ W/m^2 per year. The ratio of energy collected to the energy which falls on the collector is then $367{,}000/204 \times 365 \times 24 = 0.21$, close to that which is assumed here.

Some data are available from previous structures. In the references consulted, it is not always clear whether the values are theoretical or experimental. The type of collector is also not always described; most are probably flat-plate. Anderson (386) describes a house which collects 159 kWh/m^2 per year. He notes that the Lof house of 1945 produced 137 (in the same units); the 1960 Phoenix house produced 69; and the 1959–60 MIT house produced 171. Sabady (387) notes that a house in France produced 222; the Zero Energy House in Denmark produced 215 (389). In perhaps a later measurement, it was stated (456) that the Zero Energy House produced 174, of which 109 were "useful." Szokolay (391) indicates that the 1959 MIT house produced 185–232, somewhat different from the value mentioned in Reference 386. The Boulder house of 1945 had a solar contribution of 145 (392); the 1956–57 Denver house had 270 (393); the Nagoya, Japan house had a value of 152 (395); the Tokyo house had a value of 101 (396); the Milton Keynes house had a value of 186 (398); the Anglesey house had a value of 273 (399). The only structure listed in this reference which had a value higher than that chosen here was the Albuquerque office of 1956, which had a value of 421 (394). However, this was in an area with high solar insolation of about 2150, yielding an efficiency of 421/2150 = 20%. This is lower than the value of 23% chosen here. A Japanese collector (403) was expected to gather 189. Bush (404), in a house probably near Berkeley, California, expected to collect 318, but actually collected 145. McConnell *et al.* (408) estimate 213 for a house 170 km north of Montreal, with a vertical screen air collector. A house in Sweden collected 144 in 1975–76 (457). A well-instrumented house in Colorado (503) produced 50,600 megajoules of heat during 1976–77 from a 64-m^2 absorber area, or 50,600/64 × 3.6 = 222. A building in Spring Valley, California, supplied 257 (net) over one year recently (506). Shaw (522), under New Zealand conditions, collected 217 kWh over 0.74 m^2 in a year, for 293. Kouba *et al.* (547) collected 167 in a house in Tulsa, Oklahoma, over 1977–78. Nash and Cunningham (580), in a survey of nine U.S. hot water systems during 1978–79, found values ranging from 1.3 to 145, with an average of 62. Taylor (581) had a value of 337 for the first six months of 1979 for a water heater in New Jersey. Tipton and Rockefeller (582) measured 47 for a hot water heater in Oklahoma in August 1979. Aungst, in a comprehensive study of 20 water heaters in Pennsylvania in 1978–79 (585), had values ranging from 22 to 464, with an average of 194. Skidmore and Smok (586), in a study of five large U.S. systems during 1978–79, determined a range of 33–194, with an average of 111. Karaki *et al.* (587) found an air-heating system in Colorado utilized 251 during 1978–79 (assuming a net collector area of 51 m^2). Murray (588) found 180 removed from storage for an air-heating system in New Mexico. Cain (590) found 275 in an agricultural installation in Maryland. Nemetz (593) obtained a value of 303 for a Hawaiian hot water heater in 1978. A Colorado system (595) had a value of 294 for hot water and space heating during 1977–78. An installation in New Mexico (596) had values of 195 and 185 for delivered energy in 1976–77 and 1977–78, respectively. Leverenz and Joncich (597) obtained a value of 257 for 1977–78 for an Illinois system. A Colorado system (598) provided 144 in 1977 and 173 in 1978. Bedinger and Bailey (599) had a value of 142 during 1977–78 for a Tennessee installation. A large building in Virginia, using solar heating and cooling, had a value of 297 (600). A Texas installation (602) had a value of 241 in 1978. A New Mexico installation had an average value of 221 during March–April 1978 (603). Ward (604) obtained a value of 41 for "controlled" energy for a Colorado house during 1977–78. Boleyn (606) found 79 for one house in Oregon during the same period. A well-instrumented German system (613), using evacuated tubes and a seasonal storage tank, utilized 134 during 1976–77. A French theoretical study (356) expected 148 annually at Le Havre, and between 278 and 327 at Avignon. A St. Louis system (625) was expected to produce 194. An Arizona house praised as an "outstanding example" (658) generated 280 from storage during 1978–79 (657). A British house collected about 58 of useful energy during 1975–78 (660). Tabor notes that at a recent Canadian solar meeting the annual useful yield was said to be about 190 (661). Finally, since some of the above values may be theoretical, it is of interest to note that a value of 233 was found experimentally in what is probably the largest solar heating demonstration ever conducted in North America, with 100 units of varying types (212). Of course, it is possible that in the future the output per unit area may improve.

One possible reason why there are different estimated values of energy gathered per unit area noted in the literature, some as high as 700 kWh/m^2 (122), may be due to failure to differentiate between energy which can be gathered and that which can be used. Only the latter is considered in the present calculations. For example, Loftness (322) notes that for a location in Columbus, Ohio, the incoming radiation produced the heat equivalent of 17.9 kilowatt-years annually. He assumed that the efficiency of energy capture was 40%, higher than the 23% assumed above. However, not all of the $0.4 \times 17.9 = 7.2$ kilowatt-years assumed to be collected could be used. In the summer, heating demand was zero. (Cooling demand was significant, but cooling is not considered in this appendix.) As a result, only 3.9 kilowatt-years were used for space and water heating. The effective efficiency was then $3.9/17.9 = 22\%$. This is close to what is assumed here.

This point is also made by Mears (555), who found that the ratio of collected energy to incident energy for four different solar collectors in the United States ranged from 27% to 57%, with an average of 40%. Carscallen and Sibbitt (579) notes that for six Canadian working systems, the ratio of delivered to collected solar energy ranged from 15% to 56%, with an average of 31%. The ratio of measured solar energy used to incident energy ranged in Ref. 555 from 3.3% to 38%, with an average of 23%, close to what is assumed here. Nash and Cunningham (580), in a survey of nine U.S. water heating systems during the 1978–79 season, found total efficiencies ranging from 1.4% to 24%, with an average of 14%. Taylor (581) determined an average efficiency of 25% for a hot water heater in New Jersey for the first half of 1979. Tipton and Rockefeller (582) found an efficiency of 2% for a hot water heater in Oklahoma in August 1979. Skidmore and Smok (586) found that system efficiency (defined here as the product of collector and storage efficiency) for five large U.S. systems during 1978–79 ranged from 11% to 28%, with an average of 20%. Karaki *et al.* (587) found an overall system efficiency of 27% for a Colorado installation during the same period. Murray (588) obtained a system efficiency of 16%–22% (depending on definitions) for a New Mexico system during 1978–79. Ward (591), in an introduction to a conference on operational results, chose a presumably typical value of 24% for system efficiency. A value of 40% was assumed for collector efficiency only. Freeborne (592) found system efficiencies ranging from 0% to 29% for 16 houses during 1978, with an average of 11%. The data of Nemetz (593) can be used to deduce a value of 18% for a Hawaiian system. A hot-water installation in California had a system efficiency of 24% during 1978 (594). A New Mexico system (596) had system efficiencies of 14% for both 1976–77 and 1977–78. An Illinois installation (597) had an average of 23% for 1977–78. Eden and Tinsley (598) found values of 14% and 16% for 1977 and 1978, respectively, in Colorado. A Tennessee system (599) had an efficiency of 19% for 1977–78. A Virginia system, producing both solar heating and cooling, had a value of 21% (600). A California laundry had a system efficiency of 23% during 1977–78 (601). The value for a Texas installation in 1978 was 12% (602). The data of Murray (603) implies an average value of 14% for March–April 1978. Ward (604) found a system efficiency of 7% for a Colorado house during 1977–78, and 22% for 24 houses during the same period. Boleyn (606) found a system efficiency of 5% for an Oregon house with large storage during 1977. Hale and Murphy (612) found a value of about 17% for an Iowa school during 1977–78. A well-instrumented German system (613) with annual storage had a value of 12% for 1976–77. An installation in Colorado (358) had 29% efficiency when first built. After 15 years, this had dropped to 21%. A French study (356) expected a theoretical efficiency of 13% at Le Havre and 17%–20% at Avignon. Leach (662) found the average efficiency in a British school to be 15%. A study (614) of ten systems in four countries found an average efficiency of 22%.

A recent article (530) discussed solar collectors in Israel, where the per capita use of this type of energy is probably highest in the world. It can be deduced (530) that the efficiency of collectors there is about 8%–12%.

b. Canadian Conditions

Comparatively little data are available on Canadian experimental conditions. Lorriman (335) recently showed data for a 64.4-m^2 flat plate collector near Toronto. He had a storage tank of 18 m^3, making the ratio of storage to area 18/64 = 0.28 m, between the high and low values considered in this appendix. Lorriman found that the "solar heating system performed very well" (334), collecting 19,300 kilowatt-hours over a year. This is 19,300/64.4 = 300 kWh/m^2. However, only 6460 kWh were apparently used in space heating, and 1740 kWh for water heating, for a total of 8200 kWh. This is 8200/64.4 = 127 (in the same units). The calculation is complicated by the fact that a heat pump was in use, but the values give some idea of what would be expected under Canadian conditions.

As another example, Besant and Dumont (375) expect that evacuated tube collectors in Saskatchewan will produce 5.1 gigajoules per year, with a collector area of 17.9 m^2. No experimental values were shown. It should be noted that these are not flat plate collectors, as is assumed in this appendix. The expected energy collected is then $5.1 \times 10^6/17.9 \times 3600 = 79$ kWh/m^2 per year. Carscallen and Sibbitt (579) note that of the 14 houses in a National Research Council program, only six had data available at the end of the 1977–78 heating season. Of these, only four had data for the entire season. They had an average delivered energy of 248. Another Canadian design (85) uses a value of 207, and yet another Canadian system with a small storage tank assumes a value of 172. A Canadian theoretical study by Sasaki (357) expected about 270 for houses in the Ottawa area. An Ontario Hydro design (83, 84) using a flat plate collection assumes 233.

c. *Storage Volume*

The Ontario Hydro system noted above (83) has a storage volume of 80 m^3. This is probably enough to eliminate requirements for additional back-up and storage, although the original paper states that only 70% of total heat requirements for the building to which it is connected will be met. The ratio of volume to area for the Ontario Hydro model is 0.9 m. The Saskatchewan house mentioned above (572) has a ratio of 0.7. That of Reference 214 has a ratio of 2.2 m, indicating that the storage volume assumed here is not the highest noted in the literature. See Section vii of this appendix for a discussion of storage with a low ratio of volume to collector area.

If the Ontario Hydro model is considered, for every square meter of panel area, 0.9 m^3 of storage is required. Since the Ontario Hydro model has an area of 88 m^2, this implies a storage volume of $88 \times 0.9 = 80$ m^3.

Other installations have had large storage tanks. A house in Germany had a storage volume of 42 m^3 (388). The Zero Energy House in Denmark had a storage volume of 30 m^3 (389). The 1939 MIT house had storage of 62 m^3 (390). A house in Tokyo has a storage volume of 46 m^3 (396). The proposed design for a 1960 house in Toronto had a storage volume of 225 m^3 (397). Anderson notes (385) that "long-term storage is probably the only alternative to an auxiliary system."

There is some controversy about the appropriate ratio of storage volume to collector area. Both large and small storage volumes are considered in this appendix. Dickey *et al.* (333) note that "the larger the storage reservoir the smaller the relative heat loss" and advocate that "practical demonstration requires that at least 100,000 gallons (450 m^3) storage be considered." This is considerably larger than either of the two sizes of storage considered here.

d. *Material Requirements*

Collectors naturally vary in their weight per unit area. Rogers (216), in a survey of seven collectors, found an average weight of 37 kg/m^2. Baron (619), in a system described by Rogers, assumes 48 kg/m^2. Reference 362 estimates weights, in kg/m^2, of steel, 20.4; glass, 17.9; copper, 9.0, and aluminum, 4.7, for a total of 52 kg/m^2. The Honeywell system (82) used here weighs 34 kg/m^2. Another Honeywell system (527, 529) has a weight of 59 kg/m^2. The weight depends on the model, but the value chosen here is probably reasonable.

The Honeywell (82) system uses 31.7 kg steel, 14.5 kg glass, 4.6 kg aluminum, etc., in each 1.67 m^2 panel. Heat exchangers weighing 23 kg are not included in calculations, although they should be if proper account is to be taken of energy distribution.

If the annual energy gathered is 308 kWh/m^2, then the average power is 308,000/8760 = 35 W/m^2. A 1.67 m^2 panel then produces $35 \times 1.67 = 59$ W of power. An average power of 1000 MW then requires $1000 \times 10^6/59 = 16.9 \times 10^6$ panels. In addition, each 88 m^2 requires a storage tank of volume 80 m^3, if annual storage is assumed. (A discussion of short-term storage is held later in this appendix.)

Table G-1 shows the materials required for solar space heating based on the Ontario Hydro and Honeywell models (82, 83). This table corresponds to Table F-1 for photovoltaics and Table E-1 for solar thermal. Table G-2 shows the risk for material acquisition and construction, on a unit energy basis. Inclusion of the category "Fabricated metal products" implies that the metals used are fabricated to about the same level, in terms of stamping, forging, etc., as the average for typical metal uses. This is clearly an approximation.

Table G-1 considers two materials which have not been discussed previously, fiberglass insulation and copper, the risk of which can be analyzed in the same way as other materials. To do so, two further assumptions are

Table G-1. Material Required for 1000-Megawatt Average Thermal Power Solar Space Heating System, in 10^5 Metric Tons (d), (h)

	Steel	*Glass*	*Aluminum*	*Copper Tubing*	*Insulation (Fiberglass)*	*Miscellaneous*
Collector and Tubing (e)(g)	5.4	2.4 (i)	0.78 (j)	0.52 (a)	0.41	0.23
Storage (b)	13.7 (f),(g)				0.29 (c)	
Total	19.1	2.4	0.78	0.52	0.70	0.23

(a) After Ref. 88, since Ref. 81 also includes tubing for air conditioner and cooling tower. As a result, data for latter source are far too large.

(b) System of Ref. 81, the data from which are adapted here, had a capacity of 3.79 m^3. Assuming storage is in the form of an equisided cube, this corresponds to a side of $(3.79)^{1/3}$ = 1.56 m. Its area is then 6 × $(1.56)^2$ = 1.46 m^2, corresponding to its weight of 561 kg. The mass/area ratio is then 561/14.6 = 38 kg/m^2. The side of an 80-m^3 storage tank is $(80)^{1/3}$ = 4.3 m, and so its overall area is 6 × $(4.3)^2$ = 110 m^2. Its mass is then 110 × 0.038 = 4.26 metric tons, assuming that the steel thickness is the same for both systems. A storage tank for a Honeywell system (526) has a steel thickness of 0.25 in. If the density of steel is 7.8, this corresponds to 7.8 × 0.25 × 2.54 × $10^4/10^3$ = 50 kg/m^2. It may be deduced from Lenchek (342) that a weight/area ratio of 97 kg/m^2 is assumed for his two storage tanks.

(c) Listed as "foam" in Ref. 81.

(d) Water used in storage system not considered. Also, ducts, wiring, motors, etc. are not tabulated. Reference 85 notes that this latter group can cost 14% of the entire system. The entire table may be underestimated by this amount, assuming that cost is roughly proportional to material weight.

(e) Reference 87 estimates are higher than those shown here.

(f) If the small storage tank of 3.79 m^3 were used, the steel requirements would be much less. However, in this case substantial amounts of energy back-up would be required to maintain reliability. See Sec. vii for discussion. Reference 547 also assumes a steel tank.

(g) The ratio of storage to collector weight is about 1.4 in this table. Lenchek (342) has a value of 2.2. Heat exchangers and tubing are included in Lenchek's storage weight.

(h) Lenchek (342) also implies 0.1 (in the same units) of paint, if the energy per unit area used here is assumed.

(i) Lenchek (342) has a value of 4.7 (in the same units), if the energy per unit area used here is assumed.

(j) Lenchek (342) has a value of 3.1 (in the same units), if the energy per unit area ussed here is assumed. Cast iron weight is 3.8 for piping.

made: first, that the average grade of copper ore is 1% (90), and second, that as many man-hours go into making fiberglass from glass as for making glass from sand and other materials. See footnotes to Table G-2.

The ratio of storage weight to collector weight is about (13.7 + 0.29)/(5.4 + 2.4 + 0.78 + 0.52 + 0.41 + 0.23) = 1.4, as shown in Table G-1. Another system (308) used in Portland, Oregon, also had a ratio of 1.4 (exclusive of steel pipe).

Material requirements for heat exchangers can be estimated, although they are not included in the calculations. In the New England experiment referred to above (212, 213), an average collector area of 4.7 m^2 required 20 ft^2 (1.86 m^2) of heat exchangers. The diameter of these exchangers was not specified. Assuming it was 3/8 in. (9.5 mm) inside diameter and made of copper, the weight per unit length is 0.30 kg/m (320). The outside diameter is 1.27 cm, so the area per unit length is $3.14 \times 0.0127 = 0.04\ m^2/m$. To produce an outside area of 1.86 m^2 requires 1.86/0.04 = 46.5 m, weighing $0.30 \times 46.5 = 14$ kg. For a 0.25 in. (6.4 mm) tube, the corresponding weight is 13 kg; for 0.50 in. (12.7 mm), the corresponding weight is 16 kg. If these results relate to a panel of 4.7 m^2 with a rate of annual energy gathered of 308 kWh/m^2, then the weight of heat exchangers for 1000 MW power is $(8760 \times 10^6/4.7 \times 308)$ (13–16) kg = $(7.8\text{–}9.7) \times 10^4$ metric tons of copper. This is larger than shown for copper nonheat exchanger tubing in Table G-1. In another example, Lenchek (342) stated that steel heat exchangers for an installation in Colorado with collector area 71 m^2 would weigh 1820 kg, or 25.6 kg/m^2. This corresponds to the values of (13–16)/14.7 = 2.8–3.4 kg/m^2 estimated for the New England system. Heat exchangers are the heaviest single component in Lenchek's system.

Some estimates of risk are higher than those of Table G-2. Ramsay (492) calculates 6×10^{-4} annual injuries and illnesses and 3×10^{-6} deaths in material acquisition (including transportation) for a collector producing 1.27 kW-yr. (A calculational error was corrected in Reference 492.) This is then $(3 \times 10^{-6} \times 6000 + 6 \times 10^{-4} \times 50)/1.27 \times 10^{-3} = 38$ man-days lost/MW-yr. Transportation produces about 19 man-days lost/MW-yr, (see Section vi), so the total attributable to acquisition of aluminum, steel, and glass is 38 − 19 = 19 man-days lost/MW-yr.

e. *Lifetimes*

Results of risk calculations are shown in Table G-2. The lifetime assumed for the system is 20 years, in contrast to the 30 years for other

Table G-2. Material Acquisition and Construction Risk for Solar Space Heating (per Megawatt-Year Net Thermal Output)

	Materials (Metric Tons)	*Man-Hours (per Metric Ton)*	*Man-Hours*	*Man-Days Lost, Accident*	*Man-Days Lost, Illness*	*Deaths* $(\times 10^3)$	*Man-Days Lost, Total*
Material and Equipment Acquisition							
Iron ore mining	145	0.44	64	0.036	–	0.021	0.13
Hard coal mining (e)	142	1.2	170	0.170	0.005	0.102	0.79
Steel	95	10.0	950	0.383	0.009	0.032	0.59
Bauxite mining	16	0.50	8	0.004	–	0.002	0.02
Aluminum	4.0	12.4	50	0.016	0.001	0.002	0.03
Nonmetal mining	15	0.50	7.5	0.002	–	0.001	0.01
Flat glass (h)	15	38	570	0.176	0.049	0.020	0.30
Fiberglass insulation (c)	3.7	38	140	0.082	0.002	0.009	0.14
Copper ore mining	260	0.47 (b)	122	0.069	0.001	0.029	0.24
Copper (a), (d)	2.7	13.2	36	0.027	0.001	0.002	0.04
Fabricated metal products (g)	102	149	15,200	6.9	0.210	0.70	11.4
Total	800			7.9	0.28	0.92	13.7
Construction							
Plumbing			6000 (f)	2.18	0.075	0.72	6.54
Roofing and sheet metal			6000 (f)	5.36	0.075	1.74	15.9

Total	7.54	0.15	2.46	22.4

(a) Reference 91 states that U.S. employees of all nonferrous metals industries totaled 314,000. Nonferrous miners totaled 94,000 in 1974. Copper miners totaled 36,000 (92). Assuming that the proportion of copper ore miners to all copper workers is the same as nonferrous ore miners to all nonferrous workers, the number of copper industry workers outside mining is 36,000 × 220,000/94,000 = 84,000. They produced about 14,000,000 short tons of recoverable copper in 1974 (93). Assuming 2000 man-hours per worker, this is 84,000 × 2000 × 1.1/14,000,000 = 13.2 man-hours per metric ton.

(b) Using footnote (a), the fact that about 1% of all copper ore is recoverable, and a total of 36,000 copper miners, 0.47 man-hours per metric ton is obtained. This is comparable to iron and bauxite ore rates.

(c) Total man-days lost for accidents and illness was 122 per 100 man-years for mineral wool; distribution for accidents, illnesses, and deaths assumed to be the same as for fabricated metal products.

(d) Total man-days lost for accidents and illness was 156 per 100 man-years; distribution for accidents, diseases, and deaths assumed to be the same as for steel.

(e) Soft coal mining would change results only slightly.

(f) Some estimates are lower (87, 94), but others are comparable (85), taking labor at $10 per hour.

(g) Fabricated metal products equals sum of copper, steel, and aluminum. It is assumed that almost all metals in this system will be fabricated, in contrast to solar thermal and solar photovoltaic.

(h) Includes glass used to make fiberglass.

systems (84). Rogers (272) also assumes a 20-year lifetime. Patton (326) analyzes solar systems on a 15-year life-cycle basis. An Aerojet Company report (327) gives the life expectancy of the solar system considered as 19 years. Shamo and Fichtenbaum (406) suggest a lifetime of 15 years. Reckard (436) assumes a lifetime of 20 years. Shaw (522) does the same for New England conditions. A report by the British Building Research Establishment (501) implies a lifetime of 10 years.

The Meinels (269) state that ". . . achieving more than a ten-year operating lifetime for solar collectors is complicated by the need to use inexpensive materials and construction processes and the unreliability of inexpensive materials." This implies that the 20-year lifetime assumed here may be an overestimate.

f. Construction

Construction times are generally unknown. One of the few estimates for these times in actual systems is from an extensive solar heating demonstration in New England (212, 213). It was found that installation of about 5.7 m^2 of solar panelling and tubing, but excluding storage tanks, required about 48 man-hours. Dividing by the lifetime in order to allocate construction time over each year the system lasts, this is $48/20 \times 5.7 = 0.42$ man-hours/m^2. Dividing by the output per unit area, this becomes $0.42 \times 1000 \times 8760/308 \times 2000 = 6.0$ man-years per megawatt-year. This might be an underestimate, since installation of storage tanks has not been considered. In terms of trades, one of the few estimates states that half the time is employed in roofing and half in plumbing (95). This then implies that about $2000 \times 6/2 = 6000$ man-hours are needed in each of the two trades.

The value of 0.42 man-hours/m^2 derived above is comparable to estimates of 0.14–0.36 man-hours/m^2 predicted for space and water heating requirements in Canada for 1985 (187). Hollands and Orgill (302) estimate that 1 man-day (or 8 hours) is required to install 2.8 m^2. Dividing by the lifetime of 20 years, this is $8/2.8 \times 20 = 0.14$ man-hours/m^2. Carnegie and Pohl (611) found experimentally that 0.5 man-hours/ft^2 were required for a California agricultural installation. Allowing a 20-year lifetime, this is $0.5/0.0929 \times 20 = 0.27$ man-hours/m^2.

An indirect estimate of construction time may be deduced from the data of the Meinels (268), who found, as of 1975, "Collector costs of \$70–\$100/m^2 FOB factory; \$200–400/m^2 installed." If one assumes that the difference is due to the installation labor at about \$10 per hour, this is the equivalent of (\$130–\$300)/\$10 = 13 man-hours per m^2. Assuming a 20-year lifetime and an average of 308 kWh/m^2 collected per year, this is (13–30)/20 × 308 = 2.1–4.9 × 10^{-3} man-hours per kWh, or 9.2–21.5 man-years per megawatt-year. Even if only half the difference between parts and the final installed costs were due to installation labor, the average of the calculated range would be higher than the labor time assumed here.

A similar calculation can be made from the data of Cain *et al.* (590), who found an installation cost of \$78.60/m^2. Using the same assumptions as in the previous paragraph, this is 78.6/10 × 20 = 0.39 man-hours/m^2.

A British report (501) estimates a cost of 15 pounds/m^2 for installation in 1975. Assuming about 2 pounds per man-hour, which might be the rate of pay at that time, this is 15/2 × 20 = 0.38 man-hours/m^2, annualized over 20 years. Another part of this document (502) estimates 12.5 pounds /m^2 for installation, or 12.5/2 × 20 = 0.31 man-hours/m^2, also annualized. Recent Canadian projects found labor costs for installation of panels to be \$3800 for 31 m^2 (518); \$2440 for 37 m^2 (519); and \$10,000 for 61 m^2 (520). Allowing \$8 per hour of labor, and dividing by the estimated lifetime, this is 3800/31 × 8 × 20 = 0.77; 2440/37 × 8 × 20 = 0.41; and 10,000/61 × 8 × 20 = 1.0 man-hours/m^2. Another estimate is due to Shippee (405), who states that 2 man-days are required to install 48 ft^2, or 2 × 8/48 × 0.093 × 20 = 0.18 man-hour/m^2 over a panel's lifetime. It is not known whether this value was obtained experimentally, or if interior plumbing is included. The Hunn house in New Mexico (573) required \$255 to install 20 ft^2 of collector and associated equipment. Again allowing a 20-year lifetime, and \$10 per man-hour, this is 255/10 × 0.0929 × 10 × 20 = 0.69 man-hours/m^2.

Other indirect estimates are due to the U.S. Office of Technology Assessment and MITRE Corp. (490) who estimate costs of \$16–\$32/m^2. Allowing \$10 per hour and a 20-year lifetime, this is (16–32)/10 × 20 = 0.08–0.16 man-hour/m^2. Fraser (610) estimates a total of \$30–\$80/m^2 in 1977 dollars for collectors and support structures. Using the assumptions made immediately above, this is (30–80)/10 × 20 = 0.15–0.40 man-hours /m^2.

Indirect estimates can also be made from statements of Jordan and Liu (330), Field (329), and Ruegg (332). Jordan and Liu estimated that installation costs would be 13%–22% of total costs. Field estimated that installation would add 20% to the cost of collectors, implying that installation costs are 0.2/(1 + 0.2) = 17% of total costs. Ruegg estimated that installation costs would be 39% of total costs. A house in New Mexico

(573) had a proportion of 0.20. Applying these values to the average cost of \$2200 for the New England experiment (212), this implies that (0.13–0.39) 2200 = \$290–\$860 of the cost per installation was for labor. Allowing \$10 per man-hour, this is 29–86 man-hours for an installation of area 4.7 m^2. Dividing by the lifetime of 20 years, this is (29–86)/4.7 × 20 = 0.31–0.91 man-hours/m^2.

Reference 359 noted that costs for an installer or subcontractor would be about 26% of total cost for single or double-glazed collectors. This is in the middle of the range noted above. A higher value is due to Cain *et al.* (590), who found a proportion of 50%.

Not included here is the time to fabricate solar panels, as opposed to fabricating the metal used in them. The latter is included in Table G-2. Some estimates of the former time, expended by apparently amateur as opposed to professional fabricators, are available (517). In man-hours per m^2, they are 9.4, 6.7, 6.7, 4.2, 0.6–1.4, 10, 4.0, and 8.4, for an average of 6.3. If a panel supplies 308 kW/m^2 annually and has a lifetime of 20 years, this is 6.3 × 8760 × 1000/308 × 20 = 9000 man-hours/MW-yr.

g. *Results*

The results of Table G-2 show the risk from material acquisition and construction. The number of man-days lost might be less if it were not for the decentralized nature of the space heating system. Each house or building is assumed to have its own panels, storage, and copper tubing. As Daey Ouwens (78) points out: "material usage of large systems per kW is usually less than for small systems."

The major component of risk in Table G-2 is the use of steel for storage. For example, the categories of fabricated metal products, steel, coal mining, and iron ore mining constitute about 93% of the man-days lost by accidents in material and equipment acquisition. One might eliminate some of this risk by substituting plastics for steel, or dispensing with liquids entirely by using rock storage. However, this procedure would only substitute one type of risk for another.

Another approach would be to reduce the size of the storage tank. This produces limited decrease in steel usage because the weight does not vary directly with the tank capacity. For example, the two tanks analyzed in footnote (b) of Table G-1 have a ratio of capacities of 20, but a ratio of weights of only 7.6. See Section vii for a discussion of small storage tanks.

An important point not discussed above is possible deterioration of performance over time. There is comparatively little data on this subject, but Ward and Lof (358) found that the measured energy collected per unit area dropped by 27% over 15 years of operation of an actual collector. This corresponds to a decrease of 1.8% per year. In turn, this implies that the average energy collected per year over a 20-year lifetime would be around 84% of that collected in the first year. This possible decrease has not been taken into account above; it should be in a comprehensive analysis.

ii. Emissions

As in Appendices E and F, it is assumed that emissions incurred in material production are proportional to the weight of steel and aluminum used to fabricate the material. A total of 95 metric tons of steel per unit energy is used. This corresponds to 95(0.058–0.18) = 5.5–17 man-days lost per unit energy. Values for aluminum are 15(0.3–0.9) = 4.5–14, for a total of (5.5 + 4.5)–(17 + 14) = 10–31.

By way of comparison, Reference 284 estimates that 5700–7200 metric tons of sulfur oxides are produced from building 10,000 collectors. Each is claimed to produce 16 MWh annually. If their lifetime is 20 years, this corresponds to (5700–7200) $\times$ 8760/(16 $\times$ 20 $\times$ 10,000) = 15.6–19.7 metric tons of sulfur oxides per megawatt-year. If one ton of sulfur oxides corresponds to 6–18 man-days lost, this system produces 6 $\times$ 15.6–18 $\times$ 19.7 = 94–350 man-days lost per megawatt-year.

iii. Operation and Maintenance

There have been few attempts to estimate the time required for maintenance of collectors. Some records were kept of a recent New England experiment involving 100 collectors (213). In the first year of the experiment, there were 2.0 calls per collector of 4.4 m^2 area. The length of time per service call was not specified, but the second year of the program had 1.4 calls per unit at a cost of \$49 in all. Allowing \$10 an hour for labor costs, this corresponds to 4.9 man-hours, or 4.9/1.4 = 3.5 man-hours per call. Then the first year had operation and maintenance time of 3.5 $\times$ 2.0 = 7.0 man-hours per 4.4 m^2. The units collected 233 kWh/m^2 annually

(212). Allowing 2000 man-hours per man-year, the operation and maintenance time was $7.0 \times 1000 \times 8760/4.4 \times 233 \times 2000 = 30$ man-years per megawatt-year. In the second year of the experiment, the number of calls per unit decreased, as noted above. The area of the units increased, as did the energy gathered per unit area.

a. Indirect Estimates

One estimate (222) states that \$100–\$150 in annual costs would be required for those who purchase maintenance services. Assuming that parts are not included in this estimate (in accordance with the implied assumption in the rest of this appendix that all parts last the full lifetime of the system, or 20 years), and a labor cost of \$10 per hour, this is about (100–150)/10 = 10–15 hours annually per installation. The size of the installation to which this corresponds is not indicated. However, the same document (214) lists two apparently typical systems with annual average energy utilizations (somewhat lower than the average energy collected) of between 1.23 and 1.87 kilowatt-year. The maintenance time is then (10–15)1000/(1.23–1.87)2000 = 2.7 – 6.1 man-years per megawatt-year. This range will be used here as values for professional maintenance.

Another estimate by Ramsay (489) assumes \$40 in maintenance costs for a system producing 22 gigajoules per year. Assuming all costs are due to labor and not parts, \$10 per hour is the wage rate, and noting there are 31,500 GJ/MW-yr, labor requirements are $40 \times 31{,}500/22 \times 10 \times 2000 =$ 2.9 man-years/MW-yr.

In a related set of data, Flaim *et al.* (557) note that operations and maintenance costs for seven solar agricultural and industrial process heat systems range from \$10.87 to \$37.46/m^2/year, with an average of \$18.15. If one assumes that most of this cost is due to labor, the systems collect the energy per unit area assumed above, and \$10 per hour is the wage rate, then $18.15 \times 8760 \times 1000/308 \times 10 \times 2000 = 25.8$ man-years/MW-yr would be needed.

Reckard (436) assumes \$19 per year for O & M costs for solar water heating, and \$30 per year for solar water and space heating. The two systems have estimated capital costs of \$720 and \$4,650, respectively, and are stated to produce 11.2 and 56 million Btu per year, respectively. Assuming \$10 per man-hour, this yields $19 \times 10^{12}/10 \times 11.2 \times 10^6 \times 33.5 \times 2000 = 2.6$ and $30 \times 10^{12}/10 \times 56 \times 10^6 \times 33.5 \times 2000 = 0.8$ man-years per megawatt-year for water and space/water heating, respectively. However, the estimated energy collected per unit capital cost is apparently higher than that of some experimental data. For example, the New England Electric systems mentioned previously (212, 213) had a capital cost of \$2200. Producing 233 kWh/m^2 with an average area of 4.7 m^2 corresponds to an average power of $233 \times 4.7/8{,}760 = 0.125$ kW. The model of Reference 436 has an average power of $11.2 \times 10^6 \times 1055/3600 \times 8760 = 0.375$ kW. The relative ratio of capital costs to power for the two systems is then $(2200/0.125)/(720/0.375) = 9.2$. If the experimental values of the New England Electric systems applied, and other related factors increased in proportion, this would correspond to $9.2 \times (0.8–2.6) = 7.4–24$ man-years/MW-yr.

Another estimate (331) states that maintenance costs would be 1%–5% of the total capital costs of a system. The New England field test (212) had an average cost of \$2200, implying an annual maintenance cost of $(0.01–0.05) \times 2200 = \$22–\$110$ to be expected on a long-term basis. (The actual costs were between these values, as noted above.) Assuming a cost of \$10 per man-hour, this corresponds to $(22–110)/10 = 2.2–11$ man-hours annually per installation. Using the reasoning of the paragraph at the beginning of this subsection, this corresponds to $(2.2–11) \times 1000/(1.23–1.87) \times 2000 = 0.6–4.5$ man-years per megawatt-year. Ward (604) estimates \$80–\$90 in annual maintenance costs for a typical installation in Colorado. If the "controlled" energy production for this installation is about 6.4 gigajoules/year (604), most of the maintenance cost is due to labor, and \$10 per man-hour is assumed, this is $(80–90)3600 \times 8760 \times 10^6/10 \times 6.4 \times 10^9 \times 2000 = 20–22$ man-years/MW-yr.

b. Decentralized Maintenance

Homeowners are unlikely to be as skilled in maintenance as trained and full-time workers. It is well-known that the major location of disabling accidents is not the factory, but the home. The number of man-years per unit energy should be multiplied by a factor greater than 1 to take account of this increased risk. While the factor will not be known exactly until more maintenance experience is gained, a conservative, or lower bound, estimate could be 1.5. This factor would presumably also take into account the fact that many homeowners would probably take longer to do the job than skilled professionals, as well as incurring greater risk per man-hour.

Schurr *et al.* (434) assume the ratio of roofing deaths for "amateur" as compared to professional repairmen is between 5 and 10, greater than is assumed here.

This factor would produce an equivalent time of 1.5(2.7–6.1) = 4.1–9.2 man-years per megawatt-year. Not included in this value is the time required to operate, as opposed to maintain, the system. It is not clear how much extra risk this would add.

It may reasonably be assumed that the trade corresponding to maintaining this system is roofing and sheet metal. Using the appropriate values of Table E-2, it is found that the man-days lost to accident and illness are 3.53(4.1–9.2)2000/4100 = 7.1–15.9 and 0.0509(4.1–9.2) 2000/4100 = 0.10–0.23, respectively; the number of deaths, 1.15(4.1–9.2)2000 × 10^{-3}/4100 = 2.3–5.2 × 10^{-3}; and the total number of man-days lost, 21-47.

Some risk estimates are lower than used here. Ramsay (491) estimates 10–20 deaths and 3000–6000 disabling injuries for 25 million solar homes. Assuming (a) 6000 man-days lost per death and 50 per injury, and (b) 1.27 kW-yr annually produced per home (491), this is [(10–20) × 6000 + (3000–6000) × 50] /1270 × 25 = 6.6–13.2 man-days lost/MW-yr.

It is possible to reduce maintenance time; however, Sayigh (505) found that the amount of heat collected dropped by 30% after 3 days without cleaning, under Saudi Arabian conditions. North American conditions would produce substantially less decrease, but there would still be some loss without maintenance.

It should not be assumed that the risk per installation is high. Consider the apparently typical installation noted near the beginning of this section, producing annual energy of 1.23 kW-yr. If the number of man-days lost per megawatt-year is 21–47, the annual man-days lost per installation is (21–47) 1.23/1000 = 0.03–0.06, or 16–35 minutes.

iv. Energy Back-up

No energy back-up is assumed. This is probably an underestimation of risk, since only 70% of thermal energy required in the home is supplied by the installation considered (83). By way of comparison, Reckard (436) assumes that back-up requires 26% of total energy for water heating, and 26%–44% for solar water/space heating. The size of storage tank is not given. Flaim *et al.* (564) assume a gas or electric back-up for the systems they consider. The Meinels (267), in a comprehensive study, note that "Most solar energy applications . . . rely on electrical back-up for periods."

v. Energy Storage

The storage system described in Section i is assumed to supply enough energy for sunless days.

vi. Transportation

Again it is assumed that transportation risk is proportional to the total nonsand weight of the system, including both the primary and the finished products. For solar space heating, Table G-2 shows that this is 785 metric tons per unit energy, 0.23 times that of the 3500 metric tons for coal. The transportation risk from Table A-2 is multiplied by this factor. See also Appendix P.

Ramsay (492) estimates 3.0×10^{-5} deaths and 6×10^{-3} injuries for a solar household producing 1.27 kW-yr annually. Allowing a 20-year lifetime, this is $3.0 \times 10^{-5}/1.27 \times 10^{-3} \times 20 = 1.2 \times 10^{-3}$ deaths/MW-yr and $6 \times 10^{-3}/1.27 \times 10^{-3} \times 20 = 0.24$ injuries/MW-yr. Allowing 6000 man-days lost per death and 50 per injury, this is $1.2 \times 10^{-3} \times 6000 + 0.24 \times 50 = 19$ man-days lost/MW-yr. Note that this value probably does not take account of the raw materials used in producing the steel, aluminum, and glass, so the transportation risk may be higher than that of Ramsay.

Solar space heating is a technology which is deployed in a decentralized manner, unlike those previously discussed. It might be thought that this would reduce transportation risk, since materials might be used closer to the site where they are manufactured. However, Table G-2 indicates that a considerable part of the system's weight lies in the steel and the materials used in producing it. The steel industry is not decentralized, so transportation risk is probably close to the above estimates.

As noted in the main text of this book, some energy systems will probably be located in a noncentralized way, but all are assumed to be manufactured in the present way, i.e., somewhat centralized. If the components for decentralized systems are also manufactured in a decentralized manner, there is reason to believe that the risk per unit energy could be greater than estimated here. Labor statistics often suggest that worker safety in large establishments is better than in small ones.

vii. Small Storage Tanks

If a smaller storage tank were assumed, it might be argued that overall risk would decrease. However, in that case, there would be a need for back-up energy. For example, in what may have been the largest demonstration project of solar space heating up to that time in North America, 41% of the energy used for water heating was solar-derived, and 59% was back-up (213). Most of the systems in that experiment had relatively small storage. In addition, the smaller the storage tank, the lower the efficiency of the collectors, generally speaking.

It is difficult to give a rule for this decrease, since it depends on factors such as climate and collector type. However, the demonstration experiment mentioned above had an efficiency about 31% lower than the value assumed here. Another system, with collectors apparently similar to that assumed here and with a smaller storage system, had an efficiency about 45% less than that used in the present calculations (214).

As an example of the decrease in efficiency, a French theoretical study (356) on systems in Le Havre (with a storage tank of volume 3 m^3) and Avignon (with a storage tank of volume 4 m^3) found that the energy collected per unit area would drop by 60% as the volume/area ratio drops from 0.52 to 0.04; 39% as the ratio drops from 0.42 to 0.05; and 56% as the ratio drops from 0.50 to 0.06.

In order to give solar space heating the benefit of the doubt, the same efficiency of collection of solar energy as that of a system with large storage, i.e., about 23%, is assumed. With a smaller storage system, the materials used in the storage will decrease. The exact amount of decrease will depend on the ratio of storage volume to collector area.

As noted above, the seasonal storage assumed here had a ratio of volume to area of 0.9 m. Smaller storage systems, designed for American states bordering Canada, had a ratio of about 0.08 m (221). The area of collectors is, as before, $16.9 \times 10^6 \times 1.67 = 28 \times 10^6$ m^2. (See Sections ia and id). The volume of storage is then $28 \times 10^6 \times 0.08 = 2.3 \times 10^6$ m^3.

On the other hand, a recent study (571) of 14 Canadian solar systems shows that for the seven with liquid storage, the ratio of storage volume to collector area ranged from 0.13 to 0.44 m, with an average of 0.29 m. Leverenz and Joncich (597) had a ratio of 0.38 m for a system in Illinois. Eden and Tinsley (598) used a ratio of 0.19 m for a Colorado installation. Boleyn (606) had a ratio of 0.43 m for one system in Oregon, and 0.78 m for another. Hale and Murphy (612) describe a Virginia system with a ratio of 0.24 m. This suggests that the ratio adopted in the following may be too low for practical systems, and that the risk may be somewhat underestimated.

a. 3800-Liter Tank

Table G-1 notes that one size of a small storage tank is about 3.8 m^3. The number of these tanks required for the collector area noted above is then $2.3 \times 10^6/3.8 = 5.9 \times 10^5$. Each has a weight of 561 kg steel, so their total weight is $5.9 \times 10^5 \times 0.56 = 3.3 \times 10^5$ metric tons. This implies that the second line of Table G-1 should be multiplied by 3.3/13.7 = 0.24.

The total amount of steel is then $5.4 + 0.24 \times 13.7 = 8.7 \times 10^5$ metric tons, and the fiberglass $0.41 + 0.24 \times 0.29 = 0.48 \times 10^5$ metric tons. Results are shown in Table G-3. The materials acquisition risk for steel and related materials should then be multiplied by 8.7/19.1 = 0.46; that for fiberglass by 0.48/0.70 = 0.69. From Table G-2, this corresponds to decreases of about (1 - 0.46) (0.13 + 0.79 + 0.59) = 0.81 and (1 - 0.69) (0.01 + (0.30/4) + 0.14) = 0.07 man-days lost for (a) steel and related materials, and (b) fiberglass and related materials, respectively. The weight of fabricated metal products is multiplied by 0.46, yielding a decrease of (1 - 0.46) (11.4) = 6.2 man-days lost. Total decrease for materials acquisition is then 6.2 + 0.81 + 0.07 = 7.1 man-days lost. Construction risk remains constant, since it was assumed that this was applicable only to the collectors, as opposed to storage. The emissions risk will also decrease due to the smaller weight of steel being used. The emissions risk is multiplied by 0.46, yielding a decrease of (1 - 0.46) (10-31) = 5.4-17 man-days lost.

The transportation risk will also change. The weights of iron ore, coal, steel, and fabricated metal products are multiplied by 0.46, and the glass for fiberglass and the fiberglass itself by 0.69. This yields a decrease of 261 and 2.3 metric tons per unit energy, producing a nonsand weight of 522 metric tons, compared to the weight of 785 for seasonal storage. The transportation risk then decreases by 2.3–6.9 man-days lost.

Other sources of risk, such as energy for pumps and operation and maintenance, are assumed to remain about constant, since they generally deal with the collectors not the storage. The total decrease in risk for a small as opposed to a seasonal storage system is then (7.0 + 5.4 + 2.3)–(7.0 + 17 + 6.9) = 15–31 man-days lost. This is about 20% of the total risk. Since there are possible variations in the factors producing risk much greater than this fraction (see, for example, the discussion on collector efficiency), it could be concluded that this decrease is not significant.

On the other hand, there are areas where the risk could increase with a smaller storage tank. The largest potential increase would arise from the

Table G-3. Material Required for 1000-Megawatt Thermal Power Solar Space Heating System with Small Storage Tanks, in 10^5 Metric Tons

	Steel	*Glass*	*Aluminum*	*Copper Tubing*	*Insulation (Fiberglass)*	*Miscellaneous*
				3800 Liter Tank (81)		
Collector and tubing	5.4	2.4	0.78	0.52	0.41	0.23
Storage	3.3				0.07	
	8.7	2.4	0.78	0.52	0.48	0.23
				4000 Liter Tank (86)		
Collector, tubing, and storage (Model 1)	5.8	1.6	0	0.96	0.41	
Collector tubing, and storage (Model 2)	4.0	3.2	0.36	0.77	0.41	

back-up energy required. As mentioned in the main text, each system considered in this book is evaluated on a baseload basis, implying fairly high reliability. The two systems with small storage tanks discussed here, the New England demonstration project (212) and the Canadian system (215), required at least 40% back-up energy. Even a system with a storage tank larger than assumed here for seasonal storage required more than 10% back-up (214). Using the values of Table N-1, 40% back-up corresponds to 0.4(145 - 431) = 58-172 man-days lost. This is (58/15)-(172/31) = 3.9–5.4 times the average anticipated decrease in risk with a small storage system. Even if the back-up energy for a small storage system were only 10% of the total energy, the added risk would be 0.1(145–431) = 15–43 man-days lost, greater than the decrease calculated above.

It can then be concluded that using a small storage system as compared to a seasonal one does not decrease the risk significantly, and may in fact increase it.

b. 4000-Liter Tank

Another example using a small storage tank has a ratio of volume to area of 0.10 m (86). The storage tank has a volume of 4.0 m^3. The two models considered had weights (in kg) of steel, 1055 and 725; copper, 175 and 140; glass, 295 and 590; aluminum, 0 and 65; and fiberglass insulation, 75 and 75. Materials for a 1000-Mw system are shown in Table G-3. It was claimed that this amount of materials would supply 16 MWh per year (or 1.8×10^{-3} MW). This value is apparently a hypothetical quantity and no experimental values were shown. If this value is assumed, the weights (in 10^5 metric tons per 1000 MW) for the two models are then: steel, 1055/180 = 5.8 and 725/180 = 4.0; copper, 175/180 = 0.96 and 140/180 = 0.77; glass, 295/180 = 1.6 and 590/180 = 3.2; aluminum, 0/180 = 0 and 65/180 = 0.36; and fiberglass, 75/180 = 0.41 and 75/180 = 0.41. The values of steel, copper, glass, aluminum, and fiberglass insulation in Table G-1 are then multipliedby 5.8/19.1 = 0.30 and 4.0/19.1 = 0.21; 0.96 /0.52 = 1.85 and 0.77/0.52 = 1.48; 1.6/2.4 = 0.67 and 3.2/2.4 = 1.33; 0/ 0.78 = 0 and 0.36/0.78 = 0.46; and 0.41/0.70 = 0.59 and 0.41/0.70 = 0.59, respectively. Using these values, and the assumptions contained in Table G-2, it can be shown that the risk due to material and equipment acquisition falls by 8.2 and 9.2 man-days lost per unit energy for the two models considered. Calculations for deaths, accidents, and illnesses follow the same procedure but are not shown here.

In terms of construction, it can be assumed that the same value of risk per unit area as noted in Section i holds. Since the energy claimed to be collected per unit area for these models has a ratio of 400/308 = 1.3 compared to that of the rest of this appendix, the construction risk then drops by 22.4(1 − 1/1.3) = 5.1 man-days lost/MW-yr.

In terms of transportation, the weights of most materials transported drop by varying degrees. The nonsand weights of the two models are 411 and 332 metric tons per unit energy. Since there were 3.4–10.4 man-days lost assuming 785 metric tons transported (both per unit energy), the transportation risk drops by (785 − 411) (3.4–10.4) /785 = 1.8–5.0 and (785 − 332) (3.4–10.4) /785 = 2.0–6.0 man-days lost per unit energy for the two models.

The total decrease in risk is then 15.1–18.3 and 15.3–19.3 man-days lost per unit energy for the two models of Reference 86. It will be assumed that the risk attributable to energy for pumps (see below) and operation and maintenance remain the same in comparison to the model considered in Section i.

On the other hand, the risk can rise as well. As noted in Section ii, the man-days lost per unit energy for emissions from the two models are 94-350. The increase in emissions compared to the system with a large storage tank is then, using Table G-4, (94 - 9) - (350 - 28) = 85 - 320 man-days lost per unit energy. Also, the two models with a small storage tank supply only 50% of the total space and water heating load, so the risk attributable to this proportion of back-up, or 0.5 (170 - 470) = 85 - 235 man-days lost per unit energy, is added. This last value is obtained from Appendix N. The total increase is then (85 + 85) - (320 + 235) = 170 - 550 man-days lost per unit energy. This quantity is (170/15) - (550/19) = 11 - 29 times the estimated decrease in risk.

viii. Energy for Pumps

In the water-based system considered here, electrical energy is usually required to run pumps for circulating the fluid. Electricity is also required for fans in air systems, which are not evaluated here. A complete risk analysis should take account of this electrical energy.

In Appendix J, it is noted that the electricity requirements to produce methanol could, in principle, be supplied by burning the methanol itself. This cannot be done for solar space heating since the output is thermal not electrical.

Baron (271) stated that solar systems need the equivalent of 6.5% of their energy output to run their pumps while Rogers (270) estimated about 5%. A theoretical study (308) indicates a ratio of about 3%. A comprehensive study by Nash and Cunningham (580) for nine water heating systems during the 1978–79 season found proportions ranging from 0% (for a passive system) to 135%. If the proportions are averaged linearly, the average is 40%; if the total electrical requirement for the nine systems is divided by the total solar contribution, the average is 18.9%. Taylor (581) found a proportion of 17% for the first six months of 1979 for a water heater in New Jersey. Tipton and Rockefeller (582) found a proportion of about 100% for a water heater in Oklahoma for August 1979. This proportion was lower than that of the previous month. Fanney and Liu (583) found proportions of 8.7% and 16% for two water heaters in Maryland. Aungst (584) obtained a proportion of 19% for a hot water heater in Pennsylvania. Skidmore and Smok (586) found values ranging from 2.2% to 19.8% for five U.S. sites during 1978–79, with an average of 11%. Karaki *et al.* (587) determined a value of 7.4% for a Colorado system in 1978–79. Murray (588) found a proportion of 20% in comparing fan energy to energy removed from storage from rocks in New Mexico. Cain *et al.* (590) found a proportion of 59% for an agricultural installation in Maryland. Ward (591), in introducing a conference on operational results, assumed a theoretical value of 8.3%. Nemetz (593) has data that imply a value of 13.4%. A system in Colorado (595) had a value of 9.5% for combined hot water and space heating during 1977–78. Hedstrom *et al.* (596) found values (defined as the ratio of "parasitic power" to delivered energy) of 20% and 16% for 1976–77 and 1977–78, respectively, in New Mexico. Leverenz and Joncich (597) obtained values of 8.9%–12.8% for an Illinois installation. A Colorado system had a value of 45% (598). Bedinger and Bailey (599) found a ratio of 24% for a Tennessee system. A New Mexico system (603) had a ratio of 16%–42% during early 1978. Ward found a ratio of 9% for 24 solar homes in Colorado (604). Five other houses in Colorado (605) in 1978 had ratios ranging from 6% to 300%, with an average of about 120%. A house in Oregon (606) had a ratio of 124% during 1977. Higgin (85) estimates that pump energy for houses in Ontario will range from 12.3% to 14.7% of total energy. For a forced-air system he estimates 25.3%. McConnell *et al.* (408) estimate a requirement of 6.3%. Gilman *et al.* (454) studied an office building in Albuquerque which used both direct heat and a heat pump. They found that the compressor for the heat pump and auxiliary pumps required about 30% of

Table G-4. Solar Space Heating Risk (per Megawatt-Year Net Thermal Output) (c)

Risks	*Material Acquisition and Construction*	*Emissions (b)*	*Operation and Maintenance*	*Energy for Pumps*	*Transportation (d)*	*Total*
Occupational						
Accidental						
Death	0.0031	–	(2.1–4.7) $\times 10^{-3}$	(6–14) $\times 10^{-5}$	(3.4–10) $\times 10^{-4}$	(55–89) $\times 10^{-4}$
Injury	0.28	–	0.12–0.28	0.010	0.003–0.010	0.41–0.58
Man-days lost, excluding deaths	14.0	–	6–14	0.50	0.15–0.5	21–30
Man-days lost, total	32.6	–	19–42	0.86–1.3	2.2–6.5	55–82
Disease						
Death	(a)	–	(1.3–2.9) $\times 10^{-5}$	(5.4–86) $\times 10^{-7}$	–	(13–38) $\times 10^{-6}$
Disability	0.0082	–	(3.0–6.8) $\times 10^{-4}$	(41–72) $\times 10^{-6}$	–	(85–90) $\times 10^{-4}$
Man-days lost	0.41	–	0.093–0.21	0.005–0.055	–	0.50–0.67
Public						
Accidental						
Death	–	–	–	(4–5.9)	(16–40)	(2.0–4.6)

				$\times 10^{-5}$	$\times 10^{-5}$	$\times 10^{-4}$
Injury (b)	–	–	–	(5–35) $\times 10^{-5}$	3.3×10^{-4}	(3.8–6.8) $\times 10^{-4}$
Man-days lost	–	–	–	0.22–0.37	1.0–2.4	1.2–2.8
Disease						
Death	–	(2.4–7.5) $\times 10^{-4}$	–	(1.9–6.0) $\times 10^{-4}$	–	(4.3–14) $\times 10^{-4}$
Disability	–	1.5–4.5	–	1.0–3.1	–	2.5–7.6
Man-days lost	–	9–28	–	6.5–19	–	16–47

(a) All occupational deaths in material acquisition and construction assumed to be accidental and not disease.
(b) See footnote (t), Table A-2.
(c) As mentioned at the beginning of this appendix, all values of risk are divided by 1.11 to take account of transmission and distribution losses in nonsolar space heating systems.
(d) See footnote (p), Table A-2.

the total energy collected. The situation has been made more complicated by the heat pumps, but the auxiliary pumps alone required about 6%. Esbensen and Korsgaard (456) estimated theoretically that about 5% of solar energy collected for a house in Denmark would be required for pumps and valves while McNamara (455) estimated theoretically about 7% for a location in Chicago. A Tucson system had a ratio of operating energy to energy from storage of 29% (657, 659). The Phoenix system of 1960 (386) collected 35.3×10^5 Btu. The electricity used was stated to be 5×10^5 Btu, or 5/35.3 = 14% of total energy collected. On the other hand, another table in Reference 386 indicates that the possible electrical requirement was 8675 kWh, or the equivalent of 29.5×10^5 Btu. If this is correct, the ratio of energy for pumps, etc. to that collected is 29.6/35.3 = 84%. Dean (453) notes, "One should always be alert to the situation in which more energy is consumed by the blower than is extracted from storage at these low temperatures."

One can also make indirect calculations. In 1975, Ruegg (328) estimated that the electricity cost to drive pumps for a building with annual heating requirements of 2.8 kilowatt-years, 60% of which was supplied by solar energy, was $20. Assuming a cost of 2¢–3¢ per kWh of electricity, the ratio of electrical energy to supplied solar energy is $2000/(2–3) \times 2.8 \times 0.6 \times 8760 = 0.045–0.068$. A British report (502) estimates the electrical cost for pumps for a collector area of 4 m^2 was 5 pounds per year. At a rate of 1.93 pence per kWh, this is 5/0.0193 = 259 kWh/yr. If the collector delivers 308 kWh/m^2 annually, the ratio of pump energy to that collected is $259/308 \times 4 = 21\%$.

For systems employing rocks as a storage medium, the ratio of electrical energy used to heat collected may be high. For example, Ward and Lof (363) found that the electrical energy required in 1974–75 for a collector in Colorado was 10.7 gigajoules. The heat collected was 31.8 gigajoules, so the ratio is 10.7/31.8 = 34%. A later report (503) from the same group says that the delivery of 50,600 megajoules of solar heat over the 1976–77 season required 4690 megajoules of electric fan power, for a ratio of 4690/50,600 = 9%.

Choosing a value of 5% to give the benefit of the doubt to this form of energy, this implies that the risk corresponding to 0.05 megawatt-years of electrical energy should be added to the risk already calculated. If one uses the "weighted" risk for Canadian conditions shown in Appendix N, this corresponds to an occupational risk, in man-days lost, 0.8–1.5; a public risk of 7–22; and a total risk of 8–24.

ix. Summary

Results of calculations are shown in Table G-4. The risk of solar space heating appears to be comparable to that of other solar systems.

Some partial estimates of risk are fairly similar to those used here. Ramsay (491, 492) assumes 85–95 deaths and 18,000–21,000 injuries for 25 million solar homes annually for finished material acquisition, transportation, construction, operation, and maintenance. (A calculational error is corrected in Reference 492.) These references assume, as is done here, amateur as opposed to professional repair, but exclude nonroofing repairs. Risk attributable to storage is not included. These classifications of risk are only part of the total risk of this system. Table G-4 suggests that these components produce about one-half to two-thirds of total risk. Allowing 6000 man-days lost per death and 50 per injury, and 1.27 kW-yr produced per home annually, this is [(85–95) × 6000 + (18,000–21,000) × 50] /1270 × 25 = 44–51 man-days lost/MW-yr.

As noted in the introduction to this appendix, all values for risk are divided by 1.11 to take account of transmission and distribution losses in nonsolar space heating systems.

APPENDIX H. WIND

As nonconventional technologies which are more and more removed from present-day use are considered, estimations of material requirements and other contributions to risk become less precise. This is true for wind technology.

One of the problems in calculating windpower risk is determining the relationship between maximum power and average power. The second quantity divided by the first is the load factor. It has proved difficult to obtain verified experimental values of load factors for working windmills.

In what may be the most comprehensive analysis available (210), involving about 80 Danish windmills, a load factor of between 0.10 and 0.14 can be deduced. The largest commercial windmill ever built in North America until the 1970s, at Grandpa's Knob, had a load factor of about 0.14 (195). Winds at Grandpa's Knob exceeded 48 km/h (30 mi/h) over 70% of the time (97), hardly typical of most locations. In a place (Ottawa, Canada) with perhaps a more typical wind distribution, a load factor of 0.06 was found for an experimental windmill (98). Another estimate states: "A typical windpower generator operates at a load factor . . . only one-fourth that of a typical fossil-fuelled plant" (201). Since the load factor for coal and oil was taken as 0.70, one-fourth of that is 0.18. British windmills were expected to have an average load factor of 0.20 (305). Noll (514) estimates load factors of 0.14–0.21 for a 200-W turbine, corresponding to windspeeds of 10–14 mi/h (16–23 km/h). The Gedser windmill in Denmark had a load factor of about 0.20 in 1959–60. The Bogo mill had a value around the same time of 0.18–0.20 (255). Simmons (257) noted that an Israeli turbine which ran from 1953 to 1956 had a load factor of 0.19. Another summary of this 2.5-kW windwill estimates a load factor of 0.10 (548). Of this, only about 0.04 was usable, the rest being lost due to storage limitations. See Section v of this Appendix. The 100-kW turbine at Plum Brook, Ohio, was expected to have a load factor of around 0.20 (261), but there has been comparatively little data published on this machine. In one report (251), a 200-kW (rated) windmill in the United States was expected to have a load factor of around 0.50. However, results for the first five months of its operation in 1978 indicated a load factor of around 0.20 (249). Results for the first seven months indicated a value of around 0.16 (250). For smaller windmills, the load factor can be variable. One report (324) notes that a 50-W machine in 10–12 mi/h (16–19 km/h) winds should produce 5–7 kWh in a month. The load factor is then (5–7) 1000/50 × 30 × 24 = 0.14–0.19. For windmills of rated capacity (in kW) of 0.1, 0.5, 1, 4, 8, and 12, load factors were estimated to be, in the same wind, 0.11–0.15, 0.10–0.13, 0.09–0.12, 0.08–0.11, and 0.08–0.10, respectively. A 25-kW (rated) system in New England apparently produced 7376 kWh in 166 days (620), for a average load factor of 7376/166 × 24 × 25 = 0.07. Scheinbein, a development engineer for a wind turbine company, says (528) that all the top commercial wind turbine models perform with about the same efficiency and prospective buyers should not hold out hope that some amazing breakthrough is going to occur in windmill design.

On the other hand, some estimates of load factor have been as high as 0.63 (323), although these are believed to be theoretical. A Soviet windmill in the 1930s (341), rated at 100 kW, was claimed to have a load factor of about 0.32. However, there seems to be little information available on this or other machines for which similar claims have been made.

It is clear that the load factor will vary with location and type of machine. Until more typical data are produced, the load factor will be assumed to be 0.20, in order to give the benefit of the doubt to this system.

The "rated" maximum power of a windmill is generated at a high wind speed, such as 20–25 mi/h, although there are no fixed rules for selecting this speed. In effect, the rating is partially arbitrary, implying some degree of arbitrariness in calculating the load factor. The key physical quantity is the amount of energy that an installation produces in a given time.

i. Material Acquisition and Construction

How much material is required for wind electrical energy? One estimate states that the production of 4 MW of rated capacity requires 400 short tons of steel, 60 tons of fiberglass and plastics, and 10 tons of copper (199). This does not include the concrete in the foundations. The sum is then (400 + 60 + 10)/4 × 1.1 = 107 metric tons per MW (rated), which is used here. This weight is higher than the value of 75, in the same units, for the General Electric MOD-1 windmill built in the United States (209) but close to the estimates of 119 and 106 for experimental windmills in a recent Swedish report (99). Another estimate for a MOD-1 machine suggests a value of about 150 (252). The Clayton machine referred to above (249) has a value of around 200. Reference 78 estimates 150 of metal and 100 of concrete. A Canadian design (283) weighing 360 kg is claimed to produce 4.5–9.0 MWh annually. If its load factor is 0.20, it weighs 0.36 × 0.20 × 8760/(4.5–9.0) = 70–141. Two other Canadian designs (377) by DAF Ltd., for mills rated at 4 kW and 6 kW, had weights, including tower, of 450 kg and 900 kg, respectively. These are equivalent to 113 and 150. De Renzo (510) notes two proposed systems, rated 0.5 and 1.5 MW, have weights (including steel towers) of 75 and 113 metric tons, respectively. This corresponds to 75/0.5 = 150 and 113/1.5 = 75. If concrete towers are assumed, the weights are 480 and 236. For two other designs, rated 0.1 and 1.0 MW, he notes (511) a weight (including tower) of 80–228. A later discussion in the same document (512) estimates the weight of 0.5 and 1.5-MW (rated) turbines as 145 and 85, respectively. Lawrence (618) estimates 430–534, made up of 4–7 for aluminum, 224–253 for concrete, 2 for copper, 11–17 for fiberglass, and 189–254 for steel.

On the other hand, the value of 109 metric tons is much less than that of nine other European turbines evaluated by Vadot (101), who found an average weight of 240. The Grandpa's Knob windmill, referred to above, weighed 427 (195). This is confirmed elsewhere (78). This apparently includes the entire windmill; a previous estimate (248) of 230 apparently included only part.

For small windmills, of the order of 10 kW, the weight is about 476, in addition to about 950 of concrete (217). Simmons (256) lists a number of small windmills, ranging in rated power from 200 W to 6 kW, and notes a weight of 300–1000. These values should be used with caution, since it is difficult to tell from Simmons' description whether or not storage systems such as batteries are included with some of the windmills. Noll (514) notes a weight of 205 for a 200-W turbine. A British report (305) estimates weights of 450 for 1-kW machines; 230 for 10-kW machines; 180 for 100-kW machines; and 75 for 1-MW machines. It seems fairly clear that the smaller the windmill, the greater the weight per rated power, at least on the basis of the above data.

There have been few estimates for very large windmills. Mensforth (384), in a design for a 10-MW (rated) system, used 72, without a tower. It is likely that a tower would increase the value substantially.

It can then be seen that the estimate of weight per unit power rating used here (107 metric tons) is conservative in the sense of possibly underestimating risk. To transform this estimate into average output, the load factor must be known. It is assumed to be about 0.20 (see above). Then 1 MW (rated) produces 0.20 (average). Transforming the weights of steel, fiberglass, and copper to the usual 1000-MW average output, a total of $400 \times 1000/4 \times 1.1 \times 0.2 = 4.5 \times 10^5$ metric tons of steel, $60 \times 1000/4 \times 1.1 \times 0.2 = 0.68 \times 10^5$ of fiberglass and plastics, and $10 \times 1000/4 \times 1.1 \times 0.2 = 0.11 \times 10^5$ of copper are required. Results are shown in Table H-1.

Reference 199 did not allow for foundations. One study estimates foundations to weigh between 0.11 and 0.17 of the rest of the structure (202), although another windmill (217) has a ratio of about 2. Assuming an approximate value of about 0.2, this yields about $0.2(4.5 + 0.11 + 0.68) \times 10^5 = 1 \times 10^5$ metric tons of concrete per 1000-MW average

Table H-1. Materials Required for 1000-Megawatt Average Output Wind System, in 10^5 Metric Tons (199, 202)

	Steel	*Copper*	*Fiberglass and Plastics*	*Concrete*
Windmills	4.5	0.11	0.68	
Foundations				1
Total	4.5	0.11	0.68	1

output. This quantity is shown in Table H-1. It is less than another estimate, deducible from Daey Ouwens, of 90×10^5 metric tons (78) for windmills at sea.

As noted from the wide range of values for weight per rated power and other quantities, it is difficult to specify an "average" windmill. Part of the range is due to the fact that windmills differ greatly in size, from those in the rated kilowatt range to the 2.5-MW machines being proposed. One possible solution in terms of data is to discuss a windmill in the middle of the range of power, perhaps the 200-kW(rated) systems being evaluated in the United States. By choosing this size for discussion only, nothing is implied about its desirability or the desirability of any other size.

In terms of material requirements, the present assumption of 107 metric tons per MW (rated) and a load factor of 0.20 yields 107/0.2 = 535 metric tons/MW-y. The 200-kW (rated) turbine weighs about 200 metric tons/MW (rated) (251), so if its load factor is the same, this corresponds to about 200/0.2 = 1000 metric tons/MW-yr. If its load factor increases to 0.35 in the future, this would correspond to about 200/0.35 = 600 metric tons/MW-yr. This calculation suggests that the ratio of weight to energy production assumed here is fairly close to what might be achieved for an "average" windmill in the future.

Similar values of weight per megawatt-year are found in Reference 549, where this quantity ranged from 434 to 710 metric tons/MW-yr for the MOD-O, MOD-OA, and MOD-1 windmills. The value for the proposed MOD-2 windmill is lower, but the load factor estimated for this system (0.42) is yet to be confirmed.

a. *Lifetimes*

There is some disagreement over what is an expected lifetime of a windmill. Some estimates have been 30 years or more. However, Reference 355 states: "The assumption of a 30-year lifetime in the above studies may be dubious in view of the fact that the blade lifetime expected in the NASA windmill is only 5 years . . . the blades on the 125-foot diameter windmill comprise 36% (of the cost) of the unit." Brown and Warne use a value of 15–20 years (376). The large number of Danish windmills used during World War II had a lifetime of around 10 years (380). Kornreich and Tompkins (253) state that "We consider there is no experience in having achieved a 25–30 year lifetime."

Another reference (94) uses the value of 20 years, which is assumed here. Rayment (516) assumes the same for Britain. The weights of Table

H-1 are then divided by 20 × 1000 = 20,000 to convert them into weight per megawatt-year over the system lifetime.

This value of lifetime is probably optimistic, i.e., on the side of likely underestimating risk. For example, the windmill at Grandpa's Knob, referred to above, lasted about 3½ years (196). The Magdalen Islands installation, the largest windmill ever built in Canada, lasted less than a year (197). A French windmill installed by Ontario Hydro lasted about a month (195). An 800-kW windmill in France lasted about 1½ years (198). A British turbine (382) failed after about one-half year. A Canadian windmill was estimated to have a lifetime of 10 years (323).

It is probable from the above description that parts of the system will last for about 30 years, and other parts for less. The value of 20 years chosen here is in some sense an average of the two types of parts.

b. *Construction*

There have been a number of estimates for construction times. Reference 199 states that 112 workers are needed to produce 4 MW of rated capacity in 1977. However, it is not clear whether this applies to (a) construction, (b) materials acquisition and fabrication, or both. It has been stated (218) that construction of a 2.5-MW (rated) windmill being built at present requires about 115,000 man-hours. Using the factors noted above, this is about 115,000/2000 × 2.5 × 0.2 × 20 = 6 man-years per megawatt-year.

Another estimate can be based on Reference 321, which noted that erection cost for a 1.2-MW (rated) windmill in 1971 dollars was $102,900. Allowing $6 per man-hour (a likely rate in 1971), a load factor of 0.2, a 20-year lifetime, and 2000 man-hours per man-year, the erection time is 102,900/1.2 × 2000 × 6 × 0.2 × 20 = 1.8 man-years per megawatt-year. This can be taken as an upper bound.

To put this last value on a more understandable basis, consider the 200-kW (rated) Clayton windmill mentioned before (251). It is implied that it would require 1.8 × 2000 × 0.2 × 20/280 = 10 workers for seven weeks (280 hours) to erect it. De Renzo (509) states that times of 6–8 weeks are required to erect steel (truss) towers for large windmills. However, the number of workers is not specified. He notes that the erection time for concrete towers is 8–12 weeks.

Another estimate gives the installation costs in 1974 to be $120/kW (rated) (378). Assuming $8 per man-hour (a likely rate in 1974) with the same assumptions about load factor, lifetime and man-hours per man-year, this is 120 X 1000/8 X 0.2 X 20 X 2000 = 1.9 man-years per megawatt-year. De Renzo (511) estimates erection and installation costs per MW (rated) for 0.1 and 1.0-MW turbines as $162,000–$175,000 and $24,600–$33,700, respectively. Using the load factor and lifetime assumed above and a wage rate of $10 per hour, this is (162,000–175,000)/10 X 2000 X 0.2 X 10 = 2.0 - 2.2 and (24,600 - 33,700)/10 X 2000 X 0.2 X 20 X 0.1 = 3.1 - 4.2 man-years/MW-yr.

A General Electric study (63) yields a construction labor requirement of 2400–6400 man-hours per MW (rated) for 1.5-MW units. Using the assumptions about load factor and lifetime mentioned above, this corresponds to (2400–6400)/2000 X 0.2 X 20 = 0.3–0.8 man-years per megawatt-year. The average of these values, (0.3 + 0.8)/2 = 0.55, can be taken as a lower bound. These values are used in Table H-2, where it is assumed that half the construction time is spent on electrical work and half on roofing and sheet metal. Concrete work is left as a separate category. The lower bound corresponds to 10 workers erecting the Clayton (251) windmill in about two weeks.

Flaim *et al.* (567) list site-dependent costs (taken to be primarily labor costs) for 11 wind turbines. If the load factor is assumed to be 0.2, labor $10 per hour, and the lifetime 20 years, the manpower ranges between 0.37 and 10.2 man-years/MW-yr, with an average of 2.4 man-years/MW-yr. This is higher than the range of 0.55–1.8 man-years/MW-yr used here. Flaim *et al.* assume 100 units of production for the systems they consider, which probably produces a lower labor requirement than is presently the case.

In addition to construction time, de Renzo (508) notes that assembly times are involved. For 0.5 and 1.5-MW (rated) systems he estimates assembly costs, in 1975 dollars, of $31,400 and $35,100, respectively. Allowing $8 per man-hour and the load factor and lifetime assumed above, this is (31,400–35,100)/8 X 2000 X (0.5–1.5) X 0.2 X 20 = 0.4–1.0 man-years/MW-yr. On the other hand, de Renzo (509) also estimates 16 man-weeks for assembly of nacelles. The rating of the wind turbines is apparently not specified but may be assumed to be 0.5–1.5 MW. The assembly time would then be 16 X 40/(0.5–1.5) X 0.2 X 20 X 2000 = 0.05–0.16 man-years/MW-yr.

There is some evidence that risk of material acquisition may be underestimated (see Appendix E, Section i). Using the values of Ramsay (491), the steel and glass specified in Table H-2 corresponds to (23/0.50)5 X

Table H-2. Material Acquisition and Construction Risk of Wind Power (per Megawatt-Year Net Electrical Output)

	Materials (Metric Tons)	*Man-Hours (per Metric Ton)*	*Man-Hours*	*Man-Days Lost, Accident*	*Man-Days Lost, Illness*	*Deaths (X 10^3)*	*Man-Days Lost, Total*
Material and Equipment Acquisition							
Iron ore mining	38	0.44	16.7	0.009	–	0.004	0.03
Hard coal mining (b)	34	1.2	41.3	0.041	0.001	0.025	0.19
Steel	23 (g)	10.0	231	0.093	0.002	0.008	0.15
Copper ore mining	57	0.47 (a)	27	0.015	–	0.006	0.05
Copper (d)	0.57	13.2 (a)	7.5	0.006	–	–	0.01
Nonmetal mining	4.0	0.50	2.0	0.001	–	–	–
Flat glass	3.4	38	130	0.041	0.001	0.004	0.07
Fiberglass insulation (c)	3.4	38	130	0.007 (e)	0.002 (e)	0.008 (e)	0.13
Cement (h)	5	0.84	4.2	0.001	–	–	–
Fabricated metal products	23.1	149	3400	1.54	0.046	0.155	2.51
Total	191			1.75	0.052	0.21	3.14
Construction (f)							
Roofing and sheet metal		162	550–1800	0.46–1.52	0.007–0.022	0.11–0.37	1.2–3.8
Electrical		56	550–1800	0.14–0.47	0.003–0.011	0.05–0.16	0.43–1.4
Concrete	5	0.33	1.6	0.003	–	–	–
Total				0.6–2.0	0.01–0.03	0.16–0.53	1.6–5.2

(a) From Table G-2.
(b) Soft coal mining would change results only slightly.
(c) Plastics considered as fiberglass for computational purposes.
(d) See footnote (d), Table G-2.
(e) See footnote (c), Table G-2.
(f) See footnote (e), Table M-1, for discussion.
(g) Reference 280 states that 1000-MW (rated) windmills require 2.9×10^5 metric tons of steel. If these windmills have a lifetime of 20 years and a load factor of 0.20, they produce $1000 \times 20 \times 0.2 = 4000$ megawatt-years over their lifetime. This corresponds to 72 metric tons of steel per megawatt-year.
(h) Includes cement, sand, gravel, etc.

$10^{-9} = 2.3 \times 10^{-7}$ and $(3.4/0.36)2 \times 10^{-7} = 1.9 \times 10^{-6}$ of U.S. production, respectively. The total number of injuries attributable to the weight of steel and glass is $2.3 \times 10^{-7} \times 1 \times 10^5 + 1.9 \times 10^{-6} \times 5 \times 10^3 = 0.033$. The total number of deaths is $2.3 \times 10^{-7} \times 235 + 1.9 \times 10^{-6} \times 70 = 1.9 \times 10^{-4}$. Allowing 6000 man-days lost per death and 50 per injury, this is $6000 \times 1.9 \times 10^{-4} + 50 \times 0.033 = 2.8$ man-days lost/MW-yr.

Results are shown in Table H-2. Inclusion of the category "Fabricated metal products" implies that the metals used are fabricated to about the same level, in terms of forging, stamping, etc., as the average for typical metals. This is clearly an approximation.

ii. Emissions

The emissions due to windpower will be less than that arising from solar space heating, primarily because of the relatively small amount of coal used in the steel for building windmills. Using the assumption from Section 2I that the effects of emissions vary directly with steel use, the number of man-days lost is (0.058–0.18) × 23 = 1.3–4.1.

By way of comparison, Reference 282 estimates that 26 metric tons of sulfur oxides will be produced during the construction of 10,000 small turbines. However, this omits sulfur oxides from steel production. If a value of 0.01 metric tons of sulfur oxides per metric ton of steel is assumed (220), this would produce an extra 0.01(3000–3600) = 30–36 metric tons of sulfur oxides, based on the amount of steel assumed. This yields a total of (30 + 26)–(36 + 26) = 56–62 metric tons. It is claimed (283) that these turbines will produce 4.5–9.0 MWh per year. If their lifetime is 20 years, the 10,000 turbines will produce 20 × (4.5–9.0) × 10,000/8760 = 103–204 MW-yr over their lifetime. Then (56–62)/(103–204) = 0.27–0.60 metric tons of sulfur oxides per megawatt-year are produced. Since 1 metric ton of sulfur oxides corresponds to 6–18 man-days lost, the model considered in this paragraph corresponds to (6–18) (0.27–0.60) = 1.6–11 man-days lost per megawatt-year.

iii. Operation and Maintenance

The number of maintenance workers for windpower will probably be higher than needed in the solar technologies. The number of maintenance workers in one estimate is about 75% of construction workers (102). (This value may include engineering, administration, and surveying personnel.) Using the value of construction labor from this reference (higher than that of Section ib), 7 × 0.75 = 5.3 man-years per megawatt-year are required, in contrast to the 1.0 value for solar thermal (349). The risk values for solar thermal are then multiplied by 5.3/1.0 = 5.3, not including the climatic factor for solar thermal. It is implicitly assumed that the trades are similar.

To put these values on a more understandable basis, consider the Clayton windmill (249). If its present load factor was about 0.20, with its rated capacity of 200 kW it would produce about 200 × 0.2/1000 = 0.04 MW-yr/yr. The operation and maintenance assumed here would be about 5.3 × 0.04 × 2000 = 420 man-hours per year for this installation, or about one-fifth of a man-year. This time would include all operation, supervision, and repair of components.

A British work (313) estimates operation and maintenance costs annually to be \$21.60–\$31.20 per kilowatt installed. Assuming a load factor of 0.20, this is 5(21.60–31.20)1000 = \$108,000–\$156,000 per megawatt-year. If one allows about \$5 per man-hour, which might be the current British rate, this is (108,000–156,000)/5 × 2000 = 10.8–15.6 man-years per megawatt-year, where there are 2000 man-hours in a man-year.

Another estimate is due to Brown and Higgin in Ontario (309). They estimate that the annual maintenance cost for a 150-kW (rated) machine is \$36,100. Using a load factor of 0.20 and a cost per man-hour of \$10 (North American conditions), operation requires 36,100 × 1000/0.2 × 150 × 10 × 2000 = 55 man-years per megawatt-year. Even if half this time were spent operating the diesel equipment assumed to accompany the windmill, about 28 man-years per megawatt-year would be required, greater than is assumed here.

As another method of estimation, Brown and Warne (376) estimate that operation and maintenance costs are 7.7% of the purchase price per year for smaller windmills. Purchase prices vary strongly. An Aerowatt 4.1-kW (rated) costs around \$55,000 (400); a 6-kW (rated) machine costs around \$6000 (401). Assuming \$8 per man-hour and the usual load factor and lifetime, operation and maintenance time could be between 6000 × 0.077 × 1000/8 × 2000 × 6 × 0.2 × 20 and 55,000 × 0.77 × 1000/8 × 2000 × 4.1 × 0.2 × 20, or 1.2–16 man-years per megawatt-year. If half the cost were for parts, this could be 0.6–8 man-years per megawatt-year. In this case, risk attributable to manufacturing the parts would have to be added.

Flaim *et al.* (558) estimates annual operating and maintenance costs to range from \$14,000 to \$41,000/MW (rated) for five turbines from 0.2 MW (rated) to 2.0 MW (rated). Allowing $10 per man-hour and assuming that all the cost is due to labor, this produces (14,000–41,000)/10 × 0.2 × 2000 = 3.5–10.3 man-years/MW-yr, if the load factor is 0.2.

For small windmills, Reference 338 states, "Most common models require a complete overhaul every 1000 hours of use." However, it is not clear how much time this would take or what trades would be used.

Requirements for operation and maintenance are still somewhat speculative. In the Clayton work (249), "The operating and maintenance costs have yet to be studied fully both because of insufficient data and because it would be difficult, if not impossible, to separate 'normal' start-up costs from the ones associated with (this) machine." As more data become available, it is likely that the above values will be modified.

It has also proved difficult to obtain data on older windmills. One source (380) suggests that the Danish windmills of World War II had "relatively high working and maintenance costs." No quantitative data are given, or data on the division of these costs between parts and labor. The Danish Gedser windmill, which operated in the 1950s and 1960s, ceased operation because operating expenses were too high (381). Again, no quantitative data were given. These results suggest that the values assumed here may be low.

iv. Energy Back-up

Windpower is weather-dependent, although it may be more or less so than solar power, according to its location. It is assumed here that the back-up energy that windpower will require is about the same as solar thermal or solar photovoltaic. The risk calculated for these two technologies may then be used.

Miller writes (484): "For a larger amounts of wind-power (greater than 5%–10% of total installed electrical capacity) there would have to be a back-up system as there is no evidence that any guaranteed minimum output can be achieved from a wind generator and therefore there would be no reduction in the amount of generating plant required to meet the peak demand."

Flaim *et al.* (559), in calculating the cost of windpower, add the cost of the storage and back-up systems. Their previous publications (565, 566) also assume back-up. Simmons discusses back-up and storage in some detail, although without evaluating them quantitatively. He notes (258) that "the best solution (of wind variability) consists in having a diesel-driven generator in continuous operation."

There is no rule that can be given for the exact amount of back-up required. Reckard (436) assumes a back-up of 30%. In another estimate, Wolff and Meyer (254) note that about 22% back-up was required for a small 8-kW windmill. It can also be assumed that no back-up is required or that low-risk back-up is used. Appendix N can be used to show that these two possibilities are similar in terms of the overall risk ranking of this system.

v. Energy Storage

It is likely that the rock-and-oil storage system assumed for solar thermal electric will not be suitable as storage for windpower. Pumped storage or batteries would presumably be used. However, it is noted in Appendix E that the risk associated with storage systems other than rock-and-oil will probably be higher. In the interest of giving windpower the benefit of the doubt, it will be assumed that the risk of storage is that of rock-and-oil, with a maximum storage time of 6 hours. Flaim *et al.* (565) use a value of 10 hours, as does Lovins (577). Reckard (285) suggests 3–4 days storage. Reference 205 indicates that 38 days of storage might be necessary, although this would eliminate the back-up requirement. Littler and Thomas (504) state that for a house requiring 5 kWh/day, a probable minimum capacity would be 250 kWh, or 250/5 = 50 days. This would also eliminate back-up requirements. Daey Ouwens (78) assumes that storage is required but does not specify a value. De Renzo (513) states that 3 days storage is required for small windmills. Stewart (409) suggests 5 days storage. The combination of storage time and back-up chosen should produce an average load factor of about 0.7.

As noted in Section iv, Simmons has discussed the problem of storage, indicating that at least some storage will be required. He states (259): "Systems must operate with storage, or be interconnected to base-load power grid, to meet base-load and peak-load demands. Individual units may require up to 60 days or more of energy storage capacity, depending on region, unless supplemental energy supply is used or units are intertwined on a large dispersed network." He goes on to note (260): "It is incorrect to view and to cost out wind-powered systems solely as fuel savers. A self-contained storage system is needed, and power should be offered for sale on demand." De Renzo (507) assumes an "energy storage cycle time of about ten hours. This may not be sufficient for some wind energy base-load applications where storage times of several days may be required to compensate for short time variations in wind velocities, or where storage times of six months or more may be required to allow for seasonal variations in wind strength." Simmons (257) gives an example of the relationship between storage and generating capacity. In the Israeli windmill referred to above (a 4.3-m diameter Jacobs unit equipped with a 150 A-h battery), only about 65% of the energy produced was utilized, the rest regarded as loss during battery charging. Another estimate (548) suggests that only 43% of the energy extracted from the wind was usable. Finally, several of the smaller windmills he discusses (256) either come equipped with batteries or have some provision for storage. However, Simmons does not discuss the relationship of storage to power quantitatively.

One installation (325) of a 2-kW windmill uses batteries with an energy capacity of 130 A-h at 120 V. The energy in these batteries when fully charged is then 120 X 130/1000 = 15.6 kWh. The storage time is then 15.6/2 = 7.8 h, more than is assumed here. The batteries are claimed to have an average life of 15–20 years. De Renzo (507) assumes an average lifetime of 10–20 years for batteries and 20 years for thermal storage. In another example, Noll (514) recommends 230 A-h of battery capacity for a 200-W turbine. If the voltage is 12 V, this is 230 X 12/200 = 14 h of storage.

As another example, Justus (360) calculated the storage times required to produce a given power reliability for an average power output of 200 kW from a 1500-kW (rated) windmill, corresponding to an average load factor of 200/1500 = 0.13. He found that 24–48 h of storage would be required to produce reliability of 95% in New England, and more than 95% in the central United States. This implies that storage requirements may be substantially underestimated here. A Swedish proposal for a 1000-MW windmill suggests a storage time of 5 days, or 120 h (379).

For small windmills, Reference 339 states, "If we are to proceed to construct an operating wind power system from reliable existing components, we must content ourselves with the good old lead–acid battery." Appendix F indicates that using these batteries will likely produce substantial risk.

vi. Transportation

As before, it is assumed that the transportation risk is in direct proportion to the weight of material transported. The weight of nonsand and cement materials is 180 metric tons per unit energy, from Table H-2. To this should be added, in principle, the weight of the storage system. As mentioned above, the storage system is assumed to be rock-and-oil, although a workable system probably would be pumped storage or batteries. In order to give wind energy the benefit of the doubt, it will be assumed that the rock and oil are moved short distances, with negligible risk. If the weights, mode of transport, and distances moved for pumped storage or battery materials are determined, the associated risk should be added to the following. The transportation risk is then the coal transportation risk multiplied by 180/3500 = 0.051. See also Appendix P.

There is some evidence that risk estimated here is low (see Appendix E, Section vi). Ramsay (491) notes that 1.11 metric tons correspond to 6.4×10^{-3} trucking and warehousing accidents and 3.0×10^{-5} deaths. Using the present weight of finished materials (steel, copper, etc.) yields $(53/1.11) \times 6.4 \times 10^{-3} = 0.31$ accidents and $(53/1.11) \times 3.0 \times 10^{-5} = 1.4 \times 10^{-3}$ deaths. Allowing 6000 man-days lost per death and 50 per injury, this is $6000 \times 1.4 \times 10^{-3} + 50 \times 0.31 = 24$ man-days lost/MW-yr. Risk due to transportation of raw materials would be extra. Transportation of nonsand and cement raw materials by rail would produce another (129/3500) (15–47) = 0.6–1.7 man-days lost, for a total of 25–26.

vii. Summary

Results of the above calculations are shown in Table H-3. Much of the risk comes from the energy back-up and storage required, with another sizable contribution from material acquisition and construction. In the foregoing, bizarre accidents like windmill blades striking passers-by are not emphasized, but only the ordinary day-to-day risk in building and operating windmills. However, the risk from unusual sources should not be disregarded in future analyses. One source (281) notes "The risk to human life from blade throw and tower collapse is an important consideration especially in densely populated areas."

APPENDIX I. OCEAN THERMAL

The concept of producing energy by taking advantage of the thermal gradients in oceans is an old one, dating back to the Frenchman d'Arsonval, who proposed it in 1881. It is really a solar system that works like an inside-out refrigerator. A refrigerator uses an electric pump to produce a temperature difference between an insulated box and the room. The ocean thermal system uses the thermal gradient between the warm surface and cold lower levels of the ocean to drive a fluid, which turns a turbine

Table H-3. Windpower Risk (per Megawatt-Year Net Electrical Net Output)

Risks	*Material Acquisition and Construction*	*Emissions*	*Operation and Mainte-nance*	*Energy Back-up (d)*	*Energy Storage*	*Transpor-tation (f)*	*Total*
Occupational							
Accidental							
Death	(3.7–7.4) × 10^{-4}	–	2.6 × 10^{-3}	(14–31) × 10^{-5}	3.8 × 10^{-4}	(8–26) × 10^{-5}	(36–43) × 10^{-4}
Injury	0.05–0.08	–	0.15	0.022	0.02–0.07	(7–25) × 10^{-4}	0.24–0.32
Man-days lost, excluding deaths	2.5–3.9	–	7.5	1.1	1.0–3.5	0.035–0.13	12–16
Man-days lost, total	4.7–8.3	–	23.2	2.0–3.2	3.3–5.8	0.5–1.8	34–42
Disease							
Death	(a)	–	1.4 × 10^{-5} (c)	(1.2–19) × 10^{-6}	(4.4–11) × 10^{-6} (c)	–	(1.9–4.4) × 10^{-5}
Disability	(13–18) × 10^{-4}	–	3.2 × 10^{-4} (b),(c)	(9–17) × 10^{-5}	(1.0–2.6) × 10^{-4} (c)	–	(1.8–2.6) × 10^{-3}
Man-days lost	0.065–0.088	–	0.10	0.012–0.12	0.03–0.08	–	0.21–0.39

Public							
Accidental							
Death	–	–	–	$(8–13) \times 10^{-5}$	–	$(4–10) \times 10^{-5}$	$(1.2–2.3) \times 10^{-4}$
Injury	–	–	–	$(1.0–7.7) \times 10^{-4}$	–	0.8×10^{-4}	$(1.8–8.5) \times 10^{-4}$
Man-days lost, total	–	–	–	0.48–0.81	–	0.26–0.60	0.74–1.4
Disease							
Death	–	$(3.5–11) \times 10^{-5}$	–	$(4.2–14) \times 10^{-4}$	–	–	$(4.6–15) \times 10^{-4}$
Disability	–	0.22–0.65	–	2.3–6.9	–	–	2.5–7.6
Man-days lost	–	1.3–4.1	–	15–43	–	–	16–47

(a) See footnote (a), Table G-3.
(b) See footnote (e), Table E-4.
(c) See footnote (f), Table E-4.
(d) See Appendix N for discussion of risk assuming no back-up or low-risk back-up.
(e) See footnote (t), Table A-2.
(f) See footnote (p), Table A-2.

to produce electric power. The system is sketched in Figure I-1. Greater detail is available elsewhere (103).

Although the ocean thermal system was tested off the coast of Cuba in the early 1930s and off Africa in the 1950s, comparatively little is known about its material, labor, and other requirements. What follows can only be called a rough estimate.

i. Material Acquisition and Construction

Ocean thermal systems use some of the power they generate in order to operate, so that there is a difference between the gross output and net output. For example, Boot and McGowan (104) describe a system with

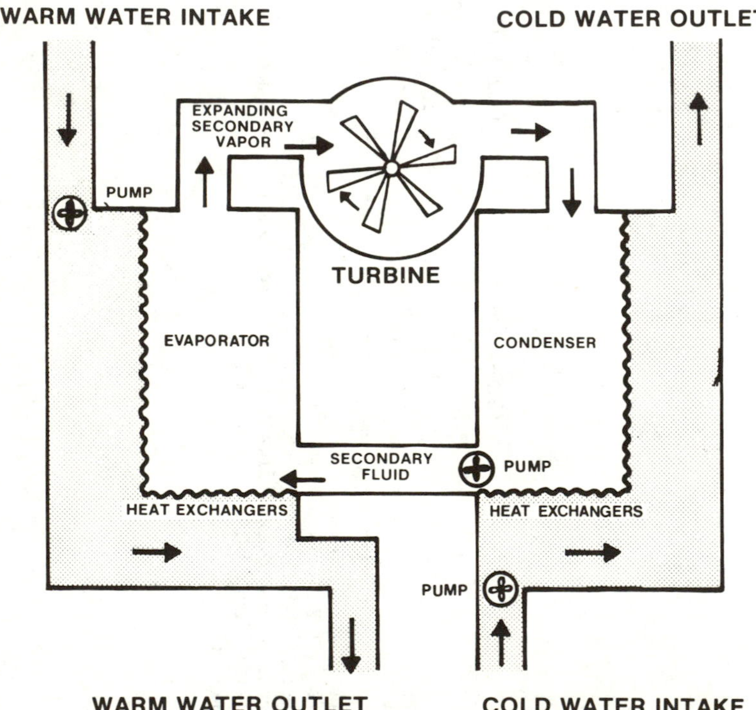

Figure I-1. *Ocean thermal energy system. The efficiency of the system will depend on the temperature difference between the surface and bottom layers, the design of the heat exchangers, and many other factors.*

100-MW gross output, but only 59.1-MW net output. It is assumed that the net output is 60% of the gross, although there are systems with as low as 54% efficiency (105). [A small system being proposed for Hawaii has 20% efficiency (607).]

Pollard (106) discusses the design of a 160-MW system, nominally rated. It will be assumed that this is net power. He notes that an authoritative report specifies the use of 242,000 metric tons of concrete, 41,600 of steel, and 1640 of titanium for main system, generators, pumps and motors, and turbines. One report (422) suggests that the aluminum requirement for 1000 MW of capacity would be 58,000 metric tons. If one assumes that this aluminum could substitute for titanium, the question of calculating risk for titanium does not arise.

There are at least two other materials (107, 112) used in ocean thermal systems. Copper is used for cables, and ammonia is used as the working fluid to transmit heat from one part of the apparatus to another. With the assumptions on net power and load factors noted above, the weight of copper and ammonia for the Pollard system would be 1840 and 2170 metric tons, respectively.

Table I-1 shows the total weight of materials required for the standard of 1000-MW net output used here. The amounts are clearly less than the other nonconventional systems considered, although they are by no means small.

For lack of evidence to the contrary, the ocean thermal station was assumed to have a lifetime of 30 years. This may be an overestimate, since operations in a hostile ocean environment, with storms and corrosion, will

Table I-1. Materials Required for 1000-Megawatt Average Ocean Thermal System in 10^5 Metric Tons (a),(c)

Aluminum	*Steel*	*Concrete*	*Copper (b)*	*Ammonia*
0.83	4.1	24	0.16	0.19

(a) Load factor of 0.7. Some estimates have been up to 0.9 (423, 424). However, they have yet to be confirmed in practice.

(b) Original data had values for extensive copper cable system for power transmission. Since transmission lines were generally not considered in previous sections, they were not included in this case.

(c) An estimate for a 160-MW (net) system, scaled up to an average 1000-MW power, is (in 10^5 metric tons): steel, 4.8; concrete, 16.9; titanium, 0.49 (301).

almost certainly reduce lifetimes considerably. Until tests are performed, it is not known how much this reduction will be. One higher estimate of lifetime has been 38 years (424), but this has not been yet confirmed in practice.

Results of the calculations are shown in Table I-2. The risk from material acquisition is considerably lower than the other nonconventional technologies considered, primarily due to the small weight of material per unit energy produced. Inclusion of the category "Fabricated metal products" implies that the metals used are fabricated to about the same level, in terms of stamping, forging, etc., as the average for typical metals. This is clearly an approximation.

Questions about the methods and risk of construction are not simply answered, because of the novelty of the system. About 5800–8300 man-years have been estimated to construct what is probably a standard 100-M net prototype (107, 498). Using a 30-year lifetime, with a load factor of 0.7, this is (5800–8300)2000/30 $\times$ 0.7 $\times$ 100 = 5500–7900 man-hours /MW-y.

The allocation of these man-hours among the building trades is also a difficult task. For simplicity, one-half is allotted to each of shipbuilding and miscellaneous contracting. Concrete construction, which in any case produces a small risk value, is put into a separate category. The exact proportion in the two main groups does not affect the results significantly.

The risk from construction is much greater than for materials acquisition and fabrication. In terms of total man-days lost, the factor is between 4.5 and 7. This is in accordance with the results of Appendix E.

There is some evidence that risk of material acquisition may be underestimated (see Appendix E, Section i). Using the values of Ramsay (491), the steel and aluminum specified in Table I-2 correspond to (12.4/0.50) $5 \times 10^{-9} = 1.2 \times 10^{-7}$ and (2.8/0.25)$6 \times 10^{-8} = 6.7 \times 10^{-7}$ of U.S. production, respectively. The total proportion for steel and aluminum is $(1.2 + 6.7) \times 10^{-7} = 7.9 \times 10^{-7}$. The total number of injuries attributable to the weight of steel and aluminum is $7.9 \times 10^{-7} \times 1 \times 10^{5} = 0.079$. The total number of deaths is $7.9 \times 10^{-7} \times 235 = 1.9 \times 10^{-4}$. Allowing 6000 man-days lost per death and 50 per injury, this is $6000 \times 1.9 \times 10^{-4} + 50 \times 0.079 = 5.1$ man-days lost/MW-yr.

ii. Emissions

As usual, the emission due to material acquisition is assumed proportional to the weight of steel and aluminum used. The latter material has a greater risk per unit weight. The estimated weight of sulfur oxides is 0.01 $\times$ 12.4 + 0.049 $\times$ 2.8 = 0.26 metric tons/MW-yr, where 12.4 and 2.8 are the weights of steel and aluminum, respectively. From Section 2I, this weight of sulfur oxides corresponds to 0.26(6–18) = 1.6–4.7 man-days lost.

iii. Operation and Maintenance

A 32–35 man crew for a 100-MW plant is assumed (113). With a load factor of 0.7, this is (32–35)2000/100 $\times$ 0.7 = 910–1000 man-hours/MW-yr net output. The closest analogy to the type of maintenance done on these plants in terms of standard industrial classifications is "water transportation services," in which a total of 506 man-days was lost to nonfatal injuries and illness per 100 full-time workers in 1974. To obtain the results of Table I-3, it is assumed that the relative proportion of accidents, disease, and death is the same as in a trade like roofing and sheet metal. From Table M-2, the ratio of deaths to man-days lost from accidents and illness for roofing and sheet metal is 0.056/175 = 3.2×10^{-4}, so the number of deaths per 100 man-years for water transportation services is deduced to be about $3.2 \times 10^{-4} \times 506 = 0.16$.

In another estimate, the operating and maintenance costs for a 160-MW (net) system was given as $940,000 per year (316). This included a service ship, minisubmarine and crew, etc., but excluded spare parts, assumed to be about 1/3 of the cost. If labor is counted at $10 per hour, with a load factor of 0.7 the number of man-hours per megawatt-year is 940,000/10 $\times$ 160 $\times$ 0.7 = 840, lower than assumed here. On the other hand, a 55-MW (net) system had estimated operating costs of $2.5 million per year (317). If spare parts accounted for 1/3 of the cost, as in the previous example, if the load factor is as above, and if labor is again counted at $10 per hour, the number of man-hours per megawatt-year is 2,500,000 $\times$ (2/3)/10 $\times$ 55 $\times$ 0.7 = 4300. This is higher than assumed here.

Flaim *et al.* (563) note that TRW and Lockheed estimates of operation and maintenance costs are 0.48 and 0.19 cents/kWh, respectively. Again, assuming that spare parts are 1/3 of the cost, and that labor is $10 per hour, this is 0.67(0.19–0.48)8760 $\times$ 1000/10 $\times$ 100 = 1110–2820 man-hours/MW-yr. In the same reference, Lavi estimated in 1978 that costs for cleaning processes only would be 0.04–0.36 cents/kWh, for an average of 0.19 cents/kWh. If all of this is due to labor, making the above assumptions again yields 0.19 $\times$ 8760 $\times$ 1000/10 $\times$ 100 = 1660 man-hours/MW-yr for cleaning alone.

Table I-2. Material Acquisition and Construction Risk of Ocean Thermal Power (per Megawatt-Year Net Electrical Output)

	Materials (Metric Tons)	*Man-Hours (per Metric Ton)*	*Man-Hours*	*Man-Days Lost, Accident*	*Man-Days Lost, Illness ($\times 10^4$)*	*Deaths ($\times 10^3$)*	*Man-Days Lost, Total*
Material and Equipment Acquisition							
Iron ore mining	20.6	0.44	9.1	0.005	0.8	0.002	0.02
Hard coal mining (a)	18.6	1.2	22.3	0.021	0.6	0.013	0.10
Steel	12.4	10.0	124	0.050	12.3	0.004	0.08
Copper ore mining	54	0.47	25.4	0.014	2.6	0.006	0.05
Copper (e)	0.54	13.2	7.1	0.005	1.4	0.001	0.01
Bauxite mining	11.1	0.50	5.5	0.003	0.5	0.001	0.01
Aluminum	2.8	12.4	34.7	0.014	3.4	0.001	0.01
Nonmetal mining (g)	80	0.55	44	0.012	3.3	0.007	0.06
Ammonia	0.6	1.1 (b)	0.66	(c)	(c)	(c)	
Fabricated metal products	15.7	149	2340	1.04	316	0.105	1.70
Total	216			1.16	340	0.14	2.0
Construction (f)							
Concrete	120	0.33	39.6	0.014	5.3	0.0065	0.058
Shipbuilding (d)			$(2.8\text{–}4.0) \times 10^3$	2.2–3.2	660–950	0.22–0.31	3.6–5.2
Miscellaneous contracting			$(2.8\text{–}4.0) \times 10^3$	1.8–2.6	250–360	0.58–0.85	5.3–7.7

4.0–5.8	910–1300	0.80–1.2	8.9–13

(a) See footnote (b), Table H-2.
(b) 12.5×10^6 metric tons were produced in the United States in 1971 (108). Of this, 70%, or 9.6×10^6 short tons, were used directly as fertilizer (109). The total use of fertilizers on U.S. farms was 35.5×10^6 metric tons (110), so ammonia constituted 24.2% of all fertilizers. Chemicals used on U.S. farms had a total value of \$3.0 billion, and fertilizers had a total value of \$2.1 billion, or 70% (110). Assuming linearity between weight and value, ammonia constituted $0.242 \times 0.7 = 0.169$ of the weight of all agriculture chemicals. The total number of agricultural chemical workers in 1970 was 41,000 (111). It is then deduced that ammonia producers numbered about $0.169 \times 41{,}000 = 6900$. The number of man-hours per metric ton is then $6900 \times 2000 \times 1.1/13.7 \times 10^6 = 1.1$. No account is taken of the primary chemicals and materials which are required to produce ammonia, so the overall risk is probably underestimated.
(c) Total man-days lost for accidents and illness were 68 per 100 man-years for agricultural fertilizers. Distribution for accidents, illness, and deaths assumed to be the same as for steel.
(d) Total man-days lost for accidents and illness were 162 per 100 man-years. Distribution for accidents, illness, and deaths assumed to be the same as for fabricated metal products.
(e) See footnote (d), Table G-2.
(f) See footnote (e), Table M-1, for discussion.
(g) Includes cement, sand, gravel, etc.

Table I-3. Ocean Thermal Power Risk (per Megawatt-Year Net Electrical Net Output)

Risks	*Material Acquisition and Construction*	*Emissions (c)*	*Operation and Mainte-nance*	*Energy Back-up*	*Energy Storage (a)*	*Transpor-tation (f)*	*Total*
Occupational							
Accidental							
Death	$(9.4–13) \times 10^{-4}$ (b)	–	$(73–80) \times 10^{-5}$	–	–	$(6–19) \times 10^{-5}$	$(17–23) \times 10^{-4}$
Injury	0.10–0.14	–	0.046–0.05	–	–	$(5.1–19) \times 10^{-4}$	0.15–0.19
Man-days lost, excluding deaths	5.2–7.0	–	2.3–2.5	–	–	0.026–0.095	7.5–9.6
Man-days lost, total	10.8–15	–	6.7–7.3	–	–	0.4–1.4	18–24
Disease							
Death	(b)	–	–	–	–	–	–
Disability	$(2.6–3.4) \times 10^{-3}$	–	5.7×10^{-4} (d),(e)	–	–	–	$(32–40) \times 10^{-4}$
Man-days lost	0.13–0.17	–	0.03	–	–	–	0.16–0.20

Public							
Accidental							
Death	–	–	–	–	–	$(31–74) \times 10^{-6}$	$(31–74) \times 10^{-6}$
Injury	–	–	–	–	–	6.2×10^{-5}	6.2×10^{-5}
Man-days lost, total	–	–	–	–	–	0.2–0.5	0.2–0.5
Disease							
Death	–	$(4–13) \times 10^{-5}$	–	–	–	–	$(4–13) \times 10^{-5}$
Disability	–	0.25–0.75	–	–	–	–	0.25–0.75
Man-days lost	–	1.6–4.7	–	–	–	–	1.6–4.7

(a) Energy storage may be required if plants are built far enough from shore.
(b) See footnote (a), Table G-3.
(c) See footnote (t), Table A-2.
(d) See footnote (e), Table E-4.
(e) See footnote (f), Table E-4.
(f) See footnote (p), Table A-2.

iv. Energy Back-up

There is no energy back-up required for this system.

v. Energy Storage

There is no energy storage required for this system, at least for sites close to land. However, as these sites are used up, new sites must be built farther from shore. This will probably require the use of intermediate forms of energy, such as methanol, and the use of tankers (112). Because these are future developments, they are not considered here.

vi. Transportation

Transportation risk is again assumed to be proportional to the weight of nonsand and cement material used in construction. It is 136/3500 = 0.039 of coal transportation risk. See also Appendix P.

There is some evidence that risk estimated here is low (see Appendix E, Section vi). Ramsay (491) notes that 1.11 metric tons corresponds to 6.4×10^{-3} trucking and warehousing accidents and 3.0×10^{-5} deaths. Using the present weight of finished materials yields $(32/1.11) \times 3.0 \times 10^{-5} = 8.6 \times 10^{-4}$ deaths and $(32/1.11) \times 6 \times 10^{-3} = 0.17$ accidents. Allowing 6000 man-days lost per death and 50 per injury, this is $6000 \times 8.6 \times 10^{-4} + 0.17 \times 50 = 14$ man-days lost/MW-yr. Risk due to transportation of raw materials would be extra.

vii. Summary

The overall level of risk in ocean thermal energy production, as shown in Table I-3, is by far the lowest of the nonconventional technologies considered here. Whether or not this is due to serious underestimates of the materials required and the load factor will not be known until operating data are acquired. One way of explaining the results is by noting that the ocean itself acts as a collector of solar energy, and so there is no need for "collectors" of large area as in the case of solar thermal and solar photovoltaic.

APPENDIX J. METHANOL

Using wood to produce wood alcohol (methanol) is another nonconventional source of energy. The actual methanol synthesis from hydrogen and carbon monoxide feedstock is a well-established technology, and the production of the liquid goes back to the destructive distillation of wood in the mid-1800s. The development of copper-based catalysts has permitted the production of methanol at medium pressures and temperatures of 250°C. Details are given elsewhere (114).

Methanol can be used in transportation as well as electricity production; the latter is considered here. It differs from the other nonconventional technologies considered here by being based on a chemical process. In addition, its end-use involves burning, as do coal and oil. From this viewpoint, methanol has characteristics of both conventional and nonconventional energy systems.

In effect, methanol production involves the use of a renewable resource —wood—in the same way that solar collectors use renewable energy from the sun. The difference lies in the fact that an effort must be made to replant and harvest more trees, whereas no such effort is required for solar and wind energy.

There are other methods for producing energy from wood or wood products, such as using milling or logging residue, burning wood directly, etc. Methanol can also be produced from agricultural and municipal wastes. These are generally not considered here.

Because of the hybrid nature of methanol production, elements of the methodology used for both conventional and nonconventional energy systems are adopted. This does not pose a barrier to analysis, since both methodologies are only slightly different ways of describing the total risk of a system.

It is not suggested here that the system outlined below will actually be used to generate baseload electricity. As mentioned above, a possible use for methanol will be transportation. However, the efficiency of converting chemical to mechanical energy is fairly low. Assuming transportation as an end-use will produce relatively high risk because of this low efficiency. The reasoning in terms of end-use employed below can be viewed as a method for minimizing the calculated risk associated with methanol.

i. Material Acquisition and Construction

For the purposes of calculation, logging and harvesting of wood is assigned to the area "Operation and Maintenance." This section is concerned only with acquiring materials for and the construction of the methanol factory.

No estimates could be found for the weight, type of materials required, and construction times to build a methanol factory. However, an oil refinery is a close analogy to this type of factory. An indirect approach will be taken in calculating these quantities. A 100 million gallon per year methanol plant costs about $94 million. Of this, $30 million is for the gasifier, $16 million for the oxygen plant, $8 million for the reformer, and $15 million for the methanol synthesis plant (291). It is shown in Section vii that this plant produces 46 Mw-yr of electricity annually, so the monetary cost is about $94,000/46 = $2040 per kW(e) (rated).

Reference 33 notes that the cost to produce 1 kW(e) (rated) for an oil refinery is $57.40. Both estimates relate to costs in the mid-1970s. If it is assumed that the material weights and construction times are roughly proportional to monetary costs, the weights and times for a methanol plant should be 2040/57 = 36 times that of an oil refinery.

Mackay and Sutherland (312) find that a methanol refinery producing 910 metric tons per day has a capital cost of $170 million (described as a realistic value), in 1976 dollars. The systems considered here produce about 1100 metric tons per day at a capital cost of $94 million. The Mackay–Sutherland model then has a capital cost per ton of methanol equal to 170 X 1100/910 X 94 = 2.2 times that assumed here. Presumably its material weight and construction times would also be greater.

Reference 33 notes that 55.5 man-hours of labor, 0.38 metric tons of metals (assumed here to be steel), and 0.42 metric tons of concrete are associated with 1 MW-yr output from an oil refinery. However, these values are based on a 30-year lifetime. The economic lifetime of a methanol refinery seems to be about 20 years (292, 293). It is not clear why the lifetimes apparently differ, but it may be due to the fact that a methanol plant handles wood, and an oil refinery handles petroleum.

As a result, the materials and construction times noted above should be multiplied by (30/20)36 = 48, yielding 2660 man-hours of labor, 18.2 metric tons of metals, and 20.2 metric tons of concrete. By way of comparison, Reference 624 estimates that 2 to 5 times more construction workers are needed to install a given capacity to burn wood fuel than are needed for the same capacity to burn coal. From Table A-1, construction time per megawatt-year output for a coal-fired plant is 429 man-hours. This implies that the wood-fired construction time is (2–5) 429 = 850–2100 man-hours/MW-yr. It is likely that building a methanol refinery would take more construction time.

In addition to this, there is a risk associated with the capital costs of harvesting wood. The costs include machinery for felling, hauling, loading, sorting, chipping, etc. The cost is about $169 million over about 20 years (294). If 46 MW-yr of electricity are produced annually from the methanol refinery being considered (see Section vii), the capital cost of harvesting per kW(e) (rated) is $169,000/46 = $3670.

It is difficult to estimate the material weights and construction times associated with this cost. However, the capital costs per kW(e) in harvesting coal and oil are about $58 and $71, respectively (9, 33). The latter two types of operation may be similar to that of methanol in terms of machinery per unit cost if not in other respects. About 1.2 metric tons of steel and 50–106 man-hours of labor per megawatt-year electricity output are associated with harvesting coal and oil, respectively. If it is again assumed that the material weight and construction times vary proportionately with costs, then about 3670 × 1.2/(58–71) = 62–76 metric tons of steel and 3670 × (50–106)/(58–71) = 3200–5500 man-hours of labor are associated with the capital requirements for harvesting wood for methanol.

Finally, there is risk associated with building the facility for converting methanol to electricity. As a rough approximation, it can be assumed that the facility will be similar to that used for burning oil. From column 3 of Table B-1, the construction labor is 250 man-hours per megawatt-year output. About 1.4 metric tons of metal and 2.4 metric tons of concrete are required. It is possible that these values may be an underestimate, since the volume of methanol used per unit electrical output will be greater than that of oil.

In total, there are 2660 + (3200–5500) + 250 = 6100–8400 man-hours of labor, 18.2 + (62–76) + 1.4 = 82–96 metric tons of steel, and 20.2 + 2.4 = 23 metric tons of concrete per megawatt-year of electricity. Using the ratio of risk to material weight and construction time (assuming it falls into "miscellaneous contracting") for the appropriate amounts of coal, iron ore, steel, fabricated metal products, cement, concrete, and construction time as shown in Table F-2, the numbers of man-days lost due to accidents, illnesses, and deaths per unit energy output are 9.9–12.3; 0.22–0.28; and 12–15, respectively. Results are shown in Table J-1. Inclusion

Table J-1. Material Acquisition and Construction Risk of Methanol (per Megawatt-Year Net Electrical Output)

	Materials (Metric Tons)	*Man-Hours (per Metric Ton)*	*Man-Hours (d)*	*Man-Days Lost, Accident*	*Man-Days Lost, Illness*	*Deaths ($\times 10^3$)*	*Man-Days Lost, Total*
Material and Equipment Acquisition							
Iron ore mining	137–160	0.44	60–70	0.04–0.05	0.001	0.017–0.020	0.13–0.16
Hard coal mining (b)	123–144	1.2	148–173	0.15–0.17	0.003–0.004	0.088–0.103	0.68–0.80
Steel (a)	82–96	10	820–960	0.33–0.38	0.008–0.010	0.029–0.034	0.50–0.59
Nonmetal mining (e)	23	0.50	12	–	–	–	0.02
Cement (c)	23	0.84	19	0.01	–	–	0.01
Fabricated metal products (a)	82–96	149	12,200–14,300	5.4–6.4	0.16–0.19	0.55–0.65	8.9–10.4
Total	470–542		13,330–15,550	5.9–7.0	0.17–0.21	0.68–0.81	10.2–12.0
Construction (f)							
Concrete	23	0.33	7.6	–	–	–	0.01
Miscellaneous contracting			6100–8400	4.0–5.4	0.05–0.07	1.3–1.7	11.6–16.0
Total			6100–8400	4.0–5.4	0.05–0.07	1.3–1.7	11.6–16.0

(a) Assuming all steel is fabricated.
(b) Soft coal mining would change total results only slightly.
(c) Includes cement, sand, gravel, etc.
(d) Equals the product of the two previous columns.
(e) Equals weight of cement.
(f) See footnote (e), Table M-1.

of the category "Fabricated metal products" implies that the metals used are fabricated to about the same level, in terms of stamping, forging, etc., as the average for typical metals. This is clearly an approximation.

By the way of contrast, Segerson (532) estimates that 528 construction workers would be required for 2–3 years to build a 50-MW (rated) wood-fired plant. Excluded from this calculation are 243 workers in unspecified supporting services. If the plant lifetime is 20 years and it has a load factor of 0.8 (Segerson assumes a higher load factor but this is as yet unverified), there are $2000 \times 528 \times (2–3)/50 \times 20 \times 0.8 = 2600–4000$ man-hours/MW-yr for construction. This is comparable to the sum of the values used here for constructing (a) the methanol refinery and (b) the methanol-to-electricity facility, excluding the construction time for the wood-harvesting equipment. Adding in the manpower in the supporting services would make the values deduced from the values of Reference 532 higher than used here. A comparable value is deducible from Reference 621, where it is stated that the capital costs of a wood-burning facility are 5–10 times that of a comparable fossil-fuel facility, per unit energy output. Assuming that construction times are roughly proportional to capital costs, and using the values from Tables A-1 and B-1, column 3 this is (5 - 10) (250 - 429) = 1250 - 4290 man-hours/MW-yr for construction of a wood-burning installation.

The end-use of methanol or the wood used to produce it can vary. It is assumed here that the end-use is electricity. This is in line with the suggestion of Reference 289, which indicated that a best use for this fuel would be in gas turbines for electrical peak power generation. The efficiency of this methanol-to-electricity conversion is about 0.33 (211). Of course, it is also possible to burn wood directly to produce electricity, but this is not the end-use completely evaluated in this appendix.

ii. Emissions

Emissions-related risk from methanol may be divided into two parts; first, emissions produced in gathering materials for factory construction, and second, emissions produced from fuel combustion. Some data, as noted below, indicate that methanol may generate pollutants in its production and combustion. In the interest of giving this system the benefit of the doubt, the second part of the emissions risk is not combined with the first.

Assuming that 82–96 metric tons of steel are used in construction, and noting from Section 2I that 1 metric ton of steel produces 0.058–0.18

man-days lost, a total of (82–96) × (0.058–0.18) = 4.7–17 man-days are lost per unit energy. These values are used here.

While wood burning (as opposed to methanol production) is not completely evaluated here, Reference 280 notes that 600–11,000 metric tons per year of sulfur oxides are produced from operations resulting in 1000-MW net power from burning wood directly. The variation depends in part on the crop rotation period. The value corresponds to (600–11,000) /1000 = 0.6–11 metric tons/MW-yr. Since it is assumed that each metric ton of sulfur oxides corresponds to about 12–36 man-days lost (see Section 2I), there are about (0.6–11) × (12–36) ≐ 7.2–396 man-days lost per megawatt-year.

In another reference dealing with the general subject of biomass, it is noted that cornstalks are 0.11% by weight sulfur, as compared to an estimate of 0.2% for oil (410). It is later stated (411) that the sulfur content of wood is 0.1%. To convert these values into weight of sulfur oxides, another table in this document (412) notes that 0.11%–0.3% sulfur weight of oil from biomass (origin unspecified) produces 0.14–0.49 lb of sulfur oxides per million Btu. Assuming a conversion efficiency of oil-to-electricity of about 0.35, this produces (0.14–0.49) × 3600 × 8760/10^6 × 1.055 × 2200 × 0.35 = 5.4–19 metric tons of sulfur oxides per megawatt-year. This corresponds to 31–340 man-days lost per megawatt-year.

In terms of general nonsulfur pollution from use of methanol, Posner (193) has noted potential problems. DiNovo *et al.* (515) have stated: "Use of this highly toxic material (methanol) in fuel distribution systems will present a host of new problems . . . spills and pipeline leakage in transport could adversely affect both humans and animals in contact. . . ."

iii. Operation and Maintenance

The operation and maintenance requirements can be divided into two parts: in the factory, and in the forest, felling timber and hauling logs.

A 100 million imperial gallon per year methanol plant produces 46 MW-yr net energy (see Section vii below). It requires 208 man-years, or about 41,600 man-hours (190). Then 904 man-hours are required to produce 1 MW-yr of net energy. Smith *et al.* (33) find that 204 man-hours of operation and maintenance in an oil refinery produces (3.7–5.6) × 10^{-5}, (3.1–4.0) × 10^{-3}, and 0.38–0.54 deaths, injuries, and man-days lost, respectively. Assuming that the risk in a methanol plant is the same per man-hour as in an oil refinery, the risk for methanol is then (1.6–

2.5) $\times$ 10^{-4}, 0.014–0.017, and 1.7–2.4 deaths, injuries, and man-days lost, respectively. This estimate may be low for three reasons. First, risk from disease has been set to zero for lack of data. Second, since methanol is produced from wood and wood chips, handling these materials may be riskier than handling oil products. Third, another estimate (23) gives an upper limit on the total number of man-days lost per unit time in oil refineries which is an order of magnitude greater than that of Smith *et al.* Further investigation is needed to resolve these points.

By way of comparison, Reference 624 finds that a wood-burning plant requires 2.7–7.5 times the operating labor of a coal-burning plant. Using Table A-1, this implies a requirement of (2.7–7.5)407 = 1100–3050 man-hours/MW-yr, higher than is assumed here.

The second part of operation and maintenance to consider is logging. A total of 730 man-years of logging is required for a 100 million imperial gallon per year plant (117). Methanol produces 21.6 kWh (74,000 BTU) of energy per imperial gallon, so the gross energy output of this plant each year is $0.0216 \times 10^8/8760 = 246$ MW-yr. What is the electrical energy? The 33% efficiency of the methanol-to-electricity process plays a part here (211). The gross electrical energy is then $246 \times 0.33 = 81$ MW-yr. It is shown in Section vii that the net electrical energy, taking account of energy requirements to produce methanol, is 46 MW-yr.

Then 730/46 = 15.9 man-years of logging per megawatt-year of useful energy are needed. A total of 296 man-days is lost due to accident and disease per 100 man-years in the category "logging camps and logging contractors" (15). If the distribution of disease, accidents, and deaths is similar to that of roofing and sheet metal (another outdoor trade), the number of man-days lost due to disease and accidents per 100 man-years would be 4.1 and 292, respectively. In addition, there would be 0.095 fatalities. This value is lower than the rate of 0.12 fatalities per 100 man-years recorded for Canadian forestry workers from 1966 to 1975 (344). There would then be $0.159 \times 4.1 = 0.65$; $0.159 \times 292 = 46$; and $0.159 \times 0.095 \times 6000 = 91$ man-days lost due to disease, accidents, and deaths, respectively. Results are shown in Table J-2.

The Office of Technology Assessment (624) finds that the logging and wood-products industry has 14 times more occupational injuries and illnesses per unit energy of fuel produced than bituminous (underground) coal mining. Most of the risk values in Table A-2 for coal mining refer to a combination of above-ground and underground mining, so they cannot be used as a direct comparison. However, Reference 22 estimates 7–20

Table J-2. Methanol Risk (per Megawatt-Year Net Electrical Net Output)

Risks	*Material Acquisition and Construction*	*Emissions (b)*	*Operation and Maintenance*	*Transportation (c)*	*Total*
Occupational					
Accidental					
Death	$(2.0\text{–}2.5) \times 10^{-3}$	–	1.5×10^{-2}	$(2.0\text{–}7.1) \times 10^{-4}$	0.017–0.018
Injury	0.20–0.25	–	1.0	$(1.6\text{–}6.8) \times 10^{-3}$	1.2
Man-days lost, excluding deaths	10.1–12.7	–	49	0.08–0.34	59–62
Man-days lost, total	22–28	–	141	1.2–4.5	164–173
Disease					
Death	(a)	–	9.0×10^{-5}	–	9.0×10^{-5}
Disability	0.0045	–	2.1×10^{-3}	–	6.6×10^{-3}
Man-days lost	0.22–0.26	–	0.65	–	0.87–0.91
Public					
Accidental					
Death	–	–	–	$(1.0\text{–}2.7) \times 10^{-4}$	$(1.0\text{–}2.7) \times 10^{-4}$
Injury	–	–	–	$(2.0\text{–}2.3) \times 10^{-4}$	$(2.0\text{–}2.3) \times 10^{-4}$
Man-days lost, total	–	–	–	0.6–1.6	0.6–1.6

Disease					
Death	–	$(1.3–4.6) \times 10^{-4}$	–	–	$(1.3–4.6) \times 10^{-4}$
Disability	–	0.8–2.7	–	–	0.8–2.7
Man-days lost	–	4.7–17	–	–	4.7–17

(a) See footnote (a), Table E-4.
(b) See footnote (t), Table A-2.
(c) See footnote (p), Table A-2.

and 0.03–0.4 man-days lost per megawatt-year for accidents and disease, respectively, in underground mining, for a total of 7–20 man-days lost. Multiplying this by the ratio of 14 noted above, this would yield 98–280 man-days lost per unit energy for logging and wood products. This may be an underestimate, since the logging industry has 1.4 times the lost workday cases of accidents and illnesses per worker (624) as compared to the combined logging and wood-products industry. If the average number of workdays lost per case is the same for (a) logging and (b) logging and wood products, the man-days lost per unit energy for logging alone is 1.4(98–280) = 140–390. It seems logical to consider logging alone, since other phases of methanol risk are evaluated elsewhere in this appendix.

The system considered here requires 2.2 metric tons of dry wood per metric ton of methanol (295). MacKay and Sutherland (306) assume a value of 3.2 metric tons of dry wood for the same output. If the collection efficiency were the same, the time per unit output from wood harvesting would have to be multiplied by 3.2/2.2 = 1.45. If the risk per unit time were the same, this implies that the present risk from this source may be underestimated.

If wood were burned to produce electricity directly, results could be somewhat different. Rennie, in Carlisle (290), finds that 12,000 oven-dried tons of wood per day are required to fire a 1000-MW wood-burning electricity plant. The 100 million imperial gallon methanol plant discussed above requires 880,000 oven-dried tons of wood per year (295), produced by 730 workers. Each then supplies 1210 oven-dried tons per man-year. Assuming that the same rate of collection applies to the wood-burning plant, a total of 12,000 × 365/1210 = 3600 man-years are required. This is 3.6 man-years/MW (rated), or about 23% of the value noted above. Reference 532 assumes a value of 3.0 man-years/MW (rated) for a 50-MW plant. On the other hand, it can be deduced from the data of Marshall (296) that 9.5 man-years/MW (rated) are needed, under somewhat different assumptions. It is not surprising that these values are lower than those for producing methanol, since a subtantial portion of the energy in wood does not appear in methanol as a final product.

Since methanol can be called a "renewable" technology, it is presumed that reforestation will be undertaken on land where wood has been harvested. Fertilizers and pesticides may be used. Risk attributable to their production and use should, in principle, be added to risk from other sources. Due to lack of data, this is not done here.

iv. Energy Back-up

Because of the continuous nature of the methanol process, no energy back-up is required.

v. Energy Storage

No energy storage is required.

vi. Transportation

Transportation risk forms a small part of total risk for most energy systems, and this also holds true for methanol. The nonsand weight of the system is about 426–496 metric tons per unit energy. The transportation risk is then the coal transportation risk multiplied by (424 - 496)/3500 = 0.12 - 0.14. Results are shown in Table J-2. Transportation risk in logging is assumed to be part of operation and maintenance. Public transportation risk in operation and maintenance, such as people killed in collisions with logging trucks, is not considered here. See also Appendix P.

vii. Electricity and Oil

To produce 1 ton of methanol requires 600 kWh of electricity (286). Another estimate (287) notes that this value could be high as 3740 kWh/ton. In the interest of giving methanol the benefit of the doubt, the lower value will be chosen. The 100 million imperial gallon plant discussed above produces 1200 tons/day (288). Then 0.6 × 1200 × 365/8760 = 30 MW-yr of electricity are required annually. The plant has a gross energy output of 81 MW-yr (see Section iii), so this corresponds to 30/81 = 0.37 MW-yr of electricity per unit energy output.

There are two ways of evaluating the electricity requirement in terms of risk. One way is to multiply the weighted risk of Appendix N by 0.37 and add it to the other sources of risk. This would produce about 63–170 man-days lost for the Canadian case and about 120–340 for the United States case.

Another assumption which could be made is that the methanol itself is used to produce the required electricity. This would be analogous to ocean thermal power, where some of the electricity produced is used in operation. It is not known whether methanol would be used to produce the electricity necessary for the refinery. If it were, the risk from all other sources would be multiplied by 1/(1 - 0.37) = 1.59, since the net output of the system would be reduced. However, this assumption would produce a lower total risk than assuming that electricity was supplied by the "average grid" of Appendix N. To give methanol the benefit of the doubt, the "self-generating" assumption will be made.

In addition, 10^6 Btu of oil are required per ton of methanol (286). If there are 6.3 X 10^6 Btu per barrel of oil, and 13,200 barrels of oil are required to produce 1 MW-yr of electricity (208), then the oil required would produce 1200 X 365 X 10^6/6.3 X 10^6 X 13,200 = 5.3 MW-yr. The risk associated with this oil is the occupational risk of gathering and handling it. From column 1 Table B-2, this is 5.3 (1.6 - 16) = 8.5 - 85 man-days lost.

If the oil is used for its petrochemical qualities, then this risk should be added. If it is used for its thermal properties, it may again be assumed that the methanol is burned instead. Allowing 74,000 Btu per imperial gallon, this would produce 74,000 X 10^8 = 74 X 10^{11} Btu (assuming perfect efficiency). There are 1200 tons of methanol produced daily, so the thermal requirements in terms of oil are 10^6 X 1200 X 365 = 4.4 X 10^{11} Btu. Then the ratio of oil thermal requirements to thermal methanol output is 4.4/74 = 0.06. If the electricity and oil inputs are supplied from the methanol, the risk from other sources should be multiplied by 1/(1 - 0.37 - 0.06) = 1.75. The net annual energy of the facility is then 81/1.75 = 46 MW-yr of electricity. Both assumptions about methanol supplanting electricity or oil produce a lower calculated total risk.

viii. Summary

Results are shown in Table J-2. Some of this risk could be reduced if wood were burned directly to make electricity. This possibility is not considered in detail here. However, the risk would still be non-negligible.

APPENDIX K. HYDROELECTRICITY

Hydroelectric stations now generate the largest proportion of Canadian electricity and are expected to do so until well past 1990. The data on hydroelectricity were not in the same form as that for other energy systems, so it was difficult to use some of it in the present analysis. As a result, the following calculations can be viewed as an estimate.

In spite of this, some quantitative observations can be made. Jassby (137) summarized the situation succinctly:

> The notion that hydroelectric power is a "clean" and dependable alternative to other, more obvious polluting energy sources is not a viable one.

i. Material Acquisition

The risk due to hydroelectricity can be divided into two parts: (a) that incurred in material acquisition, construction, and other forms of normal operation, and (b) that due to catastrophic dam failures. Information on (b) will be analyzed after the routine risk is considered.

Data on the amount of steel, concrete, and other raw materials required for a typical dam are not plentiful. Part of the problem is that there exists no "typical" dam; all are somewhat different. However, Morison of Ontario Hydro (138) has made estimates of the experience in Ontario over the past 30 years. Data are shown in Table K-1.

The materials shown can be grouped into three categories: steel, rock and earth, and concrete. Albers *et al.* (140) have indicated that materials like aluminum, chromium, copper, manganese, mica, and nickel will also be required. However, the amounts specified are small and can be neglected without altering the results significantly.

Table K-1. Materials Required for 1000 Megawatt Capacity Hydroelectric System, in 10^5 Metric Tons (138) (c),(d),(f)

Reinforcing Steel	*Structural Steel*	*Rock (a),(g)*	*Earth (a),(g)*	*Concrete (b),(e)*
0.14	0.068	117	44	27.5

(a) Density assumed to be 1.9 metric tons/m^3 (142).
(b) Density assumed to be 2.4 metric tons/m^3 (139).
(c) A Bechtel Corporation report (143) lists rock and earth requirements approximately seven times as great.
(d) Turbines, transformers, and other electrical equipment may not be included here. It is implicitly assumed that they are a small fraction of steel requirements.
(e) In a previous era, the Hoover Dam required 4.4×10^6 cubic yards of concrete for a dam rated at 1345 MW (546). The weight for 1000 MW (rated) was then, using footnote (b), $44 \times 2.4/1.345 \times (1.09)^3 = 60 \times 10^5$ metric tons.
(f) Load factor assumed to be 0.6; see text for explanation.
(g) This may include material only excavated, rather than used in construction.

The range of values for material requirements can be illustrated for some dams in the western United States. Ranging in capacity from 40 MW to 1060 MW, their steel and concrete requirements for an average 1000-MW power (in the same units as Table K-1) were 0.13–4.3 and 6.7–340, respectively (314). The values of Table K-1 are somewhere in between.

Table K-2 relates these quantities to the unit energy of 1 MW-yr over the lifetime of the system. Morison implicitly assumes a lifetime of 50 years with a load factor of 0.60, which means that each megawatt of installed capacity produces 50 × 0.6 = 30 MW-yr over its lifetime. Inclusion of the category "Fabricated metal products" implies that the metals used are fabricated to about the same level, in terms of stamping, forging, etc., as the average for typical metal uses. This is clearly an approximation.

Since the lifetime of a dam and its load factor play a part in subsequent calculations, a brief discussion is in order. Estimates of lifetimes vary considerably. Thomas (141) has summarized them:

> 60 years . . . is considered to be the life of a dam evaluation of worth . . . a life of 100 years may be given to a dam when assessing the unit cost of electricity generated at the site.

While Morison's estimate is somewhat lower, the small risk values of Table K-2 indicate that the precise lifetime chosen will not make a substantial difference in calculations. As examples, Wardrop (353) estimates a station life of 50 years at a capacity factor (load factor) of 0.7. This would mean that each megawatt of installed capacity would produce 50 × 0.7 = 35 MW-yr over its lifetime. Reference 664 assumes a capacity factor of 0.56 and a lifetime of 60 years, so that each unit of capacity produces 0.56 × 60 = 34 units of energy.

It should not be assumed that dams last forever. They may be abandoned for economic or other reasons. For example, it has been shown (147) that at least 120 dams in the United States were abandoned from 1964 to 1974.

The load factor, or ratio of energy produced to the maximum which could be produced, is lower for hydroelectricity than for most of the other energy systems considered. In terms of world hydroelectricity production, for example, the factor for 1974 was about 0.49 (144). In 1975, the factor for the United States was 0.54 (145). For Canada in 1976, it was 0.62 (146). (On the other hand, Reference 533 shows an average value for Spain for 1960–1967 of 0.33.) The factor of 0.60 chosen for computational purposes is then reasonable for North America, although it may be different for other countries.

Table K-2. Material Acquisition Risk of Hydroelectricity (per Megawatt-Year Net Electrical Output) (a)

Material and Equipment Acquisition	*Materials (Metric Tons)*	*Man-Hours (per Metric Ton)*	*Man-Hours*	*Man-Days Lost, Accident*	*Man-Days Lost, Illness*	*Deaths (X 10^3)*	*Man-Days Lost, Total*
Iron ore mining (c)	1.15	0.44	0.51	–	–	–	–
Hard coal mining (c)	1.04	1.2	1.2	0.001	–	0.001	0.01
Steel	0.69 (f)	10.0	6.9	0.003	–	–	–
Nonmetal mining (b)	628 (g)	0.50	324	0.089	0.002	0.055	0.42
Cement	92 (g)	0.84	77	0.027	0.001	0.003	0.04
Fabricated metal products (d)	0.69	149	103	0.046	0.001	0.005	0.07
Total	723			0.166	0.004	0.064	0.55

(a) See text for discussion of conversion factors from Tables K-1 to K-2.
(b) Includes rock, earth, and concrete materials.
(c) Assuming 1.67 kg of iron ore and 1.5 kg of coal per kg of steel.
(d) Assuming the weight of metal products equals the weight of steel.
(e) Reference 664 specifies 2.6 metric tons of copper, brass, and bronze, 1.3 of aluminum and castings, 0.8 of manganese, etc.
(f) Reference 664 specifies 126 of steel and castings.
(g) Reference 664 specifies 1340 of concrete.

ii. Construction

The problem with evaluating construction risk in the same way as for other energy systems is that the number of man-hours required for each task in hydro building is generally not known. That is, the hours spent in plumbing, cement work, etc. are not available for each dam. As an example from another era, the Hoover Dam required about 5000 men working for 5 years in the 1930s to build a dam of 1345-MW capacity (546). This is 5000 X 5/1345 = 19 man-years per MW capacity. If the death rate shown in Table M-2 for "miscellaneous contracting" is assumed, i.e., 0.042 deaths/100 man-years, this corresponds to 0.042 X 19/100 = 0.008 deaths/MW capacity. It is not known how representative this dam was of all dams. Reference 546 does not indicate the actual number of deaths or injuries. A similar approach is taken by Reference 664, which indicates that 4500 man-hours are required per MW-yr output.

A more direct approach based on experience in large utilities can be used. Morison (138) notes that there have been about 0.025 fatalities in Ontario per installed megawatt of capacity over the last 30 years. Potier (148), dealing with France, notes that the average number of deaths per year between 1953 and 1967 in hydroelectric construction was 20.6. The construction produced an installed power of 5315 MW, so there were 0.058 deaths per installed megawatt. Potier continues:

> Hydroelectric power is then more dangerous than thermal production. For the same period, the latter had 114 deaths for an installed capacity of 8666 megawatts, or 0.013 deaths per megawatt.

As noted in the previous part, 1 MW installed corresponds to about 30 MW-yr of energy produced over a dam's lifetime, so the Morison data correspond to 0.025/30 = 0.00083 deaths/MW-yr, and the Potier data to 0.058/30 = 0.00193. If 6000 man-days lost per death are allowed, this corresponds to 0.00083 X 6000 = 5.0 and 0.00193 X 6000 = 11.6 man-days lost, respectively, for the Morison and Potier estimates.

Potier notes that the average number of injuries in French hydroelectric construction was 3676 from 1956 to 1961. If we (a) assume this rate was similar for the 15 years over which 5315 MW were installed, (b) allow 50 days lost per injury, and (c) follow the reasoning in terms of dam lifetime and load factors used previously, the man-days lost to construction injuries are 3676 X 50 X 15/5315 X 50 X 0.6 = 17.3 per megawatt-year.

Assuming that the man-days lost due to injuries is the same for Ontario and France, the total man-days lost due to construction is then (17.3 + 5.0)–(17.3 + 11.6) = 22–29.

On the other hand, a calculation dealing with the Churchill Falls hydro dam in Canada yields estimated construction risk lower than this (425). Further analysis is required.

iii. Emissions

The weight of steel used in material acquisition was calculated to be 0.69 metric tons per unit energy, as noted in Table K-2. Using the estimated effects on public health per unit quantity of steel from Section 2I, the number of man-days lost per unit energy is 0.69(0.058–0.18) = 0.04–0.12. It is noted in Table K-1 that all the steel may not be accounted for.

iv. Operation and Maintenance

The overall risk from this source has been estimated. One source (149) indicates that occupational injuries in 1972 at hydroelectric sites in the United States occurred at about one-half the frequency and one-tenth the severity (in terms of man-days lost) of the average for all electric generating plants. However, an analysis of the same data for the years 1969 to 1972 indicates that the severity rate of accidents in hydroelectric plants was about 20% higher than for fossil-fired plants (150).

Potier (151) indicates that the death rate for operation and maintenance of hydroelectric dams in France is comparable to that of fossil-fired plants. For the former system, there was one death per 1580 MW-yr between 1953 and 1967. For the latter system, there was one death per 1160 MW-yr, or about 27% more. There are then 1/1580 = 0.00063 deaths per unit energy in hydroelectric operation and maintenance, corresponding to 6000 × 0.00063 = 3.8 man-days lost.

Bertolett and Fox (150) show that for operation and maintenance activities in the United States totaling about 7 million man-hours, there were 21 accidents per fatality. Each accident produced 72 man-days lost. Assuming that the ratio is the same in France, this implies 21 × 0.00063 = 0.0133 accidents per megawatt-year, and 72 × 0.0133 = 0.95 man-days lost. There is then a total of 3.8 + 0.95 = 4.8 man-days lost per unit energy due to death and accidents.

v. Energy Back-up and Storage

There is no back-up required for hydroelectricity. Risk due to storage is included in the normal material acquisition and construction categories. On the other hand, Lopez-Cotarelo (533) notes that for hydro dams in Spain the total storage capacity in 1978 was 1960 MW-yr. However, the actual stored energy ranged from 830 to 1600 MW-yr from 1973 to 1978. This suggests that back-up, storage, or both may be required for some hydro dams.

vi. Transportation

It is assumed that transportation risk is proportional to the weight transported, excluding cement and sand. The latter two quantities are probably moved only short distances, with small risk. As shown in Table K-2, the weight is 3.6 metric tons per unit energy. This yields a total number of occupational man-days lost between (3.6/3500) (10–35) = 0.01–0.04, assuming risk is proportional by weight to that of coal. Public man-days lost are even smaller. See also Appendix P.

There is some evidence, however, that risk estimated here is low. Ramsay (491) notes that 1.11 metric tons corresponds to 6.4×10^{-3} trucking and warehousing accidents and 3.0×10^{-5} deaths. Using the present weight of finished products (steel and fabricated metal products) yields (1.4/1.11) $\times 6.4 \times 10^{-3} = 8.1 \times 10^{-3}$ accidents and (1.4/1.11) $\times 3.0 \times 10^{-5} = 3.8 \times 10^{-5}$ deaths. Allowing 6000 man-days lost per death and 50 per injury, this is $6000 \times 3.8 \times 10^{-5} + 50 \times 8.1 \times 10^{-3} = 0.64$ man-days lost/MW-yr.

In terms of excluding sand and cement, this assumption may have to be reconsidered in the future, since dams are being built in increasingly remote locations. The James Bay development in Canada is an example.

vii. Dam Failures

Thousands of dams have been built over the past century all over the world. However, many are used for nonpower purposes, such as irrigation and navigation. Dams which have failed are allocated into two categories: those wholly or partly related to hydroelectricity production, and those which are not. For example, the well-known Johnstown flood of 1889 had nothing to do with hydroelectricity.

Because there have been many partial or total failures in the past, a decision was made to concentrate on those which caused the largest number of deaths. A list is shown in Table K-3.

In addition to this list, there have been other disasters about which detailed information was not available. This includes Hyokiri (Korea, 1961) with 250 dead, and Khadakwasla (Panshet, India, 1961) (157) with up to 250 dead. Further research is needed on these and other past dam failures.

Table K-3 indicates that between 4255 and 5729 people have died in dam disasters likely associated with hydroelectricity, since production

Table K-3. Major Disasters Likely Associated with Hydroelectricity Generation (c)

Place	*Date*	*Number Dead*	*References*
Vajont, Italy (a)	1963	2600–3000	153, 154 (b)
Gleno, Italy	1923	600	155, 156
St. Francis, United States	1926	426–450	155, 156
Kiev (Babi Yar), U.S.S.R.	1961	145	161
Koyna (Shivaji Sagar Lake), India	1967	180	158
Vega de Tera, Spain	1959	123–150	153, 155, 156
Sella Zerbino, Italy	1935	100–111	155, 156
Oros, Brazil	1960	30–1000	153, 155, 156
Coedty, Wales	1926	20–60	155, 156
Teton, United States	1977	9–11	159, 160
Bhakra, India	1959	10	155
Colorado Dam, Texas, United States	1900	8	155
Necaxa No. 2, Mexico	1909	4	155
		4255–5729	

(a) Dam did not fail, but was overtopped.

(b) Reference 155 indicated 5000 dead, but this may be in error.

(c) Since data on hydroelectricity production are considered only to the end of 1978, the Morvi dam accident of August 1979 in India is not included. This dam had a hydro component (531). A late report (535) placed the number of dead at 1335. This would increase the total associated with hydroelectricity at around 25%. On the other hand, a statement by the president of a local municipality near the dam accident places the number of missing as high as 25,000 (534). In either case, this accident is either the largest or second largest disaster in history associated with energy systems.

began in the late 19th century. While dams built many years ago are probably less well-designed than modern ones, their failure probably affected fewer people. For example, the Vajont disaster, the largest ever recorded in terms of deaths, occurred in 1963.

Some people may be injured as a result of a dam failure in addition to those who are killed. Most data on dam failures list only the number of dead. It proved possible to obtain the number of injured for only four failures. This information is shown in Table K-4.

It should be noted that three of the four cases in this table are non-hydroelectric dams. The implicit assumption is that the ratio of injured to dead is approximately the same for hydroelectric as opposed to non-hydroelectric (i.e., navigation, irrigation, etc.) dams. This is a reasonable assumption. There are many other factors, such as the amount of water behind the dam, where the victims live with respect to the dam, and the degree of preparation for emergency, which affect both the total number of victims and the ratio of injured to dead more than the type of dam.

The data of Table K-4 is quite limited. However, as an approximation the ratio of injured to dead varies between 1 and 8. This implies that the number of injured in hydroelectric dam failures throughout history is (4200–5700) (1–8) = 4200–46,000. Of course, some people may have been injured in dam failures in which nobody died.

The number of deaths and injuries must be related to the total amount of hydroelectricity produced. Since the calculated deaths and injuries refer to the world-wide experience, the data on hydroelectric energy must do so as well. Darmstadter *et al.* (152) have collected information on this subject and results are shown in Table K-5.

In order to convert these yearly quantities into a grand total, some assumptions must be made. One can assume (a) that production of hydroelectricity on a significant scale started around 1890 and increased linearly to 1925; (b) a linear increase between each following pair of years shown in Table K-5; and (c) 1978 is the last year considered. These assumptions tend to slightly overestimate the amount of hydroelectricity produced. The actual type of increase was probably exponential over most of the period under consideration, rather than linear, which would produce a slightly lower quantity than that estimated. In turn, this slight overestimate will produce a slight underestimate of the risk per unit energy.

Using the assumptions noted above, the total hydroelectricity produced from 1890 to 1978 was 3.67 million MW-yr. This information can be used to calculate the rate of death and injury per unit energy generated.

Table K-4. Ratios of Injured to Dead in Dam Failures

Place	*Date*	*Number Dead*	*Number Injured*	*Ratio of Injured to Dead*	*References*
Koyna (Shivaji Sagar), India (hydroelectric)	1967	175–180	1500	8.3–8.6	158, 162
Kelly Barnes, United States (nonhydroelectric)	1977	37	47	1.3	163
Baldwin Hills, United States (nonhydroelectric)	1963	3–5	15	3–5	155, 156
Mohegan Park (Spaulding Pond), United States (nonhydroelectric)	1963	6	6	1	155, 156

The estimated deaths per unit energy is (4255–5729)/3,670,000 = 0.0011–0.0016. The number of injuries is (4255–46,000)/3,670,000 = 0.0011–0.0128. It should be noted that these values are an average over

Table K-5. World Hydroelectricity Consumption for Selected Years (in Thousands of Megawatt-Years) (152), (a)

Year	*Quantity*
1925	9.0
1938	20.8
1950	39.6
1955	54.1
1960	78.5
1965	102.8
1967	115.1
1968	120.6
1980	216.4 (projected)

(a) Assuming one million metric tons coal equivalent equals 913 megawatt-years.

many countries and extend back considerably in time. It is not certain that they apply to present-day Canadian or North American experience.

While this report is concerned with average risk, not maximum risk, it is of interest that hydroelectricity may have the highest risk associated with a single accident (124).

viii. Summary

Results of calculations are shown in Table K-6. There are three major sources of risk: construction, operation and maintenance, and dam failures. The largest contribution is from construction. Other sources have either zero or negligible risk.

Hydroelectricity and nuclear power are two systems considered which have had major concern expressed over risk to the public in terms of disasters, although oil and liquefied natural gas tanker explosions can also create public disasters. It is of interest to compare the two sets of public risk for hydroelectricity and nuclear power.

In terms of total public risk, hydroelectricity has about 3–13 times the number of deaths of nuclear power. The ratio for disabilities and injuries is between 0.33 and 0.40. Finally, the ratio of total man-days lost in hydroelectricity as compared to nuclear is between 2.6 and 8.6.

Not all public risk is derived from real or potential catastrophes. For example, hydroelectricity has risk from transportation, and nuclear has risk from transportation and waste management. For a direct comparison of risk attributable to disasters, the public risk of the columns entitled "Electricity Production" in Table D-2 (for nuclear) and "Dam Failures" in Table K-6 can be considered.

When this pair of categories is evaluated, the ratios for hydroelectricity to those of nuclear are between 7 and 37 for deaths; between 490 and 1570 for injuries and disabilities; and between 7 and 37 for total man-days lost. It is then likely that hydroelectricity has both a higher overall and disaster-related risk than nuclear power.

APPENDIX L. TRANSMISSION LINES

For most of the systems considered in this book, the risk associated with the construction, installation, operation, and maintenance of electrical transmission lines will be low. As mentioned in Section 3A, most of the technologies considered here are or will be located close to where the energy they produce is used. That is, the length of transmission lines will be short.

By "transmission lines" is meant the lines from the energy source to the main distributing station or stations. Distribution lines are those that lead from this station to homes, factories, and offices.

In this appendix, only the risk associated with transmission lines is considered. There are, of course, many other considerations in constructing these lines. For example, their aesthetics have been carefully studied. References 164 and 165 discuss these factors in detail.

The costs of transmission lines can be substantial. Hydro-Quebec is building five 735-kV transmission lines, with a total length of about 4800 km (3000 mi), to link James Bay electrical production with the existing grid (166). The cost will be about $4 billion, which is approximately $780 per meter ($250 per foot). This value comprises capital cost, right-of-way, and installation. Of course, financial costs are not necessarily directly related to risk.

i. Material Acquisition and Construction

There are no units of transmission lines directly corresponding to that of 1000 MW-yr used elsewhere in this report. For simplicity, the unit to be used will be 1000 km in length.

In estimating the material requirements of transmission lines, it is important to consider the voltages and number of circuits. Over the years, the voltages have increased markedly, since a higher voltage can carry more power. For example, a 500-kV line can carry from 4 to 7 times the amount of power that can be carried on a 230-kV line, each with the same number of circuits (167). Generally speaking, the higher the voltage, the greater the material requirements and related costs.

All 765-kV systems use one-circuit towers (168). As mentioned above, Hydro-Quebec plans to use a 735-kV system in the James Bay project, and this will be assumed to be roughly equivalent in material requirements to the 765-kV system. Ontario Hydro has concluded that 500 kV is the appropriate voltage level for future bulk power transmission in that province (169). These high voltages are either being introduced now or are designed for the future. At present, most high-voltage lines in Ontario operate at 230 kV (170).

A brief description of typical transmission lines is now given (171). A 230-kV, two-circuit system has each of its circuits in three separate phases on each side of the vertical tower. The tower is 46.3 m high and 14.3 m

Table K-6. Hydroelectric Power Risk (per Megawatt-Year Net Electrical Output) (a)

Risks	*Material Acquisition and Construction (f), (g)*	*Operation and Maintenance*	*Transportation*	*Dam Failures*	*Total*
Occupational					
Accidental					
Death	$(8.9–20.0) \times 10^{-4}$	6.3×10^{-4}	–	–	$(1.5–2.6) \times 10^{-3}$
Injury	0.34	0.013 (e)	–	–	0.35
Man-days lost	22–29	4.4	–	–	26–33
Disease					
Death	(b)	–	–	–	–
Disability	8×10^{-5} (c)	–	–	–	8×10^{-5}
Man-days lost	0.004	–	–	–	0.004
Public					
Accidental					
Death	–	–	–	$(1.1–1.6) \times 10^{-3}$	$(1.1–1.6) \times 10^{-3}$
Injury	–	–	–	$(1.1–12.8) \times 10^{-3}$	$(1.1–12.8) \times 10^{-3}$
Man-days lost	–	–	–	6.6–10.0	6.6–10
Disease					
Death	–	–	–	–	$(1.1–3.2) \times 10^{-6}$ (d)
Disability	–	–	–	–	0.006–0.019 (d)

Man-days lost	–	–	–	–	0.04–0.12 (d)

(a) No energy back-up required; energy storage risk included in material acquisition.
(b) See footnote (a), Table E-4.
(c) Assuming 50 man-days lost per disability.
(d) From emissions due to steel production. See footnote (t), Table A-2.
(e) Reference 150 notes 72 man-days lost per injury.
(f) Reference 234 estimates 0.1–1.0 deaths per year for construction of a 1000-MW hydro plant. However, this is based on a 100-year lifetime. If this is converted to the 50-year lifetime assumed here, this is (100/50) (0.1–1.0) = 0.2–2.0 deaths. Reference 234 also assumes a 0.1–1.0 value for load factor. If the load factor of 0.6 used here is assumed, this implies that 0.6 (0.2–2.0)/(0.1–1.0) = 1.2 deaths per year are required to produce 600 MW-yr, or 1.2/600 = 2×10^{-3} deaths/MW-yr. Reference 234 incorporates disabling accidents with deaths, so allowing 6000 man-days lost per death, the result is $6000 \times 2 \times 10^{-3}$ = 12 man-days lost/MW-yr.
(g) Reference 235 indicates a value of about 0.025 deaths/MW of installed capacity at the Cabora Bassa dam in Mozambique, assuming a capacity of 2100 MW. This is similar to the values assumed here.

wide. Each phase consists of one 4.1-cm diameter aluminum conductor, steel reinforced. The 500-kV, two-circuit line is similar in general configuration to the 230-kV line. Each phase consists of four 2.4-cm diameter conductors forming a 51-cm square. The tower is 49.4 m high and 25.9 m wide. The 765-kV, one-circuit system has a tower 40.3 m high and 42.7 m wide, thus using more material in this part of the system than do the other two voltages.

Table L-1 shows the weight of materials required for each voltage per 1000 km. Data were not available for the 765-kV system, so its requirements were estimated indirectly by the following method. Reference 174 notes that the cost of a unit length of 765-kV, one-circuit line is about 1.1 times that of the same length of 500-kV, two-circuit line. Assuming that the cost is roughly proportional to the weight of material used, the second row of Table L-1 is multiplied by 1.1 to obtain the third row.

Table L-2 shows the material acquisition risk per km of line, using the same methodology as elsewhere in this book. The table evaluates the risk for two voltages: 230 and 500. The materials required and calculated risk for the latter voltage are placed in brackets below the corresponding values for 230 kV. To determine corresponding quantities for 765 kV, the 500-kV data may be multiplied by 1.1. Inclusion of the category "Fabricated metal products" implies that the metals used are fabricated to about the same level, in terms of stamping, forging, etc., as the average for typical metal uses. This is clearly an approximation.

The number of deaths per 1000 km of line varies between 0.4 and 1.2. The number of man-days lost per km, including nonfatal accidents and disease, varies between 6.3 and 17.5.

What is the construction risk? Smith (175) estimates that 4650 man-hours (or 2.33 man-years) of design and construction time are required per km of 500-kV line. It is assumed that this value is about the same for the other two voltages being considered, for lack of information to the contrary.

Data were not available on the proportion of time spent on either design or construction, nor on the trades involved in construction. As a very rough approximation, the total time can be divided into four equal parts: design, electrical construction, roofing and sheet metal, and miscellaneous contracting. While no roofing is done in transmission line construction, the risk in this trade probably approximates that incurred in working on and around high towers.

It is assumed that the design phase has little or no risk, yielding (3/4) × 4650 = 3490 man-hours equally distributed among the three trades

Table L-1. Materials Required for 1000-km Transmission Line, in 10^5 Metric Tons

Type of Line	*Steel*	*Aluminum*	*Porcelain (c)*	*Concrete*
230 kV, two-circuit (172)	0.32	0.18	0.023	1.30 (a)
500-kV, two-circuit (173)	1.06	0.32	0.040	4.14 (a)
765-kV, one circuit (b), (d)	1.18	0.36	0.045	4.61 (a)

(a) One cubic meter (1.31 cubic yards) weighs about 1.61 metric tons.
(b) See text for explanation.
(c) Used as insulation.
(d) Value for aluminum is close to that of Donakowski (233); values of steel and porcelain (assumed equivalent to glass) are about half that of Ref. 233; value of concrete is about 2.9 times that of Ref. 233.

mentioned above. Results of risk calculation are shown in Table L-2. The total number of man-days lost/km in construction, 6.1, is comparable to or lower than that of material acquisition. Even if all of the 4650 man-hours/km had been spent in roofing or sheet metal, a relatively hazardous trade, the number of man-days lost/km would have been only about 2.95 × 4 = 12.

ii. Emissions

As in previous appendices, emissions due to production of steel and aluminum must be considered in evaluation of risk. Using the reasoning noted in Section 2I and the values of Table L-1, the number of man-days lost/km for 230, 500-, and 765-kV lines are, respectively, 32(0.058–0.18) + 18(0.3–0.9) = 7.3–22; 106(0.058–0.18) + 32(0.3–0.9) = 16–48; and 118 (0.058–0.18) + 36(0.3–0.9) = 18–54.

iii. Operation and Maintenance

There is comparatively little published information available on operation and maintenance requirements for high-voltage transmission lines. One source (176) divides maintenance into three aspects: running, major,

Table L-2. Material Acquisition Risk of Transmission Lines (per Kilometer)

	Materials (Metric Tons)	*Man-Hours (per Metric Ton)*	*Man-Hours*	*Man-Days Lost, Accident*	*Man-Days Lost, Illness* $(\times 10^3)$	*Deaths* $(\times 10^3)$	*Man-Days Lost, Total*
Material and Equipment Acquisition (b)							
Iron ore mining	53 [177]	0.44	23 [78]	0.013 [0.044]	0.3 [0.8]	0.005 [0.018]	0.045 [0.153]
Hard coal mining	48 [159]	1.2	58 [191]	0.058 [0.191]	1.5 [5.0]	0.035 [0.115]	0.268 [0.883]
Steel	32 [106]	10.0	320 [1060]	0.129 [0.426]	3.2 [10.7]	0.011 [0.037]	0.199 [0.659]
Bauxite mining	72 [128]	0.50	36 [64]	0.020 [0.036]	0.4 [0.7]	0.008 [0.015]	0.071 [0.126]
Nonmetal mining	132 [418]	0.55	73 [230]	0.021 [0.066]	0.6 [1.8]	0.013 [0.040]	0.096 [0.303]
Porcelain	2.3 [4.0]	57 (a)	131 [228]	0.041 [0.071]	1.1 [2.0]	0.005 [0.007]	0.069 [0.121]
Concrete	130 [414]	0.84	109 [348]	0.037 [0.117]	0.1 [0.2]	0.004 [0.014]	0.060 [0.191]
Fabricated metal products	50 [138]	149	7450 [20,600]	3.3 [9.20]	101 [278]	0.335 [0.927]	5.44 [15.0]

Total	519 [1544]	8420 [23,200]	3.71 [10.3]	110 [330]	0.42 [1.19]	6.37 [17.7]
Construction (c)						
Electrical		1160	0.312	0.007	0.014	0.95
Miscellaneous contracting		1160	0.752	0.010	0.243	2.21
Roofing and steel metal		1160	0.999	0.014	0.325	2.95
Total			2.06	0.03	0.67	6.1

(a) Risk per unit time for porcelain is assumed to be the same as for flat glass. The number of man-hours required per metric ton for porcelain as compared to that of flat glass was multiplied by 1.5 to take account of the former material's greater complexity of shapes.

(b) First entry for each material denotes 230-kV line; second entry, 500-kV line. To obtain data for 765-kV line, multiply 500-kV data by 1.1.

(c) See footnote (e), Table M-1, for discussion.

and emergency. Running maintenance is divided into routine and diagnostic, the latter including a walking physical examination of the line.

Assuming an average velocity of 1 km/hr over comparatively rough terrain, this implies 1 man-hour/km for this part of maintenance. This does not include actual repairing of the line, replacement of parts, or major and emergency maintenance. As a rough estimate of total maintenance time, the above value should be multiplied by 10, yielding 10 man-hours/km per year. It should be recognized that this is a crude value, subject to alteration if and when more precise data are available.

The next question is, Which trade or trades does operation and maintenance comprise? Little information is available to answer this question. It is clear that maintenance includes activities that are comparatively risk-free, such as walking, and others that have more risk, such as helicopter patrols and replacement of wires on high towers. This mixture of activities is taken to be comparable in risk to that incurred in miscellaneous contracting, a construction trade. This trade includes various activities, and its risk per man-hour lies between those of hazardous trades, such as roofing and sheet metal, and safe trades, such as electrical work.

Using these assumptions and Table M-2, the number of man-days lost per km per year due to maintenance accidents is $130 \times 10/2000 \times 100 = 0.006$, and the number due to maintenance-induced disease is $1.8 \times 10/2000 \times 100 = 9 \times 10^{-5}$. The number of deaths is $0.042 \times 10/2000 \times 100 = 2.1 \times 10^{-6}$, and the total number of man-days lost is $6 \times 10^{-3} + 9 \times 10^{-5} + 6000 \times 2.1 \times 10^{-6} = 0.019$.

iv. Transportation

Table L-2 shows that the weight of nonconcrete materials for the 230- and 500-kV systems, in metric tons/km, are 260 and 710, respectively. Assuming that these materials are moved about the same distances as coal (see Table A-2), the values of deaths and injuries can be calculated as in previous appendices. Results are shown in Table L-3.

v. Summary

In summarizing the risk due to transmission lines, it should be remembered that all of the categories mentioned above, with the exception of operation and maintenance, refer to risk incurred in putting the lines into place.

To place this risk on an annual basis, one divides by the lifetimes of the lines. Written estimates of these quantities are difficult to obtain. It can be conservatively assumed that the lines have a lifetime, before replacement, of about 30 years, in analogy with the values used elsewhere in this book. This then implies that the risk from (a) material acquisition and construction, (b) emissions, and (c) transportation should be divided by this value.

Results are shown in Table L-3. Note that the results refer to 1000 km, not 1 km as in Table L-2. The values refer to 230-, 500-, and 765-kV systems.

The risk is fairly low, even when 1000 km are considered. Most of the occupational risk is from material acquisition and transport. The public risk is dominated by emissions from coal used to make the necessary steel. The occupational risk in man-days lost per 1000 km over the system lifetime is between 180 and 250 for a 230-kV line; between 450 and 630 for a 500-kV line; and between 490 and 690 for a 765-kV line. Putlic risk, in the same units and for the respective voltages is 260 - 770, 560 - 1680, and 630 - 1880.

These values of risk are of interest, but they are not on the same basis as those in the other appendices. The units used throughout this book are man-days lost per megawatt-year of energy produced. Transmission lines do not produce energy, but they do carry it.

In order to relate transmission line risk to energy carried over the line, the size of the power source, its estimated lifetime, and the length of the lines should be known. This clearly will vary with each system. However, a rough estimate of these quantities for a case that is well-known can be made. The James Bay development will generate about 10,200 MW, and its transmission lines will have a length of over 4800 km (3000 mi) (166).

While the lifetimes of hydroelectric dams are not known perfectly, Appendix K notes that an implicit lifetime of 50 years has been assumed by Morison (138). He also assumes a load factor of 0.6. The combination of these two quantities is equivalent to an effective lifetime of 30 (= 50 X 0.6) years at a load factor of 1.0. This value of 30 years is the same as that assumed for the lifetime of transmission lines, so the rated capacity can be used in the following calculation.

The James Bay development uses 735-kV lines, which can be taken as approximately equivalent in risk to the 765-kV lines for which risk has been estimated. Using Table L-3, the occupational risk, in man-days lost for the entire line, is (490 - 690) (4800/1000) = 2400 - 3300. The corres-

Table L-3. Transmission Line Risk (per 1000 Kilometers per Year Over System Lifetime) (a)

Risks	*Material Acquisition and Construction*	*Emissions (b)*	*Operation and Maintenance*	*Transportation*	*Total*
Occupational					
Accidental					
Death	0.014	–		0.004–0.012	0.020–0.028
	[0.039]		0.0021	[0.011–0.034]	[0.052–0.075]
	[0.043]			[0.012–0.037]	[0.057–0.082]
Injury	1.33	–		0.032–0.12	1.4–1.5
	[3.7]		0.065	[0.09–0.32]	[3.9–4.1]
	[4.1]			[0.10–0.36]	[4.3–4.5]
Man-days lost, total	124			32–105	180–250
	[340]	–	20	[80–260]	[440–620]
	[370]			[90–290]	[480–680]
Disease					
Death	(c)	–	(c)	–	(c)
Disability	0.067 (d)				0.069
	[0.184]	–	0.0016 (d)	–	[0.186]
	[0.202]				[0.204]
Man-days lost	3.7				3.8
	[10.1]	–	0.088	–	[10.2]
	[11.1]				[11.2]

Public					
Accidental					
Death	—	—	—	0.002–0.005 [0.005–0.013] [0.006–0.014]	0.002–0.005 [0.005–0.013] [0.006–0.014]
Injury	—	—	—	0.004 [0.011] [0.012]	0.004 [0.011] [0.012]
Man-days lost, total	—	—	—	12–29 [34–79] [37–86]	12–29 [34–79] [37–86]
Disease					
Death	—	0.006–0.02 [0.014–0.042] [0.016–0.046]			0.006–0.02 [0.014–0.042] [0.016–0.046]
Disability	—	39–116 (e) [85–255] [95–280]	—	—	39–116 [85–255] [95–280]
Man-days lost	—	250–740 [530–1600] [590–1790]	—	—	250–740 [530–1600] [590–1790]

(a) The first set of data in each entry refer to 230-kV, two-circuit; the second, to 500-kV, two-circuit; and the third, to 765-kV, one-circuit.

(b) See footnote (t), Table A-2.

(c) All deaths assigned to accidents.

(d) See footnote (e), Table E-4.

(e) See footnote (f), Table E-4.

ponding public risk is (630–1880) (4800/1000) = 3000–9000 man-days lost.

By way of comparison, Donakowski (233) finds 40,700 man-days lost in total occupational risk to build a 500-mi, 765-kV line. Putting this on a 1000 km and annual basis to correspond to Table L-3, this is 40,700 × 100/500 × 1.6 × 30 = 1700 man-days lost.

These values are then divided by the average energy produced per year, taking acount of lifetime and load factor. As mentioned above, this is 10,200 MW. The occupational risk per megawatt-year over the lifetime of the lines, in man-days lost, is (2400 - 3300)/10.200 = 0.24 - 0.32. The corresponding public risk is (3000–9000) /10,200 = 0.29–0.88 man-days lost. The total risk, both occupational and public, is then (0.24 + 0.29) - (0.32 + 0.88) = 0.5–1.2 man-days lost. The total risk for transmission lines then adds about 2%–3% to the total risk of hydroelectricity. It can then be concluded that while the risk of transmission lines is not negligible, it appears to form a small part of overall risk.

APPENDIX M. MATERIAL AND LABOR STATISTICS

i. Productivity

Labor productivity is an area where estimates can differ because of varying definitions and assumptions. For example, it is not always clear if nonproduction employees are included in manpower totals. In the following, productivity estimates using References 416 and 417 and 1972–73 data are outlined where they differ by more than 25% from those shown in Table M-1. Nonferrous smelting and refining has a productivity of 18 man-hours/metric ton (excluding nondomestic shipments), contrasting with 12.5 in Table M-1 for only the aluminum sector of this industry. Copper ore mining had a value of 0.34, in contrast to 0.47 in Table M-1. The value of 0.84 for cement in Table M-1 contrasts with 1.1 for hydraulic cement. Flat glass has a value of 16 compared to 38 in Table M-1.

Even if the above productivity values of References 416 and 417 were used instead of those in Table M-1, the overall risk of material acquisition would not change substantially. One category that could affect the calculated risk significantly is fabricated metal products. Table M-1 shows a value of 149. From References 416 and 417 one can deduce values of 83 for fabricated metal products, 64 for fabricated structural steel, 197 for

Table M-1. Man-Hours Required per Metric Ton (e)

Material	*Production, 1973 (Million of Metric Tons) (a)*	*Manpower (Thousands)*	*Man-Hours (per Metric Ton) (b)*	*References*
Steel (f)	120	603	10	17
Aluminum	4.97	31.1	12.5	178
Cement	76.7	32.6	0.84	18
Hard coal (anthracite)	6.4	3.8	1.2	179
Iron ore	68.4	15	0.44	177
Bauxite	2.0	0.5	0.5	180
Copper	12.7	84	132	91–93
Ammonia	12.5	6.9	1.1	108–111
Fabricated metal products	20	1490 (c)	149	181, 182
Glass	1.36	26	38	181, 182
Fiberglass (d)			38	
Copper ore		36	0.47	92
Nonmetal mining			0.50	183

(a) Statistics for the United States.
(b) Assumed 2000 man-hours per man-year.
(c) Misprint in original.
(d) In addition to that required for making glass. Numbers of man-hours required to make fiberglass is assumed to be the same as that to make glass.
(e) It has been suggested that, for a variety of reasons, some man-hours per metric ton shown here may be overestimated and others may be too low, especially those related to construction (425).
(f) As an example of the approximations used, the products of the steel industry in 1973 were, in the same units, pig iron 92, iron slag 26, raw steel 137, steel mill products 101, steel slag 8.8, iron castings 16, steel castings 1.7, etc. Some of these products are intermediate, i.e., partly or wholly consumed by the industry (416, 417).

engine and turbine manufacture, 146 for electrical distribution equipment, and 175 for electrical lighting and wiring manufacture. The category "Fabricated metal products" used in this report is an average over the last five categories. Until more detailed models are available, it is not clear if the present value is an over or an underestimate.

The subject of occupational risk is a large one. This appendix provides only limited information, primarily on U.S. conditions, as taken from Bureau of Labor Statistics data and other sources.

ii. Risk Severity

One area of interest is how different these data are from that of other sources. As noted in footnote (h) of Table M-2, risk per unit time clearly varies from one country to another. Within a country, these data can also vary. The National Safety Council in the United States gathers data on a somewhat different basis from the Bureau of Labor Statistics. The severity index is here defined as the total man-days lost per 100 man-years, counting deaths as 600 man-days lost. The index for NSC data is now compared to those used in this book.

For metal mining, the severity index used here is that of the NSC, since the severity calculated from BLS data by Reference 16 yields results substantially lower than NSC data. Since the NSC severity index is apparently not divided into deaths and injuries, the 1972–74 ratio of deaths to injuries in the United States, 0.023 (420) and the value of 55 man-days lost per injury in coal mining (22) is used here. This yields about 112 man-days lost due to accidents per 100 man-years, and 0.047 deaths. It should be noted that this value applies only to surface mining, which probably includes iron, copper ore and bauxite. The present severity index of surface mining is 394; the 1970–72 NSC value for underground metal mining is 964 (419). For hard coal mining, the present index is 924; the 1970–72 NSC index for underground coal mining is 1003. For nonmetal mining, the present index is 262; the 1970–72 NSC indexes are 514 for sand and gravel quarrying, 528 for stone quarrying, 394 for surface mining, and 662 for underground nonmetal mining. For steel, the present index is 124; the 1970–72 NSC index is 137. For aluminum, the present index is 138; the 1973–75 NSC index for alumina is 113 and for aluminum reduction is 155. For copper, the present index is about 230; the 1973–75 NSC index for nonferrous smelting and refining is 219–232. For glass, the present index is 106; the 1973–75 NSC index for flat glass is 94. For cement, the present index is 110; the 1973–75 NSC index is 230, comprising 208 for mills and 420 for quarries. For fabricated metal products, the present index is 146; the 1973–75 NSC index for iron and steel products is 186, comprising 135 for boiler shops, 98 for heating equipment, and 328 for structural steel fabrication. For fiberglass, the present index is 866; the 1973–75 NSC index is 668. In terms of construction the present index for roofing and sheet metal is 511; the 1973–75 NSC index for structural and ornamental metal work is 389. For concrete, the present index is 290; the 1973–75 NSC index for concrete, bridge, and dam work is 367. For miscellaneous construction, the present index is 384; the 1973–75 NSC index for heavy construction is 303, for public utility

Table M-2. Occupational Health and Safety Statistics, per 100 Man-Years (f),(v)

Industry	*Man-Days Lost, Accidents (15) (o)*	*Man-Days Lost, Illness (15),(o),(p)*	*Man-Days Lost, Total (15),(a)*	*Deaths (g),(q)*
Material Acquisition				
Metal mining (i),(h),(r)	112	2.0	114	0.047
Hard coal mining (h),(r)	199.8 (w)	5.2 (x)	205	0.120 (h)
Nonmetal mining (h),(r)	56.7	1.5	58.2	0.034 (n)
Steel (y)	80.4 (u)	2.0	82.4	0.007 (ac)
Aluminum	81.6	2.1	83.7	0.009
Copper (b)	152.2	3.8	156	0.013
Glass (z)	62.0	1.7	63.7	0.007
Cement	66.5	1.8	68.3	0.007 (ab)
Fabricated metal products (k)	89.3	2.7	92	0.009
Fiberglass (c)	118	3.6	122	0.012
Ammonia (d), (l)	66.3	1.7	68	0.006
Logging (e)	292	4.1	296	0.095 (s)
Construction (j), (t)				
Plumbing	70.0	2.4	72.4	0.023
Electrical (aa)	53.9	1.2	55.1	0.018
Roofing and sheet metal	172.3	2.4	174.7	0.056
Concrete	95.5	2.7	98.2	0.032
Miscellaneous contracting	129.7	1.8	131.5	0.042
Shipbuilding (c), (m)	157	4.8	162	0.016

(a) Does not include allocations for deaths.

(b) Distribution in first three columns assumed to be the same as that for steel.

(c) Distribution in first three columns assumed to be the same as that for fabricated metal products.

(d) Distribution in first three columns assumed to be the same as that for steel.

(e) Distribution in first three columns assumed to be the same as that for roofing and sheet metal.

(f) Statistics for United States.

(g) From Ref. 184; calculated from ratios of nonfatal man-days lost to deaths for broad occupational categories.

(h) Death rate in Norwegian coal mining is 0.18; in Poland, 0.018, in the same units (242). The death rate for all mining in the Federal Republic of Germany was 0.059 in 1970 (235). Britain had a death rate of 0.031 for 1970–75 (430). South Africa had a death rate for all mining of 0.084 for 1965–70 (539). France had a death rate of 0.05 for 1968–72 (540). On the other hand, a rate of 0.26 is estimated for Britain in 1974 (496), made up of 0.02 for accidental

Table M-2 continued

and 0.24 for occupational deaths. The rate will depend on the grade and type of coal, safety precautions, and other variables (242).

(i) See text of this appendix for derivation of values.

(j) Construction in Britain had an average death rate of 0.019 for 1971–75, in the same units (430). Construction deaths in the Federal Republic of Germany in 1970 were about 0.028 in the same units (235). South African death rate for construction was 0.077 for 1965–70 (539). France had a death rate of 0.049 for 1968–72 (540). On the other hand, Bell (497) estimates a value for Britain of 0.12.

(k) The death rate for a similar category (iron and other metals) in the Federal Republic of Germany in 1970 was 0.012 (235). Of this, about 4% was due to disease. The death rate in the United Kingdom for 1960–72 was 0.014 (235). France had a death rate of 0.012 for metalworking in 1968–72 (540).

(l) The death rate for the chemical industry in the Federal Republic of Germany in 1970 was 0.009; in France for 1968–70 it was 0.017; in the United Kingdom for 1960–72 it was 0.009 (235).

(m) The death rate in the United Kingdom for 1960–72 was 0.016 (235).

(n) The death rate for quarry workers in the United Kingdom was 0.073 for 1967–69; for France in 1968–70 it was 0.037 (235). However, this category does not correspond exactly to "nonmetal mining."

(o) The ratio of working days lost to occupational diseases to that of accidents (both groups nonfatal) in the United Kingdom was 3.5% (235). This is similar to most of the ratios in this table.

(p) Average rate for all industry in France in 1976 was 2.6 (485).

(q) Average rate for disease (excluding accidents) in France in 1976 was 0.0015 (485).

(r) Canadian death rate for 1966–75 was 0.13 for all mining (344). Ontario death rate for 1976 for all mining was 0.08. Ontario accident rate for 1975 for all mining, allowing 50 man-days lost per accident, is 680 man-days lost.

(s) Canadian death rate for 1966–75 was 0.12 for forestry workers (344).

(t) Canadian death rate for 1966–75 was 0.044 for all construction (344). Ontario death rate for 1976 for all construction was 0.028. Ontario accident rate for 1975 for all construction, allowing 50 man-days lost per accident, was 570 man-days lost.

(u) Ontario accident rate for 1975 for steel and metal industry workers, allowing 50 man-days lost per accident, is 395 man-days lost (344).

(v) Although using agricultural wastes to produce fuel is not considered here, Britain had a death rate of 0.011 in 1974 for farming employees only, and 0.019 for employees and employers (496). Allowing 50 days lost per accident, man-days lost for accidents was 94 for employees only in Britain. Man-days lost due to accidents in the U.S. for 1973–74 was 64–70 for agricultural production (15).

(w) British rate in 1974 was 960, allowing 50 man-days lost per accident (496).

(x) British rate in 1974, including pneumoconiosis, was 40, allowing 50 man-days lost per illness (496).

(y) South African death rate for iron and steel was 0.022 for 1965–70 (539).

(z) South African death rate for glass, bricks, and tile was 0.034 for 1965–70 (539).

(aa) South African death rate for electricians was 0.033 for 1971–73 (539).

(ab) Ferguson (628) estimates 0.04–0.4 deaths for one million tons of concrete produced in Britain, including transportation. Assuming, as in Table M-1, that 0.84 man-hours are required per metric ton of cement, this implies that 10^6/

$0.84 \times 2000 = 600$ man-years are required to produce 10^6 metric tons of cement. It is realized that most of concrete is not cement, but Ferguson suggests that the risk per unit weight for the noncement components will be similar to the cement. Then there are (0.04–0.4)/6 = 0.007–0.07 deaths per 100 man-years.

(ac) Ferguson (630) estimates 1.5–2.3 deaths per 10^6 metric tons of steel in Britain. If it is assumed from Table M-1 that 10 man-hours are required per metric ton, this is $(1.5\text{–}2.3)\ 2000/10^6 \times 10 \times 100 = 0.030\text{–}0.046$ deaths per 100 man-years.

Table M-3. Activities Required for Final and Intermediate Materials

Final Material	*Intermediate or Primary Materials (a)*	*Activity (b)*
Steel (c), (e)	Iron ore, coal	Steel, metal mining, hard coal mining
Aluminum (d)	Bauxite	Aluminum, metal mining
Concrete	Sand	Cement, nonmetal mining
Glass	Sand	Glass, nonmetal mining
Copper	Copper ore	Copper, metal mining
Fiberglass	Glass, sand	Fiberglass, Glass, nonmetal mining
Ammonia		Agricultural fertilizers
Fabricated metal products	Steel,	See above
	aluminum,	See above
	copper	See above
Earth, rock		Nonmetal mining

(a) From Table M-1.

(b) From Table M-2.

(c) A more sophisticated model (120) shows that to produce 1 kg of steel requires 12 g of ferroalloys, 40 m^3 of oxygen, 393 g of scrap iron, 34 g of mill cinder, 556 g of coking coal, 896 g of iron ore, 69 g of dolomite, 182 g of limestone, and 4 g of flurospar.

(d) One kg of aluminum produced in Canada requires up to 10 g of calcium fluoride, 100–200 g of pitch, 430–600 g of carbon, 80–150 g of soda ash, 100–120 g of lime, 10–70 g of fluorspar, 20 g of cryolite, 10–30 g of aluminum fluoride, and 1.91 kg of alumina or 3.8–6.0 kg of bauxite (418).

(e) Ferguson (629) states that for Britain, 1 metric ton of steel requires 1.6 of iron ore, 0.9 of coal, and 0.2 of oil.

construction 239, and for general building construction 153. The NSC also has labor categories which are not considered in the simplified model used here. The values of the indexes are as follows: airframe manufacturing, 41; communications equipment, 21; heavy electrical equipment, 62; engines and turbines, 34; general industrial machinery, 56. These last categories apply to manufacturing.

iii. Materials

Producing final materials requires raw and intermediate materials. Table M-3 shows the major assumptions used in calculating the risk of final materials. It is recognized that this table is far from complete.

APPENDIX N. ENERGY BACK-UP SYSTEMS

Some nonconventional energy systems require back-up for the times that the sun or wind is unavailable. What type of back-up should be used? As pointed out in Section 2H, one discussion of solar energy has proposed coal as a back-up for solar thermal electricity. It might be contended that because coal is a relatively high-risk energy source, its use might make the overall risk for energy systems requiring back-up excessively large. As stated elsewhere in this report, the benefit of the doubt has been given to nonconventional energy systems, wherever possible, in the direction of estimating lower risk. As a result, coal is not assumed as a back-up.

i. Weighted Back-up

Nonconventional energy systems requiring back-up have not been built in sufficient quantities to determine what type of back-up will be used. A fair assumption might then be that the back-up would be a weighted average of the electricity now used. For 1975, Canadian generating capacity was, in percent, hydroelectricity, 58; nuclear, 5.1; coal, 20; oil, 12.5; and natural gas, 4.1 (192). These proportions will probably change somewhat in future years, but a key assumption of this report is that present-day conditions prevail. It is unlikely that the proportions outlined above would change enough over the next few years to alter the results of this report significantly. Values for the end of 1977 were hydroelectricity,

58%; nuclear, 7%; and coal, oil, and gas, 36% (277). It can be seen that proportions did not change sharply between 1975 and 1977.

To calculate the weighted risk, 58% of the hydroelectricity risk, 5.1% of the nuclear risk, and so on, is added. Results are shown in Table N-1. Most of the overall risk is attributable to coal and oil. The values in the sixth column are used as the back-risk for solar thermal electricity, solar photovoltaic, and windpower, taking account of the fact that back-up forms only a small part of overall energy requirements for these systems.

Proportions of electricity generated by energy sources will vary from country to country. Back-up risk can be recalculated accordingly. For example, United States production by source in 1975 was, in percent (capacities not available): hydroelectricity, 15.8; nuclear, 8.9; coal, 44.7; oil, 15.1; and natural gas, 15.7 (191). Calculations using these proportions, showing United States back-up risk, are shown in the right-hand column of Table N-1. They are found by multiplying the first column by 15.8/58 = 0.27; the second, by 8.9/5.1 = 1.75; the third, by 44.7/20 = 2.24; the fourth, by 15.1/12.5 = 1.21; the fifth, by 15.7/4.1 = 3.83; and summing. Results are broadly similar to those in Canada, except for the greater public risk from disease. This is attributable to the greater fraction of coal used to generate electricity in the United States.

ii. Low-Risk Back-up

The use of relatively low-risk nuclear power or natural gas as back-up will lower the risk attributable to this aspect of energy. While risk may be lowered, the use of nuclear back-up will produce considerable waste of energy. Since nuclear power cannot be turned off and on as quickly as can coal or oil power, when it is not required for back-up the nuclear energy will not be used.

It can be shown that the total risk for systems that may require back-up is almost the same if no back-up, as contrasted with low-risk back-up, is assumed. In the following discussion, it should be remembered that no back-up (instead of low-risk back-up) can be assumed for the calculations.

Back-up is assumed only for wind, solar photovoltaic, and solar thermal electric. The exact proportion of back-up is still subject to controversy. A value of 10% of total energy is assumed here.

The risk per unit energy for nuclear or natural gas, as shown in Appendices D and C, respectively, is apparently low. If they are to be used as

Table N-1. Weighted Back-up Risk (per Megawatt-Year Net Electrical Output)

Risks	*Hydro-electricity (X 0.58)*	*Nuclear (X 0.051)*	*Coal (X 0.20)*	*Oil (X 0.125)*	*Natural Gas (X 0.041)*	*Total (Canada)*	*Total (United States)*
Occupational							
Accidental							
Death	(0.87–1.5) $\times 10^{-3}$	(15–38) $\times 10^{-6}$	(0.50–1.3) $\times 10^{-3}$	(0.24–2.4) $\times 10^{-4}$	(0.66–2.0) $\times 10^{-5}$	(1.4–3.1) $\times 10^{-3}$	(1.4–3.8) $\times 10^{-3}$
Injury	0.20	(0.82–1.5) $\times 10^{-3}$	(1.1–1.7) $\times 10^{-2}$	(0.23–1.75) $\times 10^{-4}$	(3.7–14.8) $\times 10^{-4}$	0.22	0.077–0.099
Man-days lost	15.1–19.1	0.12–0.31	4.0–10.2	0.41–2.4	0.06–0.20	20–32	14–32
Disease							
Death	–	(12–41) $\times 10^{-6}$	(0–1.5) $\times 10^{-4}$	–	–	(0.12–1.9) $\times 10^{-4}$	(1.9–41) $\times 10^{-5}$
Disability	4.6×10^{-5}	(1.1–2.0) $\times 10^{-5}$	(0–1.6) $\times 10^{-3}$	–	–	(0.9–1.6) $\times 10^{-3}$	(1.8–3.7) $\times 10^{-3}$
Man-days lost	0.0023	0.082–0.26	0.04–1.0	–	–	0.12–1.2	0.21–2.6
Public							
Accidental							
Death	(0.64–0.93) $\times 10^{-3}$	6.1×10^{-7}	(1.6–3.8) $\times 10^{-4}$	–	–	(0.80–1.3) $\times 10^{-3}$	(5.3–11) $\times 10^{-4}$
Injury	(0.64–7.4) $\times 10^{-3}$	5.6×10^{-6}	0.32×10^{-3}	–	–	(1.0–7.7) $\times 10^{-3}$	(0.89–2.7) $\times 10^{-3}$

Man-days lost	3.8–5.8	0.41×10^{-2}	1–2.3	–	–	4.8–8.1	3.2–6.8
Disease							
Death	$(0.6–1.9) \times 10^{-6}$	$(3.6–27) \times 10^{-6}$	$(3.4–12) \times 10^{-3}$	$(8–21) \times 10^{-4}$	$(1.0–2.9) \times 10^{-7}$	$(4.2–14) \times 10^{-3}$	$(8.8–29) \times 10^{-3}$
Disability	0.003–0.011	$(1.0–3.2) \times 10^{-3}$	19–56	4.4–13	$(0.5–1.6) \times 10^{-3}$	23–69	48–140
Man-days lost	0.02–0.07	0.03–0.18	118–350	27–81	0.003–0.011	145–430	300–880

back-up for systems like the three mentioned above, the risk due to back-up is small compared to that from other aspects of the systems. For example, the total man-days lost per unit energy of nuclear power is about 5–15. The number of man-days lost for solar thermal electric from non-back-up sources is 43–57. In consequence, risk from nuclear back-up is 0.10(5 - 15) = 0.5 - 1.5, and its proportion is (0.5/43) - (1.5/57) = 1% - 2.6%. Proportions for natural gas would be even smaller. If no back-up were assumed, results would be similar to the low-risk case.

Results are shown in the concluding graphs of the main text. The man-days lost with nuclear or natural gas back-up are shown as jagged lines. In Figure 10, showing occupational risk, the values for wind man-days lost with nuclear or natural gas as opposed to energy-weighted back-up are only slightly changed. This difference is too small to show graphically.

The relative rankings in Figure 10 do not change with low-risk back-up. The relative rankings in Figure 12, showing the public risk per unit energy, change somewhat with low-risk back-up.

Figure 14 shows the total risk. The relative ranking of energy systems changes only slightly compared to "normal" back-up. While the absolute value of man-days lost per unit energy can drop substantially as a result of substituting nuclear or natural gas instead of a system weighted by energy use as a back-up, especially in terms of public risk, the relative rankings of the energy systems in terms of risk remains almost constant.

APPENDIX O. MAN-DAYS LOST PER DEATH

There are a number of measures by which risk can be evaluated. The most obvious is the number of deaths. However, nonfatal accidents and disease also occur. These may be considered on their own, or combined mathematically with deaths if so desired. Of course, there is no way to combine nonfatal accidents and disease with deaths in a physical manner.

i. Sensitivity Analysis

A key assumption in combining deaths, accidents, and disease is the number of man-days assigned per death. As noted in Section 2F, this report assumes a value of 6000. How sensitive is either the absolute or relative value of man-days lost per unit energy to this value? The following is a simplified sensitivity analysis of this question.

Occupational risk is divided into four categories: accidental deaths and injuries, and disease-related deaths and disabilities. Public risk has the same four categories. It will be assumed in the following that the number of man-days lost for nonfatal accidents and illnesses are known, although many of the references cited indicate that the values for these categories are not certain. This leaves the four categories of death: occupational accidental, occupational disease, public accidental, and public disease.

It seems clear that occupational accidental deaths will produce about 6000 man-days lost (311). A simple calculation suggests this. If the average working life is from 18 to 65, the average worker's age is around 41. If death occurs at that age, the number of working days lost is about (65 - 41) X 250 = 6000, where 250 is the number of working days in a year. It is true that younger workers may be more reckless and careless, exposing themselves to greater chances of deaths, but on the other hand they will have better reflexes.

Public accidental risk will be from accidents at railway crossings, natural gas explosions, hydroelectric dam failures, etc. As a first approximation, it might be assumed that the age at death would be the average for the population as a whole—perhaps 30–35 years. If the average life expectancy is about 73 years at present, this implies a loss of 38–43 years of life.

At this point, the nature of the choices to be made in terms of calculations becomes apparent. If one considers the total number of lost days of life, the result for public accidental deaths are between 38 X 365–43 X 365 = 13,900–15,700. If one considers only the potential working days lost, the value is (65 - 43) 250 - (65 - 38) 250 = 5000 - 6800. A case can be made that the number of man-days lost for public accidents assumed in this report should be higher than 6000 per death. What would be the implications of such a change?

For most of the energy systems considered, public accidental deaths form only a small proportion of total deaths, and similarly form only a small proportion of total man-days lost. The proportion of total deaths is zero for some systems like oil-fired electricity (or very close to zero), and range up to 5% in the case of natural gas. Proportions for other systems are generally in the range 2%–4%. As a result, if 15,000 man-days lost were used for public accidental deaths instead of 6000, the absolute values of risk might increase by no more than (15,000/6000)5% -5% = 7½%. It is possible that the rankings of some systems in terms of overall risk might change but no analysis of this question was done.

One exception to this rule is hydroelectricity. Public accidental deaths form about 40% of all deaths for that system. If these deaths are assumed to be equivalent to 15,000 man-days lost, the total number of man-days lost would increase by about 37%. However, its relative rank in total risk apparently would not change.

In terms of risk from radiation, the assumption made in footnote (l) of Table D-2 and used in this report implies that 1 man-rem of dose is equivalent to 6000/10,000 = 0.6 man-days lost. Gonen (538) has estimated 0.55 man-days lost/man-rem for Israel, close to the value assumed here.

The next category to be considered is deaths from occupational diseases. The most prominent energy-related disease causing death might be lung and related disease from underground coal mining, but there is also lung cancer from mining uranium and other industrial diseases. It has proved difficult to estimate the number of man-days lost for each of the industrial diseases that can cause death. It is unlikely that the number of occupational man-days lost for a disease would be much above 10,000, since that would imply workers dying around the age of 25. On the other hand, there is evidence that for some industrial diseases, the occupational man-days lost could be less than the 6000 assumed here. For example, it is implied in Appendix D that lung cancer due to radiation may be delayed 2–30 years from the onset of radiation. If the number of man-days attributed to occupational disease deaths in nuclear power were 3000 instead of 6000, the estimated number of man-days lost per unit energy would drop by 0.7–2.4. This would decrease the total man-days lost by 100 (0.7/4.9–2.4/15) = 14%–16%. The relative rank of this system would remain the same. For other systems, the relative change in total man-days lost produced by altering the days assigned to each death is even smaller. For example, if the value for coal-related disease deaths were doubled to 12,000, the occupational man-days lost would increase by 0–5. However, since the total man-days lost per unit energy for this system ranges from 610 to 1900, the relative change is not great. Much the same conclusion holds for the other systems discussed in this book.

The final category to be considered is public disease-related deaths. As noted above, a case can be made for including all days lost by the public, as opposed to only working days. However, this still leaves open the question of how many days should be allocated to each death. It seems likely that the number will depend on the type of disease. In terms of coal, public disease-related deaths constitute about 17% of total man-days lost. If the number of days assigned to each death were 1000 instead of 6000, the total number of man-days lost would drop by about (5/6)17 = 14%. About the same conclusion could be drawn for oil-fired electricity.

For nuclear power, the number of man-days lost due to public disease-related deaths excluding air pollution effects is 0.4–3.1. This constitutes about 9%–21% of the total number of man-days lost per unit energy. If the latent period noted above caused the man-days lost per death to be 3000 instead of 6000, the total risk would drop by 0.5(9%–21%) = 5%–11%.

In summary, this crude sensitivity analysis indicates that the total number of man-days lost per unit energy is usually not strongly affected by the exact value chosen for man-days lost per death. Where there is a strong effect, the relative rank of systems in terms of total risk is generally not changed. Of course, the absolute values of risk components, such as public accidental or occupational disease-related risk, can change substantially if the number of man-days lost per death is altered.

No attempt has been made here to perform a comprehensive sensitivity analysis, in which the number of man-days lost assigned per death would be varied for each component of total risk and for each energy system. This work should be done in the future.

ii. Justification of Combining Deaths with Accidents and Disease

The question still remains, How valid is it to combine deaths with non-fatal accidents or disease? There is no way to answer this question objectively, and the combination done here is a philosophical assumption, nothing more.

However, this assumption is made in a recent publication of the International Commission on Radiological Protection (311), where the mathematical discussion is preceded by the warning that "no single index can be scientifically defensible." This reference puts forth a suggestion for a possible "index of harm." Rather than calculating the total man-days lost for a given unit of energy, this index of harm applies to accident and disease rates for particular industries and is unitless. The index is made up of components representing deaths, accidents, and disease. Public risk is not included.

The index of harm proposed in Reference 311 uses the fact that for many industries, the injury rate apparently increases as the square root of the accidental death rate. It takes the contribution of industrial disease as about 5% that of deaths and accidents. Using these assumptions, it can be shown that the index of harm equals $x(0.0315\ x + 0.042)$, where x is the square root of the death rate per 1000 employees in a given industry. The accident and disease rate are not shown explicitly because the former is assumed to vary with the death rate, and the latter is assumed to vary linearly with a combination of the death and accident rate.

While not apparent from the above algebraic term, it can be shown that the effect of deaths with respect to accidents in computing the index of harm is less than if 6000 man-days lost had been assigned to each death. The question deserves further consideration.

In summary, the combination of deaths, accidents, and diseases used in this report is made solely for the convenience of readers. There has been at least one other attempt to make this combination, and perhaps others. It is clear that some time will elapse before there is general agreement on whether the combination should be done at all, and if so how it should be handled mathematically.

APPENDIX P. TRANSPORTATION

As mentioned in Section 2C, the risk of transportation is generally calculated in this study by assuming all nonsand and noncement materials are moved with the same risk per ton-mile as coal. This is clearly a crude model, and a more sophisticated one can be outlined. Because of its complexity, it is not used here, however.

Many commodities are moved by nonrail means, such as truck, water, or air. (It is assumed that almost all coal is moved by rail.) The first step in devising a more realistic calculation is to estimate the number of tons of each commodity that is transported. The total number of tons moved in the United States in 1972 was 1.12×10^9 (115). This was made up of 37.8% rail, 35.7% motor carrier, 21.3% trucking, 0.1% air, 4.6% water, and 0.7% other. In the following, motor carrier and trucking are combined as "trucking." Tons ($\times 10^8$) carried by rail, truck, air, and water are then $11.2 \times 0.378 = 4.2$; $11.2 \times 0.57 = 6.4$; $11.2 \times 0.001 = 0.011$; and $11.2 \times 0.046 = 0.52$, respectively.

Each mode of transport has a different average distance for each trip. This distance has to be taken into account in risk calculations. Suppose commodity X, when it is moved by truck, travels only half the average distance of all commodities which move by this mode. It is then reasonable to assume that the risk per ton of this commodity moving by truck is half that of all other commodities. Appropriate ratios have to be calculated for each commodity and mode.

The total number of ton-miles in the United States in 1972 was 4.45×10^{11} (115). Of this, 55.9% was rail, 37.2% truck, 0.2% air, 6.1% water, and 0.7% other, producing, in 10^{10} ton-miles, $0.559 \times 44.5 = 24.9$; $0.372 \times 44.5 = 16.6$; $0.002 \times 44.5 = 0.089$; and $0.061 \times 44.5 = 2.7$, respectively. The miles traveled for each mode is then ton-miles divided by tons, or $2490/4.2 = 590$; $1660/6.4 = 260$; $8.9/0.011 = 810$; and $270/0.52 = 520$, for rail, truck, air, and water, respectively.

The next step is determining the risk, both occupational and public, per ton transported. For rail transport, the values of Table A-2 can be used. Assuming about 3500 metric tons of coal and other materials carried by rail for every MW-yr of electricity, the occupational risk per metric ton for rail is $(4.5–14) \times 10^{-7}$, $(3.7–14) \times 10^{-6}$, and $(2.8–10) \times 10^{-3}$ deaths, accidents, and man-days lost, respectively. The public risk is $(2.3–5.4) \times 10^{-7}$, 4.4×10^{-7}, and $(1.4–3.3) \times 10^{-3}$ deaths, accidents, and man-days lost, respectively.

For trucking and warehousing, the number of man-days lost to accidents per 100 man-years in 1974 was 152 (100). The total workforce was 1.19×10^6 in that year (350), so the total man-days lost to accidents was $152 \times 1.19 \times 10^6/100 = 1.81 \times 10^6$. If the number of tons carried was about 6.4×10^8 (see above), the man-days lost per metric ton due to accidents was $1.81 \times 10^6/6.4 \times 10^8 \times 1.1 = 2.6 \times 10^{-3}$.

Similar calculations can be carried out for water and air transportation. The number of man-days lost to accidents per 100 man-years in 1974 for water transport was 267 (100). The total workforce in that year was 2.04×10^5 (350), so the total man-days lost due to accident was $267 \times 2.04 \times 10^5/100 = 5.4 \times 10^5$. If the number of tons carried was about 5.2×10^7 (see above), the man-days lost per metric ton due to accidents was $5.4 \times 10^5/5.2 \times 10^7 \times 1.1 = 9.4 \times 10^{-3}$.

The number of man-days lost due to accidents per 100 man-years in 1974 for air transport was 77.6 (100). The total workforce was 3.68×10^5 (350), so the total man-days lost due to accident was $77.6 \times 3.68 \times 10^5/100 = 2.86 \times 10^5$. If the number of tons carried was about 1.1×10^6 (see above), the man-days lost per metric ton due to accidents was $2.86 \times 10^5/1.1 \times 10^6 \times 1.1 = 0.24$. However, most of air transport is concerned with moving people, not goods. The problem is allocating accidents between these two categories. The references consulted apparently do not make this allocation. As a crude approximation, it will be assumed that one-tenth of air risk is attributable to commodity transport. Calculated occupational risk is then divided by 10 in Table P-1.

Table P-1. Approximate Risk per 10^6 Metric Tons Transported

	Rail	*Truck*	*Air*	*Water*
Average distance, in miles [km]	590 [940]	260 [420]	810 [1300]	520 [830]
Occupational				
Man-days lost, accidents	340–1280	2600	24,000	9400
Deaths	0.46–1.4	0.37–0.54 (a), (b)	7.4–17	1.0–1.3
Total man-days lost ($\times 10^{-3}$)	3.1–9.7	4.8–5.8	68–126	15.4–17.2
Public				
Man-days lost, accidents	43	87–330	–	–
Deaths	0.23–0.54	0.19–0.21 (a), (b)	–	–
Total man-days lost	1420–3300	1230–1600	–	–

(a) Ferguson (627) estimates that the ratio of public to occupational deaths is about 10 in the mid-1970s in Britain, rather than the value of 0.4–0.5 used here.

(b) Ferguson (627) estimates 1–10 total deaths per 10^9 metric ton-km for Britain in the mid-1970s. Taking account of the average distance noted above, this corresponds to 0.42–4.2 total deaths per million metric tons travelled an average distance, if average British distances are the same as in North America. It is likely they are shorter.

The next step is to estimate the deaths per ton transported. The U.S. National Safety Council (264) estimates a trucking death rate of 0.021 per 100 man-years. If the total truck workforce was about 1.2×10^6 and the weight transported about 6.4×10^8 tons, the deaths per metric ton were about $0.021 \times 1.2 \times 10^4/1.1 \times 6.4 \times 10^8 = 3.7 \times 10^{-7}$. On the other hand, the U.S. Bureau of Labor Statistics (543) estimates about 380 deaths in trucking and warehousing in 1976. If the weight transported is about 6.4×10^8 tons, the deaths per metric ton are about $380/1.1 \times 6.4 \times 10^8 = 5.4 \times 10^{-7}$.

Deaths from air accidents in transportation and public utilities were about 90 (543). Of course, not all deaths of this type will be associated with air transport. If the weight transported was about 1.1×10^6 tons (see above), deaths per metric ton were about $90/1.1 \times 10^6 \times 1.1$ 74 $\times 10^{-6}$. The factor of one-tenth can be applied here. The National Safety Council (264) notes a death rate of 0.0051 per 100 man-years. If the workforce was about 3.7×10^5, the number of deaths was $3.7 \times 10^5 \times 0.0051/100 = 19$, substantially lower than that of Reference 543. Reasons for the difference are not clear. The death rate would then be $19/1.1 \times 10^6 = 1.7 \times 10^{-5}$ per metric ton.

The National Safety Council (264) estimates a death rate of 0.0285 per 100 man-years in water transport. If the total workforce was 2.04×10^5 and the tons transported about 5.2×10^7, the deaths per metric ton were about $0.0285 \times 2.04 \times 10^3/1.1 \times 5.2 \times 10^7 = 1.0 \times 10^{-6}$. This implies total deaths were $2.9 \times 10^{-4} \times 2.0 \times 10^5 = 58$. Actual deaths in 1976 were about 70 (543), yielding a death rate of $70/5.2 \times 10^7 = 1.3 \times 10^{-6}$ per metric ton.

Results are shown in Table P-1. Air transport has the highest risk per weight transported (although this may be overestimated), followed by water. Since comparatively little freight is moved by air, the exact value of the risk per unit weight for this mode will not strongly affect overall transport risk.

By way of contrast, Myers and Vant-Hull (615) assume distances traveled by rail to be 5000, 4200, and 5500 km for glass, steel, and motors, respectively. These values are considerably higher than those of Table P-1. They assume that truck distances are 480, 160, and 800 km for cement, aggregate, and polyurethane, respectively.

There is also public risk for these modes, just as there is for rail transport. For example, trucks collide with passenger automobiles. However, data on public risk due to transportation other than rail seem to be unavailable. As a crude approximation, it will be assumed that the ratio of public to occupational risk for trucking is similar to that of rail transport. From Table A-2, the ratio for deaths, accidents, and man-days lost is 0.38-0.50, 0.03-0.12, and 0.33-0.50, respectively.

Now a particular commodity can be considered, its risk calculated by this more detailed method, and this value compared to that found by assuming it travels only by rail.

Steel works and rolling mill products (to be referred to as "steel") had 11.4×10^7 tons transported in 1972 (116), of which 43.7% was by rail, 50.8% by truck, and 5.5% by water. The weight transported by rail, truck, and water, in 10^7 tons, was $11.4 \times 0.437 = 5.0$, $11.4 \times 0.508 = 5.8$, and $11.4 \times 0.055 = 0.63$, respectively.

The ton-miles transported for this commodity were 31.7×10^9, of which 51.6%, 40.1%, and 8.4% were by rail, truck, and water, respectively (119). The rail, truck, and water ton-miles were, in units of 10^9, $31.7 \times 0.516 = 1.64$; $31.7 \times 0.401 = 12.7$; and $31.7 \times 0.084 = 2.7$, respectively.

The distances traveled for this commodity are the ton-miles divided by the weight, by mode of transport. The distances for rail, truck, and water are 1640/5.0 = 330; 1270/5.8 = 220; and 270/0.63 = 430 miles, respectively. Since, from Table P-1, the average distances for these three modes are 590, 260, and 520 miles, the risk per ton for rail, truck, and water must be multiplied by 330/590 = 0.56, 220/260 = 0.85, and 430/520 = 0.83, respectively.

Suppose 100 metric tons of steel travel by the usual modes and the usual distances. The total risk would then be $(100/10^6)$ $(0.56 \times 0.437 \times A + 0.85 \times 0.508 \times B + 0.83 \times 0.055 \times D)$. The first value in each term is the ratio of distance that steel travels in each mode to the average distance for that mode; the second value is the proportion of travel for each mode; and the third value is the first, second, and fourth column of Table P-1. The occupational risk, in man-days lost, is 10^{-4} [0.24(3100–9700) + 0.43 (4800–5800) + 0.046 (15,400–17,200)] = 0.35–0.56. Public risk would be 10^{-4} [0.24(1420–3300) + 0.43 (1230–1600)] = 0.08–0.15 man-days lost. Total risk is then (0.35 + 0.08)–(0.56 + 0.15) = 0.43–0.71 man-days lost.

If the materials moved only by rail, the total risk would be (100/3500) (15–47) = 0.43–1.3 man-days lost.

In summary, the above methodology would be a more systematic way to calculate transport risk. Because the commodities used in energy systems are often known only crudely, there seems to be little reason to apply this level of sophistication.

APPENDIX Q. COMMENTS ON AECB 1119

The publication in 1978 of the report on which this book is partly based, AECB 1119, stirred considerable interest. Perhaps the word "interest" is too mild; "controversy" might be better. It is the object of this appendix to give some idea of the comments aroused by the previous report: what was said, and what replies were made. In effect, the tradition of *The Double Helix*, in which the reader was given a picture of the sometimes seething currents behind the solid facade of science, is carried on.

Many of the comments made about AECB 1119 were highly complimentary. In the interest of modesty, few are contained in this appendix.

What remains is mostly criticism. Some of the materials have appeared in public print before, but many have not. Their grouping in this way has not appeared in one place previously.

While this appendix has been assembled primarily for the benefit of readers interested in relative risk, sociologists of science may also take note. It has been contended that science advances not by small accretions of knowledge, but by debates over larger issues. If this is true, perhaps the entire controversy has helped science in some as-yet-unperceived way.

Because of the larger number of comments that were made on AECB 1119, only a small selection is offered here. As noted above, the selection is unrepresentative, being weighted toward the critical.

Some of the comments made on AECB 1119 were indeed valid and have been taken into account in this book. In consequence, readers may see a statement of the type, "Fact *X* is wrong in Appendix *Y*" in a letter, check this appendix, and not find the offending fact. This indicates that it was removed upon consideration. In order to keep the debate as fair and historically accurate as possible, no revisions were made to either the original letters or the responses. The comments are grouped by where they originated, not strict chronological order.

There is no justification here of what was said or not said by anyone connected with the debate. In the long run, the letters, articles, and the research itself have to speak for themselves.

The British Journal *New Scientist*

A brief article based on the report was published in the British magazine *New Scientist*. This elicited comments by Colin Moorcraft, Clive G. Davies, and M.T. Aitken. Because no response was received from Moorcraft and Davies for permission to reprint their "Letters," they are paraphrased here. A short reply was made by the author. In another issue of that journal, letters from M.J. Platts and Peter Musgrove of a perhaps more objective basis than the previous three were printed. A reply to their points follows.

New Scientist
1 June 1978

Dr. Inhaber's argument, based on "risk accounting", is remarkable not only for its assumptions, but also for the risks to life and health it apparently omits.

The risk of plutonium being obtained by terrorists, for example, of which Sir Brian Flowers, when chairman of the Royal Commission on Environmental Pollution, said: "I do not believe it is a question of whether someone will deliberately acquire [plutonium] for purposes of terror or blackmail, but only of when and how often".

The risk, also, of nuclear weapons proliferation—recognised by the Australian government inquiry report of 1976, which observed that "the nuclear power industry is unintentionally contributing to an increased risk of nuclear war".

These dangers are not somehow made irrelevant to the problem of safety by their having a political, and therefore unquantifiable, component. Nor will they go away if we define the problem of safety as a technical one. They are posed by the use and development of nuclear power, and have no equivalent for any other energy source—including solar energy.

If Dr. Inhaber wishes us to believe that "neither [he] nor the Atomic Energy Control Board proposes the use or non-use of any particular energy system", he should, as a first step, abandon the special pleading for nuclear power implicit in his argument.

M. T. Aitken

103 Huddleston Road
London N7

Moorcraft said "Ugh" to the statement that solar power is more dangerous than nuclear power. He said that the article is based on a pseudoscientific methodology. Complaining that all qualities of risk are reduced to the one parameter of duration, he stated that one person-year of terminal cancer is equivalent to one person-year of twisted ankles. Calling this so much drivel, he said that death apparently lasts only 16½ years in Canada. He asked if like was compared to like in the articles, saying that at least three end-points were involved in the calculations. These endpoints were electricity, solar heat and methanol. He went on to say that three of the energy systems considered were irrelevant but other options were excluded without explanation. He gave as examples "energy conservation measures, non-methanol biomass systems, and passive solar heating systems." In addition, he complained that little was said about the materials used in the systems. Moorcraft stated that the high risk of methanol due to logging was rubbish, saying that this system could have low risk if coppicing were used. He felt that this was a blatant example of distortion on the author's part.

Davies said that the article considered only risks for energy options inside Canada. The Indian atomic bomb, according to him, showed that the peaceful and military uses of nuclear power are intertwined. He felt that as more countries use nuclear power, the chances of nuclear war become greater.

New Scientist
27 July 1978

Risks Explained

Ugh. Rubbish. Special pleading. Drivel. These comments, culled from the correspondence on my article ("Is solar power more dangerous than nuclear? 18 May, p 444; Letters, 1 June, p 614) are, unfortunately, the type of response that all too often passes for debate on the merits of energy systems.

The article is no "special pleading" for nuclear power. One might as well claim that I had been bought off by the natural gas interest, since it was this form of energy that had the lowest apparent risk, rather than nuclear. To avoid the suggestion of pro-nuclear bias, I used the estimates of John Holdren, a well-known critic of atomic power, for a number of nuclear risk categories.

Let me deal briefly with the other points in the letters. It is true that plutonium is produced in nuclear reactors, but it is by no means certain that it can or will be used by terrorists. The risks I calculated for all the systems based on occupational and related statistics, are almost certain to occur. The question of plutonium and the bomb (how could a mention of nuclear power get by without that four letter word?) is, in essence, a red herring. Some of the steel used for windmills could also be employed to make rifles and knives. I didn't chalk that up against windpower.

I did use a fixed value for average man-days lost per fatality. Many risks analysts, as well as the American National Standards Institute, have used the same value to mathematically combine deaths with non-fatal diseases and illnesses. The exact value is not of great consequence, as my graphs showing only deaths per unit energy output are quite similar to those shown in *New Scientist.*

Mr. Moorcraft says that I have deliberately excluded energy conservation, non-methanol biomass and passive solar heating systems. The first is not an energy producer, and I confined myself to them. Data on the latter two proved to be elusive.

He also says that little is mentioned about the materials required to build the energy systems. He has missed Figure 2 of my article which shows both material and construction requirements. Further details are contained in my report, referred to in the article.

Finally, he says that the relatively high risk of methanol could be reduced by coppicing. Even after looking this up in my dictionary, I remain unconvinced. There is no point in searching after the Holy Grail of energy—a risk-free system. As I demonstrated in my paper, this is physically impossible. The only question is, how big is the risk?

H. Inhaber

Technology Impact Division
PO Box 1064
Ottawa, Canada

New Scientist
8 June 1978

Energy Risks

Dr. Herbert Inhaber ("Is solar power more dangerous than nuclear, 18 May, p 444) begins admirably, but ignores his own advice. His first paragraph points to the complexity of assessing risks, and suggests that "yes"/ "no" comparisons should be replaced by "maybe" statements, but he then goes on to compare various energy technologies as if the figures he has calculated are exact. A more appropriate way of expressing his concern would have been to include error bands in his figures. Then the picture might have looked different.

I cannot give an example directly related to risk accounting, but my point is illustrated by an interchange that took place a while ago about applying energy accounting to wave energy systems. Two people of opposite opinion did back-of-envelope calculations. One took the most pessimistic assumptions he could and showed that wave energy devices might not pay for themselves (in energy terms) in 60 years. The second, referring to the most optimistic estimates, showed that they might pay for themselves in less than six weeks.

As Dr. Inhaber points out, accident ratings are closely related to quantities of materials used, and to a first order, if a wide range of estimates applies in energy accounting, the same wide range will apply to risk accounting (and to financial accounting). If Dr. Inhaber's histograms showed error bands, or confidence bands, all the newer energy technologies would be seen to have wider bands on them than the better developed technologies. The results would show general implications but would not give anything as apparently explicit as Dr. Inhaber's figures.

To extend the problem, it is future projected accident ratings that have to be considered, not historical ones. Coal mining has a bad history, but is steadily becoming safer and cleaner. The nuclear industry, in contrast, is apparently moving from simpler technology towards a more complex and risk prone technology. Time projected estimates of risk are what matters.

Of course, to provide a more complete picture, it is also important to include some social accounting in assessing the various technologies. A world more full of relatively conventional industry producing solar panels, fitting them to houses and so on, might be a world which provides a spread of employment for all manner of people. A more nuclear orientated world might conversely offer more selective employment for an increasingly specialist group of scientists and engineers, feeding power to an increasingly redundant group of dependents. Such social implications are of major importance.

Work such as Dr. Inhaber's is most valuable, but it is not enough. What is inadequate is that these important matters of national policy are only touched upon in a subsidiary way by people advocating particular technologies. Policy needs to be discussed and studied in an organised manner, at national level this is not happening.

M. J. Platts

Wavepower Ltd.
Ringwood Road
Southampton SO4 2HT

Dr. Herbert Inhaber's conclusions are surprising. For my own specialty, wind energy, they are also quite wrong. Studies in the US and the UK (for example the Department of Energy's *Energy Paper* 21) have consistently shown that megawatt windmills in typical locations have energy recovery periods of less than one year, and this is significantly less than any nuclear reactor system. Corresponding to this short energy recovery period, the materials requirements for wind energy systems are nearer 10 tonnes per megawatt-year than the 1000 tonnes per megawatt-year that Dr. Inhaber misleadingly suggests. And the calculated "risk per unit energy output" should be reduced proportionately. Correcting this error invalidates Dr. Inhaber's argument–that nuclear power is safer than any renewable energy source.

Peter Musgrove

Department of Engineering
Reading University

14 July, 1978

Dr. Dixon
Editor
New Scientist
King's Reach Tower
Stamford Street
LONDON SE1 9L5
United Kingdom

Dear Dr. Dixon:

The comments of June 8 to my article ("Is solar power more dangerous than nuclear", May 18, 1978) deserve a reply. One writer says that his energy analysis of windmills shows that they have energy recovery periods which are low. He extends this to imply that the risk associated with windmills should also be low. However, one cannot assume a direct correspondence between the recovery periods indicated by energy analysis and risk to human health. For example, air pollutants from coal burning electric power stations are a significant part of the overall risk from this source, but do not affect the energy input-output ratios.

The writer also questioned the data sources used. An important one was the Project Independence studies in the U.S. A quick glance through these studies shows that the authors were highly favourable to windpower. In the case of wind, as other non-conventional systems, I often gave it the benefit of the doubt. For example, while an LOCA (Loss of Coolant Accident) may cause risk to humans in a nuclear reactor, I did not assume a parallel LOBA (Loss of Blade Accident) for windmills.

The writer says that my main argument is that nuclear power is safer than any non-conventional energy source. However, natural gas had the lowest risk of the 11 sources, and little space was devoted to nuclear.

Another author says I should have used error bands in my figures. This was done. The caption to Figure 1 states that "each system has a range of values. The maxima are the top of the bars, the minima are the horizontal dotted lines". He says that the newer (or non-conventional) energy systems should have wider error bands than the older ones. There is no inherent reason for assuming this. In the study, coal and oil have the widest bands, due to air pollution effects not being known exactly.

It is stated that future projected accident rates should be considered, not historical ones. I used present-day data, not historical. What future rates will be is not clear to me, nor anyone else. However, we should not assume that rates per unit energy output will always fall in years to come. For example, the risk per unit energy of newly-discovered oil like that of the North Sea is much higher than that of Saudi Arabia.

Finally, the author says that calculations for energy payback times for wave energy devices can vary widely. He gives as an example the differing conclusions reached as a result of back-of-the-envelope calculations. But my "envelope" contains about 150 pages and 160 references. The unavoidable variation among lifetimes, materials and construction requirements, and other risk-related aspects of energy systems is precisely the reason why error bands were included. In spite of this variation, it has been shown clearly that non-conventional energy systems can have substantial risk per unit energy output.

Yours sincerely,

Dr. Herbert Inhaber
Scientific Adviser
Technology Impact Division

Atomic Energy Control Board
Assessment and Research
Directorate
P.O. Box 1046
Ottawa, Canada
K1P 5S9

Amory Lovins and the Canadian Atomic Energy Control Board

The first letter in this section was written by A. Lovins, of Friends of the Earth, to A.T. Prince, President of the Atomic Energy Control Board at the time it published AECB 1119. Lovins took issue with the report, and asked that it be withdrawn. Prince declined, and Lovins rewrote (enclosing a copy of one of his articles, not reprinted here). Prince again declined to withdraw the report.

28 August 1978

Dr A T Prince, President, AECB
PO Box 1046
Ottawa, Ont. K1P 5S9, Canada

Dear Dr Prince,

Though a colleague who asked AECB for a copy on my behalf was told it was not available "due to extraordinary demand from around the world", I recently succeeded in borrowing a copy of "Risk of Energy Production" by Herbert Inhaber (AECB 1119, March 1978). Its subject is one with which I am acquainted [see e.g. "Cost-Risk-Benefit Assessments in Energy Policy", 45 Geo. Wash. L. Rev. 911-943 (August 1977)].

I think it only right to let you know before commenting publicly on the Inhaber study that it is the most unimpressive technical paper I have read in many years notable for exploration of the murkier backwaters of the literature.

Your Board's reputation for integrity, impartiality, and competence--on which I express no opinion--cannot fail to be tarnished by your publication and, apparently, your enthusiastic dissemination of tendentious work that cannot, in my view, withstand even cursory scrutiny. (I suspect, too, that many thoughtful advocates of nuclear power are already finding the Inhaber report an acute embarrassment.) Scientists publishing work so far below accepted professional standards bear a heavy responsibility to their peers and to the public. For a Board charged with regulating nuclear power to generate and publish such work, carrying a message that is bound to be construed as self-serving, suggests a certain insensitivity both to that responsibility and to the need to justify public confidence in the fairness and quality of the Board's work. I should not be eager, in your place, to try to defend the way the Board has sustained these duties in this instance.

The practicalities of the Board's reputation, your own, and the difficulties of nuclear advocates are not my problem. As one deeply concerned, however, with the more abstract problem of the perceived legitimacy of the regulatory process, permit me to suggest that you publicly withdraw the Inhaber report from circulation and, should further reflection leave you in doubt about its technical quality, commission its prompt, detailed, critical, and independent review. I would emphasize the importance of review that is not only independent in fact but seen to be independent in principle.

Should you not withdraw the report, I should be glad to have a copy for my library (c/o FOE Ltd, 9 Poland St, London W1V 3DG, England). In any event I should be grateful to have your views and to know what you propose to do.

Sincerely,

Amory B Lovins

c/o FOE
602 C St SE
Washington, DC 20003

15 September 1978

Mr. Amory B. Lovins,
c/o Friends of the Earth
620 C Street, S.E.
Washington, D.C. 20003
U.S.A.

Dear Mr. Lovins:

I note your interest in the Board report, "Risk of Energy Production" (AECB-1119), as expressed in your letter of August 28. I am only sorry that you were not disposed to make any substantive comments on it.

Detailed comments were received from many people around the world and incorporated in the second edition, published in May. Dr. Inhaber, the report's author, is gathering more for a possible third edition. He tells me, however, that the corrections and amendments he has made or will make do not strongly alter his overall conclusions. These conclusions are, of course, his to defend, not mine.

You raised two points in your letter on which I would like to comment. First, is the question of review. The report was reviewed extensively within the Board, by two outside federal government energy scientists, and by a private consultant. I am pleased to enclose a copy of the last-named review.

In the long run, however, neither what you or I think or say about the report is of great consequence. What will count is whether experts in risk analysis can use the information and whether they find it has fairly treated the subject. In this context, we have been gratified by the favorable opinions of experts highly respected in their fields, such as F. Niehaus of the International Atomic Energy Agency, and Sir Edward Pochin of the British National Radiological Protection Board. You may wish to write Dr. Niehaus for a copy of his detailed and exhaustive analysis.

Your second point, concerning a possible withdrawal of the report, has me mystified, quite frankly. Over the past few years, the Board has been making efforts to become more open to the public in terms of explaining what we know. Bill C-14, the Nuclear Control and Administration Act, now before the Canadian Parliament, would formalize and enlarge this trend toward more openness. Yet you are asking us to reverse this trend by suppressing one of our own documents. I can only interpret your suggestion by assuming you are unaware of Canadian conditions, a situation that is not unknown among foreign nationals.

I am even more mystified by your subsequent request for your own copy. Are you suggesting that we suppress the document for everyone except certain individuals, including yourself?

With respect to copies of the report, owing to unprecedented demand, they are in fact temporarily out of stock. This is not, as you suggest, due to "enthusiastic dissemination" on our part. Rather, requests for it grew out of its presentation as but one useful piece of testimony at the hearings of the Royal Commission on Electric Power Planning in Ontario last February. Its popularity is primarily a testimony to its topicality. When the next printing permits, we would be glad to send you a copy.

Your letter raises significant points of public policy. As part of our moves to release more information on the Board's workings, we would like your permission to release both your letter and our reply for public scrutiny.

Yours sincerely,

A.T. Prince

Encl.

Atomic Energy Control Board
Office of the President
P.O. Box 1046
Ottawa, Canada
K1P 5S9

1 October 1978

Dr A T Prince, President
Atomic Energy Control Board
PO Box 1046
Ottawa, Ont K1P 5S9, Canada

Dear Dr Prince:

Thank you for your reply of 15 September to my letter of 28 August about the Inhaber report, AECB-1119. I should of course be willing for you to release our complete correspondence, including this letter, and indeed I have taken the liberty of doing so myself. Because of your interest in risk analysis I also enclose a copy of the George Washington Law Review article mentioned in my previous letter, in case you have had trouble obtaining it.

I have read the cursory comments on AECB-1119 by Lemberg Consultants Ltd. which you kindly sent to me as AECB-1131. I assume that as part of your "moves to release more information on the Board's workings" you will also be releasing the complete texts of the reviews made "extensively within the board, by two outside federal government energy scientists," and by Dr Niehaus and Sir Edward Pochin (together with all papers bearing on the selection of these referees and of Lemberg Consultants Ltd, and with all the other comments the Board has received on the report).

None of these reviews, however, can substitute for the prompt, detailed, critical, and independent review suggested in the fourth paragraph of my letter of 28 August--a proposal on which I regret you have not commented. Neither could my own opinions fill this role, since I have not had time to do a review sufficiently detailed to be considered definitive even in identifying all major faults, and experience with past reviews has taught me that to publish any less than that is a disservice to the public (and perhaps to the author too). It is because I did not wish to review incompletely that I deliberately refrained from outlining my substantive criticisms. This must be done properly, not cursorily.

I am surprised that you think "what will count is whether experts in risk analysis...find [the report]...has fairly treated the subject." What I think will count is how much confidence the Canadian public feels in the way the Board is discharging its duties. Expert opinions of the Inhaber report will influence public opinion but not determine it, especially as many of the report's defects will be obvious to non-expert readers of critical mind. The burden of the third paragraph of my last letter is that the Board is storing up for itself precisely the sort of trouble the US Nuclear Regulatory Commission has got itself into through its sponsorship of the Rasmussen report as ammunition for nuclear industry lobbyists--a sponsorship that is backfiring badly now that the NRC's own review panel, chaired by Professor Harold Lewis, has discredited Professor Rasmussen's conclusions--especially as misrepresented in the widely circulated Executive Summary--and has largely vindicated his critics.

(The Lewis Committee, by the way, lacked the necessary appearance of independence: though Professor Lewis's integrity is beyond question, it is wrong in processual principle, as the current NRC would probably agree, that the NRC should have appointed, funded, and staffed an inquiry into the validity of a controversial NRC report which the NRC earlier had enthusiastically pressed on the public. I trust you are also not under the illusion that Dr Niehaus's IAEA or Sir Edward Pochin's NRPB are generally considered to be independent institutions save by persons with close affinities to the nuclear industry.)

As for your fifth paragraph, I am well aware of Canadian conditions as a result of frequent work with or for the federal and provincial governments. I am therefore familiar with the institutional traditions that lead you to the bizarre conclusion that publicly withdrawing from official sponsorship and circulation a discreditable and misleading report, for the cogent reasons adduced in the third paragraph of my last letter, would "reverse" the "trend toward more openness" by "suppressing" a Board document. For the Board to cease to disseminate tendentious error under its imprimatur is as much a part of its duties as to begin to disseminate the many presumably accurate internal studies and memoranda (such as the Bruce Safety Notes) that it continues to suppress. Were the Board rather to decide, in the spirit of your letter, that error useful to nuclear promotion is to continue to be broadcast with official sanction while truth inimical to it is to remain locked in the files, a most unfortunate impression of the Board's impartiality--an impression your letter does nothing to dispel--could hardly fail to be created in the minds of Canadians.

Your sixth paragraph misreads my fifth. I asked for (and look forward to receiving) a copy of the report only if you do _not_ withdraw it--which I now gather you do not propose to do. Nor is "withdrawing" a report--which means announcing it is no longer considered of a quality worthy of publication by the Board and will not be given further currency--the same as "suppressing" it, which means it is not and has never been available to outsiders. The former means withdrawing an implicit (even if incomplete) endorsement and ceasing distribution by the Board; unlike the latter, it leaves open the option of limited availability for interested scholars, particularly historians of the Board's affairs, who do not propose to use the fact of the report's publication by the Board as an instrument for promoting its acceptance by the credulous. (No doubt you are aware that the report is being widely abused in this way by promoters of commercial nuclear interests in Canada and abroad.) I trust that readers of this correspondence will have less trouble than you seem to in drawing the distinction between taking one's colours off a horse and smothering it at birth.

I am sorry that I have evidently failed to convey to you the depth of my concern at your Board's conduct in this matter--concern reinforced by the terms of your letter. While it is only proper that public confidence should be withdrawn from institutions undeserving of it, this process tends to rub off on other and worthier institutions too. The resulting widespread loss of legitimacy which I now perceive as a growing risk in Canada and elsewhere is likely to last, if it once occurs, for generations, and would greatly reduce the ability of advanced societies to cope with increasingly complex problems. It is this broad concern with the perceived legitimacy of the regulatory process and of governance itself, not only a longstanding interest and involvement in Canadian energy affairs, that moved me to write to you, and that now moves me to ask you to reconsider your response most earnestly.

Sincerely,

encl (original only)   Amory B Lovins

c/o FOE Ltd, 9 Poland St
London W1V 3DG, England

cc: Rt Hon Alastair Gillespie (EMR, Ottawa)
Professor Arthur Porter (RCEPP, Toronto)
Dr. Ray Jackson (SCC, Ottawa)
Dr. David Brooks (FOE, Ottawa)
Dr. Gordon Edwards (CCNR, Montreal)
Energy Probe (Toronto)
Dana Silk (MEC, St John NB)

4 December, 1978

Mr. Amory Lovins
c/o Friends of the Earth Limited
9 Poland Street
LONDON WIV 3DG, ENGLAND

Dear Mr. Lovins:

Thank you for your letter of October 1 concerning the AECB report "Risk of Energy Production". Your opinions will be considered along with the hundreds of comments that Dr. Inhaber has received since the document was first published. I look forward to any substantive remarks which you may care to make when you receive your copy of the latest edition, expected from the printers in approximately two weeks.

I very much regret that you have presumed the AECB launched this report as a vindication of the nuclear industry, or that it has in any way encouraged its use as a tool to be used by proponents of nuclear energy. I can assure you that this is not the case, and had the study showed nuclear power at any other point in the relative scale, the Board would still have published the report. This leads to some interesting conjecture as to who would then have been critical of the document, and who would have used it to support their objectives.

Though I have no reason to believe that the report has as many flaws as you and few others have suggested, I take a great deal of satisfaction from the fact that it has fostered discussion and debate in both the scientific community and the public sector. This was one of the purposes of the document, a point I explained in my preface to it.

Unfortunately, neither the nuclear proponents nor the critics can view the report objectively. If we could find that elusive "independent" person, I am sure he or she would see the document as no more nor less than food for thought. Being non-judgemental, the report was never intended to be anything else.

Yours sincerely,

A.T. Prince

Atomic Energy Control Board
Office of the President
P.O. Box 1046
Ottawa, Canada K1P 5S9

cc: Hon. Alastair Gillespie (EMR, Ottawa)
Professor Arthur Porter (RCEPP, Toronto)
Dr. Ray Jackson (SCC, Ottawa)
Dr. David Brooks (FOE, Ottawa)
Dr. Gordon Edwards (CCNR, Montreal)
Energy Probe (Toronto)
Dana Silk (MEC, St. John, NB)

John Holdren and *Nuclear News*

A letter by John Holdren of the University of California, Berkeley appeared in the March, 1979 issue of *Nuclear News* to which the author replied. Holdren wrote again in the April, 1979 issue and a second letter in response was published. At the end of the set is a personal letter from the author to Holdren inviting him to settle any differences with respect to the report. No answer was received. Because Holdren refused to give reprint permission for the two letters, they are paraphrased here.

In his first letter, Holdren said that he had only recently seen the item on the Atomic Energy Control Board report as noted in the July, 1978 issue of *Nuclear News*. Since he had been quoted in the publication, he felt that he should set the record straight. He began by saying that although he had written on nuclear power, he had not estimated the health impact of long-term storage or disposal of radioactive waste. His estimates of occupational and public risk from nuclear waste were only for transport and reprocessing.

Secondly, he said that data from his report had been "misunderstood, misrepresented, and misused" in Inhaber's report. He said that some errors included confusing thermal and electrical energy, exchanging megawatts with megawatt-years, making arithmetic errors, double counting labor and back-up energy requirements and introducing arbitrary correction factors.

Thirdly, Holdren said that Inhaber's values for material requirements for wind are made up of a remarkable combination of errors. He stated that these included pounds-to-tons mistakes of a factor of 2000, and a countervailing error of a factor of 20 by confusing the energy output per year with the energy output over the lifetime of the system. He stated that the net error of a factor of 100 is in all the conclusions about wind.

Fourthly, he stated that Inhaber's values for biomass are too high by a factor of 8.33. This factor supposedly corrects for an assumed 12% efficiency of converting the chemical energy in methanol to the mechanical energy in vehicles. Holdren said that Inhaber did not account for inefficiencies of end-use devices for conventional energy sources that he considered.

Fifthly, Holdren said that these errors and inconsistencies render Inhaber's values unusable, either absolutely or as measures of comparative risk. The risks of nonconventional energy sources are worthy of study but it should be done objectively, by "someone who knows a thousandfold error when he sees one."

Nuclear News
March 1979

Response

Let me first say a word about timing. Holdren was one of the first to receive a copy of the report, last April. I am sorry that it has taken ten months to receive comments from him. In the interval, I have received private comments from many individuals, and have incorporated many into two subsequent editions which have been printed. I would like to eliminate any errors or misstatements in the report, but it should be done in a rational and even-tempered manner. The scientific method of analysis, critique and re-analysis produces the most accurate work.

I did cite Holdren's work 13 times (not 30) in my report. However, I also used about 180 other references, so it is no wonder that some of my values differ from his. This is surely the nature of a document which employs many data sources.

With respect to radwastes, the data in my report was taken directly from Holdren, who, as he notes, took it from report WASH-1224. Footnote (d) of my table D-1 indicates that I copied Holdren's original footnote on the risk from radwaste almost *verbatim,* noting that it included only shipping and reprocessing. I do not know what more could have been done.

With respect to the question of long-term storage of high-level wastes, no explicit statement was made in my report to the effect that it was included. An argument can be made that while inadequate storage of high-level wastes can cause environmental damage, the likelihood of actual harm to human health is very small. Of course, environmental damage cannot be equated to human risk. In any case, if Holdren or others have data estimating risk of high-level radwaste storage, I would be pleased to incorporate it in any future edition.

The second numbered paragraph of Holdren's letter makes a great many statements about errors in my report. If they deal with matters other than those raised in other paragraphs, I cannot deal with them for the simple reason that the "errors" are not specified. Let me put the charges in some perspective. The document has been publicly available for almost a year, and I am told has been carefully read by many people, especially those who are not happy with its conclusions. Professor Holdren is the first, to my knowledge, to make such sweeping public statements about the mathematics. Can he be the only one to discover such a series of errors?

I do plead guilty to one of the statements made in Holdren's second numbered paragraph, however, I used upward "correction factors" multiplying his data—but only in the case of nuclear power. As mentioned in my text, I chose the highest values for risk of this system's subcomponents from either Holdren's work or that of a comprehensive review article by Comar and Sagan of the Electric Power Research Institute (*Annual Review of Energy,* Vol. 1, 1976, pp. 581-99). This was done for no other conventional (coal, oil, natural gas) system. The object of this procedure was to remove, as far as possible, any inadvertent pronuclear bias.

As an example of this, the values I used for risk from nuclear electricity generation, including potential catastrophy, were between 17 and 40 times that used by Holdren. If this is unfair upward adjustment, I stand convicted.

With respect to wind, I would make the general statement that the literature dealing with nonconventional energy systems is often confusing, even, I suspect, to its authors. However, I flatly deny the statement that I was off by a factor of 2000. Because of the poorly written nature of some information on wind power, I did make an error in one aspect, the ratio of windmill weight to energy produced. This lowered the over-all risk of energy from wind power by between 20 percent and 50 percent, not the factor of 100 which Holdren implied. Prof. P. Musgrove of England had previously brought my attention to the wind data. This change will be incorporated in any future editions of the report, and was incorporated at the galley stage of a forthcoming article in *Science.*

While the error noted above is significant, it does not substantially change the risk ranking of wind with respect to the other ten sources. Because data dealing with energy risk is not known as precisely as we would like, more attention should be paid to the orders of magnitude of risk or the relative ranking rather than the exact numerical values. This point is made several times in my report, and is a procedure usually adopted by responsible commentators on it.

With respect to methanol (or biomass), the value of 8.33 which Holdren quotes is not to be found in section J-i of the third and latest edition. Perhaps he is using an earlier version. An error relating to possible double-counting of conversion losses was noted some months ago, and was corrected in the third edition. This change lowered methanol risk, but did not change its over-all ranking. Incidentally, this mistake was caught by someone working at the Whiteshell Nuclear Research Establishment of Atomic Energy of Canada, Ltd. If nothing else, this shows that people working in nuclear research are not always prone to overestimating the risk of nonconventional or "renewable" energy sources.

I tried to make the end uses as comparable as possible among the systems considered. In the case of methanol, I assumed "that the methanol used is equivalent in terms of mechanical energy to the electricity that could have been used to drive autos and buses." I submit that this is a fair way of considering different end uses. While fossil fuels and nuclear power also have conversion losses, most of it occurs in the generating plants, and was taken account of in my calculations.

Holdren's fifth numbered paragraph is a repeat of the general statements made in his second one. I would like to deal with them, but cannot due to the lack of detail. As I stated above and have demonstrated by specific example, I have tried to eliminate any errors brought to my attention. These alterations have not, however, changed the report's overall conclusions, that nonconventional energy sources can have substantial risk compared to some conventional sources such as natural gas and nuclear power.

Herbert Inhaber
Associate Scientific Adviser
Atomic Energy Control Board
Ottawa, Canada

Holdren's second letter began by saying that Inhaber's response included denials of errors that do exist and in essentially the magnitude Holdren said they did. He felt that Inhaber thinks readers of *Nuclear News* will not take the trouble to look at his report themselves. He went on to "spell out the blunders" using direct reports from the first and third editions of AECB 1119.

In terms of wind, Holdren said that there had been an error by confusing tons with pounds. This would produce an error of a factor of 2000. Then lifetime output was confused with annual output, another factor of 20, for a net error of a factor of 100. The basis of Holdren's statement of a factor of 2000 was the following passage on page H-1 of AECB 1119:

> "One estimate states that the production of 250 MWh of electrical energy requires 400 short tons of steel, 10 of copper, and 60 of fiberglass and plastic (99). Note that this is not the rated capacity, but the total energy produced. Translating this to the usual base of 1000 MW years net output, this is 12.6 million metric tons of steel, 0.32 million of copper and 1.9 million of fiberglass and plastic."

Reference 99 was to the record of hearings of a U.S. Senate committee, but it is difficult to trace. Holdren, in doing the calculations himself, used the capacity factor of 0.34 and plant lifetime of 20 years given by Inhaber. He assumed a windmill of 4 MWe rated capacity. In its lifetime, such a windmill would produce 4 MWe x 8760 h/yr x 0.34 x 20 yr = 238,000 MWh. Its steel content would then be 238,000 MWh x (400 tons/250 MWh) = 381,000 tons. According to Holdren, this would weigh more than 1,000 Boeing 747s. He felt that the error was not a simple typographical one. This is because the same ratio of mass to energy was repeated in the third sentence of the quoted passage. Holdren's calculation was (400 tons/250 MWh) x 8760 h/yr x 0.907 tonnes/tons = 12.7 x 10^6 tonnes/10^3 MW-yr.

Holdren said that he thought that the error represented a confusion of tons with pounds because the numbers in the literature for a 4 MW windmill are from 340,000 to 825,000 pounds. He felt that Inhaber's error of a factor of 20 came when he divided 12.7 x 10^6 tonnes by 20,000 instead of by 1,000.

Holdren said that these errors continued into the third edition of AECB 1119. He felt that Inhaber was unable to figure out what the problem was and made an attempt to reduce the alleged discrepancy by reducing his original capacity factor of 0.34 by a value of 0.06.

Holdren stated that the third edition has the missing details on Reference 99–that the reference was the Project Independence Task Force Report on Solar Energy. This was reprinted as an appendix to hearings of the Senate Committee on Small Business. Holdren said that Inhaber found the material in his source rather confusing. He also pointed out that the Project Independence report stated that 400 tons of steel will produce 4 MWe of wind capacity. The capacity factor of these machines would be 0.34. Each machine takes 20 tons of steel and produces 600 MWh/y or 12,000 MWh in the 20-year lifetime assumed. All the machines produced 12,000 MWh/y or 240,000 MWh over 20 years. Therefore he felt that Inhaber's error in the quoted passage is a factor of about 1,000. He then said that reversing the initial confusion between lifetime and annual energy output generates a reverse factor of 20, so that net error is about a factor of 50. Holdren said that this error was not made by confusing pounds and tons, as he had supposed it was initially.

Holdren described the work as being one of "persistent incompetence." He discussed a second incident where he said that Inhaber had denied the existence of a mistake. He writes "Inhaber's biomass figures are uniformly inflated by a factor of 8.33 (on top of various other mistakes) because he multiplied by that factor (1/0.12) to 'correct' for an assumed 12% efficiency at conversion of chemical energy to mechanical work delivered to the wheels of methanol-burning vehicles." Inhaber replied: "With respect to methanol (or biomass) the value of 8.33 which Holdren quotes is not to be found in section J-i of the third and latest edition." Holdren said that he did not have a copy of the third edition at the time he wrote his first letter. In his copy of the third edition he notes on page J-2: "Avoiding double counting of conversion losses, the third multiplicative factor which must be applied to the original data is then 0.36/0.12 = 3.0." Holdren felt that the factor of 8.33 is still there but is multiplied by a factor of 0.36, the assumed thermal-to-electric efficiency of an oil-burning power plant. He said that this corrects the error of confusing thermal with electricity in the oil fuel cycle in the first edition. Methanol can be used as easily as oil to produce electricity. Holdren said that Inhaber's calculation then is equivalent to assuming electricity from methanol would be 1/3 as efficient for running automobiles as electricity from oil. This error of 3 multiplies two other errors, producing a total error of a factor of 9.

Holdren stated that other statements in Inhaber's response were as "preposterous" as the calculations in his report. He said that the report had also been found to be preposterous by many others besides himself. His letter is signed "John P. Holdren, Professor in Energy and Resources, University of California, Berkeley."

Inhaber defends results

Professor Holdren has reopened the debate on AECB-1119, "Risk of Energy Production," by writing a second letter on the subject, as printed in your last issue [p. 32]. His letter was in response to mine answering his first, in your March issue. In his second letter, he does not answer a single one of the detailed denials and explanations I made in my statements.

With respect to his personal accusations against me, I do not feel they should be dignified by an answer.

(1) In terms of wind, the error that I admitted changed the total risk from this energy source by only 20 to 50 percent. This is not denied.

(2) The first edition of the report was the basis for his claims. Many changes and corrections have been made in the subsequent two editions, and a two-minute

phone call could have brought the latest version.

(3) The bar in Fig. 3 of my *Science* article showing the construction times for windpower is about 25 percent lower than it should be on the basis of the revised Appendix H (on wind) I sent Holdren on February 15. (The final risk values took appropriate construction times into account). But that's on the other side of the issue—a case of potential *under*counting of wind risk rather than *over*counting. Holdren ignores this.

(4) The mistake in materials requirements for wind was first pointed out to me in November by a researcher at Harvard and an energy expert in Sweden. They did not blanket the world with letters criticizing me, nor write my employer demanding that the report be suppressed. There are a number of statements in Holdren's writings over the years that I could take strong issue with, but the thought of writing a letter of complaint to his ultimate employer, the president of the University of California, simply never crossed my mind. Perhaps I am merely behind the times.

(5) There was, as I admitted in my last letter, an error in the materials requirements for wind, corrected in my recent *Science* article. It arose from a misunderstanding about these requirements on an annual and a lifetime basis, and produced an overestimation by a factor of 20, the approximate lifetime in years of the system. The error was of no other origin.

Holdren originally said I was in error by a factor of 2000, but now has retracted this. "It was not made by confusing pounds and tons," he says. He settles for a factor of 50. I now estimate a total materials requirement of 109 metric tons (or 120 oridinary tons) per MW (rated capacity). If Holdren is correct, we should be able to build a 1-MW (rated) windmill using 2.2 metric tons of materials. I do not believe this has been done.

He quotes the Project Independence report as saying that 400 tons of steel are required per 4 MW (rated), or 91 metric tons per MW (rated). When copper and other matierals are added in a total of about 110 metric tons per MW (rated) are required. I simply do not know where his factor of 50 comes from.

(6) His original statement that I was off by a factor of 8.33 in materials requirements for methanol has now been changed ("I did not have a copy of this third edition") to a factor of 9. In the third edition, I assumed a thermal-to-end-use efficiency of about 0.33, close to that which is achieved in thermal power plants. If Holdren is correct, this efficiency should be nine times as high, or about 3.0. I do not believe that efficiencies greater than one are generally achieved.

(7) Other matters: (a) the 1000 Boeing 747's he uses as an example of my materials requirements are inappropriate for at least two reasons. First, aircraft are generally not used for generating primary energy. Second, if I had taken 1 gigawatt-year instead of 1 megawatt-year as unit energy, the number of aircraft would have been 1000 times big, but still prove nothing in particular. The license of this airline should be revoked. (b) "Many others have found (the report's calculations) preposterous" (in Holdren's last paragraph). Who are they, and what did they say? This unattributed statement has no part in a scientific dialogue. (c) Holdren says that reference 99 in my report was "difficult to trace." It reads "Senate Committee on Small Business, 94th Congress. First session, May 13–14, 1975. U.S. Government Printing Office, Washington, D.C., 1975, p. 3932." I had no difficulty in obtaining it in a Canadian library.

In summary, I feel I have refuted each and every point Holdren made in his two letters. The main conclusions of AECB-1119 remain unchanged. Holdren's (a) intemperate language and (b) secrecy (I still await his "detailed and fully documented analysis" of the report that I requested on February 15)—these show how *not* to go about doing a job.

Herbert Inhaber
Scientific Adviser
Technology Impact Division
Atomic Energy Control Board
Ottawa, Canada

15 February, 1979

Prof. John Holdren
Energy & Resources Program
University of California
Room 100, Bldg. T-4
Berkeley, California
U.S.A.
94720

Dear Prof. Holdren:

John Payne, the editor of Nuclear News, was kind enough to send me a copy of your letter which will appear in the March issue of that journal. I have, in turn, sent him a response to your comments on AECB-1119, "Risk of Energy Production".

In the last line of your letter, you said that you had more detailed comments on my report. I would appreciate being sent a copy.

Without seeing these comments, I of course cannot say how many are valid. However, I might point out that since the first edition was published last April, I have received many constructive comments from around the world. A considerable number have been incorporated in the two subsequent editions. As a result, some comments which might have been applicable to the first edition would not be valid for the third edition.

I say this in the light of the comments contained in your letter to Nuclear News. For example, the error dealing with methanol was removed by the third edition. There was an error in the section on wind, although much smaller, in terms of overall risk from this energy source, than you had stated. As a result, the entire appendix H, dealing with wind, has been rewritten. I enclose a copy. It will be incorporated in any subsequent editions. You will note that in addition to changes being made to the materials requirements, the backup source has now been assumed to be a weighted average of the energy systems now used in Canada. The windpower risk drops as a result of these changes, but its relative rank remains about the same.

While there have been various changes to AECB-1119 in the past ten months, none have altered the overall conclusions of the report. As is mentioned a number of times in the text, more attention should be paid to the relative rankings than the absolute values of risk.

Yours sincerely,

HI/bd
Encl.

Dr. Herbert Inhaber
Scientific Adviser
Technology Impact Division

Atomic Energy Control Board
Assessment and Research Directorate
P.O. Box 1046
Ottawa, Canada
K1P 5S9

cc: Dr. A. Buhl
Nuclear Regulatory Commission

The *Science* Controversy

An article based on AECB 1119 was printed in the journal *Science* in February, 1979. This evoked a number of letters.

The first letter on the subject is from P.H. Abelson, the Editor, dated April 25, 1979. He enclosed letters from Rein Lemberg and Richard Caputo on the article, and stated that any response by the author would not be printed. It will be noted that this decision was made before the author had a chance to formulate a reply.

Lemberg's and Caputo's letters which appeared in *Science* follow. A response dated May 3, 1979 to these letters was made. Publication was declined, based on the policy stated in Abelson's letter of April 25.

Meanwhile, the author appealed Abelson's decision in a letter dated June 4, 1979. Abelson declined to change the policy in his letter of July 6.

Holdren *et al.* made a lengthy comment on the article in the May 4, 1979 issue. A long response was made on June 11, but its publication was declined due to the editor's policy.

Finally, a letter from D. Okrent summarizes some of the debate.

25 April 1979

Dr. Herbert Inhaber
Technology Impact Division
Assessment and Research Directorate
Atomic Energy Control Board
P.O. Box 1046
Ottawa, Ontario
Canada K1P 5S9

Dear Dr. Inhaber:

I believe that you have performed a useful service by pointing out that every form of energy including solar has risks associated with it. However, the circumstances surrounding publication of your views on these risks have posed unusual problems for us.

Rarely do we find ourselves one of a number of journals printing much the same material. We note that your views have been criticized elsewhere and that you have responded extensively to your detractors. On balance your position has enjoyed considerably more space and spotlight than have the views of your critics.

Under these circumstances we do not deem it equitable to provide you further space in our publication for discussion of this particular topic.

Sincerely,

Philip H. Abelson
Editor

SCIENCE
American Association for the
Advancement of Science
1515 Massachusetts Avenue, NW
Washington, DC 20005

PHA/jk

Enclosures

Science Vol. 204, p. 454, 4 May 1979

Energy: Calculating the Risks (I)

My attention has been drawn to the 23 February issue of *Science* and the article "Risk with energy from conventional and nonconventional sources" by Herbert Inhaber (p. 718). I was commissioned by the Atomic Energy Control Board of Canada to review Inhaber's original report (*1*) after it had been sent out to be printed. My review was constructively critical and is available as AECB Report 1131, dated 27 March 1978.

My overall impression of Inhaber's work at the time was as follows:

> . . . the author did not challenge his own assumptions in the report as to how his conclusions may be altered. Nor were any alternative interpretations of the methodology presented. In this regard, the report may become subject to criticism, especially since the conclusions depict conventional energy systems to be less risky than the non-conventional ones. As this review will show, other interpretations of the methodology of risk accounting can lead to the opposite conclusion.

In the year since my review, Inhaber's report has been widely circulated and has been summarized, excerpted, and quoted as an authoritative study. But, is it really?

Before starting my review, I asked Inhaber to tell me how much effort went into the study. He replied that the report had been prepared during a 3-month period and required a total of 3 to 4 man-months of effort by Inhaber and a research assistant. Inhaber has published revised versions of his initial report, but the revisions have all been in the area of correcting data and calculations. There have been no additional revisions of improvements of his risk-accounting methodology.

There are several serious problems with Inhaber's methodology:

1) Inhaber includes *all* of the risks associated with materials acquisition, component fabrication, and on-site construction of energy facilities. This implies that every industry making or transporting anything connected with the facility would not be doing anything else if that facility had not been built. I submit that only the *incremental* risks in constructing any energy system should be measured, not the gross.

2) Inhaber's "nonconventional" energy systems include an energy back-up in the form of conventional energy. This might be acceptable if the risk contribution of the back-up system were small in proportion to that of the nonconventional system. But is it? If one looks at figure 7 of Inhaber's original report (figure 4 of his article), one can readily see that, for wind, solar thermal, and solar photovoltaic, the energy back-up systems contribute the majority of risk! Therefore, in view of the overwhelming risk contribution of conventional back-up systems to the so-called nonconventional systems. Inhaber is not truly comparing conventional with nonconventional.

3) If one uses Inhaber's data as is, removal of the risks of creating an energy facility and the risk due to the back-up system has the effect of *reversing* his conclusion. That is, nonconventional systems (which they now are because back-up has been removed) are less risky than conventional systems. This demonstrates how *sensitive* Inhaber's methodology is to the validity of the assumptions upon which it is based.

The nuclear industry has made wholehearted reference to the Inhaber report as proof positive that nuclear energy systems are safer than the nonconventional systems. There appears to be no questioning at all of Inhaber's surprising "pro-nuclear" conclusions. This can only serve to diminish the credibility of the nuclear industry.

Rein Lemberg
Lemberg Consultants Limited,
1150 Cynthia Lane,
Oakville, Ontario L6J 6A6, Canada

References

1. H. Inhaber, *Risk of Energy Production* (AECB 1119, Atomic Energy Control Board, Ottawa, Canada, 1978).

I found Inhaber's article to be surprisingly at odds with my own similar study (*1*) of electric energy systems. About half of his source material and the methodology he claimed as his own is taken from work I technically directed or had contracted at the Jet Propulsion Laboratory (JPL) (*1, 2*). Thus, I feel knowledgeable about the information and approach Inhaber used in his study.

When I received his late 1978 report (*3*), which the *Science* article summarizes, I found remarkable disagreement between results I obtained when I used the JPL study team data and the results Inhaber derived. For example, his estimates of total health risk (*4*) compared to those in the JPL final report were (i) a factor of about 15 greater for coal; (ii) a factor of about 100 greater for solar thermal electric; and (iii) a factor of about 100 greater for solar photovoltaic. However, his results were about the same for the health risk from a nuclear plant.

I notified him immediately, pointed out these enormous differences, and asked what the nature of the disagreement might be. He indicated that he had added a few things that were left out of the JPL analysis but did not identify even in a general way what these left-out factors might be. Since I had spent 3 years developing the data and had had the assistance of about 20 professionals, I expressed skepticism and advised him not to publish any further without checking his analysis. When I noticed his article about a year ago in *New Scientist* (*5*) without any substantial changes, I wrote to each member of the Canadian Atomic Energy Control Board warning them of potential inaccuracies in Inhaber's work. However, they continued to support him.

I believe the review process used by the scientific community in this case was inadequate. I am open to suggestions as to how this can be avoided in the future.

Richard Caputo

Jet Propulsion Laboratory,
Pasadena, California 91103

References

1. R. Caputo, *An Initial Comparative Assessment of Orbital and Terrestrial Central Power Systems* (Report 900-780. Jet Propulsion Laboratory, Pasadena, Calif., March 1977).
2. K. R. Smith, J. Weyant, J. P. Holdren, *Evaluation of Conventional Power Plants* (Report ERG 75-5. Energy and Resources Program, Univ. of California Press, Berkley, July 1975).
3. H. Inhaber, *Risk of Energy Production* (AECB 1119, Atomic Energy Control Board, Ottawa, Canada, 1978).
4. Total health risk in units of man-days lost per unit of electric energy generated (megawatts electric times the number of years) due to disease, accident, and death over the entire life cycle of the energy system.
5. H. Inhaber, *New Sci.* 78, 444 (1978).

3 May, 1979

Philip H. Abelson, Editor
Science
1515 Massachusetts Avenue, N.W.
Washington, D.C.
U.S.A.
20005

Dear Dr. Abelson:

The following is in response to letters from Lemberg and Caputo on May 4 dealing with an article I wrote in Science on February 23.

I am surprised that Lemberg calls the article "pro-nuclear" when in his review (AECB 1131) of the report on which it is based, he states that I avoid pro-nuclear bias. I do not know how these two viewpoints can be reconciled.

With respect to his point about back-up energy, I might note that this was assumed for only three of the six "non-conventional" energy systems considered, not all, as Lemberg implies. For the three systems with back-up, I showed in Fig. 5 of the article what the results would be if either relatively high-risk or low-risk back-up were employed. In this sense, I have followed Lemberg's suggestion that I "challenge (my) assumptions". I am mystified in being told I have not done so when Fig. 5 shows (and Lemberg notes this) that I did.

Lemberg states that removing back-up reverses my

conclusions, stated to be that non-conventional systems are riskier than conventional ones. This conclusion is to be found nowhere in my article. It is clear from Fig. 5 that coal and oil, two conventional systems, have larger risk than non-conventional systems, regardless of what back-up is assumed.

Lemberg argues for calculating occupational risk on an "incremental" basis, rather than an absolute basis. This is a philosophical argument, and its validity cannot be proved or disproved. I chose the absolute basis of calculating occupational risk for two major reasons: (a) It is not clear what risk coal and uranium miners, steelworkers, windmill operators, etc. and all others engaged in an energy system would incur if they were not so engaged. Some might be unemployed; others might be in riskier (or less risky) industries. Estimates of incremental risk would be highly speculative. (b) The methodology of absolute risk has been employed by all analysts of which I am aware: Comar and Sagan, Rose _et al_, Gotchy, Smith _et al_, Hamilton, Morris, etc. (references available on request). I saw no reason to alter a well-established tradition.

Lemberg notes that the report on which the article is based may have been misused by the nuclear industry. I deplore misuse by any group, nuclear or otherwise. Both the article and the report are studded with _caveats_, noting that risk data on all the systems considered is not known perfectly, that different models of energy systems will yield different results. These _caveats_ are ignored at the user's peril. The article is far from the last word on the subject.

Caputo notes differences between his calculations of risk for three energy systems considered in a report he prepared (1) and my article. In terms of coal, he apparently assumed a low BTU gasification with combined cycle. I believe this system is not in widespread use at present. My article was concerned with energy systems as they exist now, so I assumed a conventional system with lime scrubber flue gas desulfurization, noted in the report. The difference in oxides of sulfur produced by the two systems, and related air pollution risk accounts for the difference between Caputo's values and mine.

With respect to the differences on solar thermal electric and solar photovoltaic, I confess that I had difficulty in interpreting Caputo's study (1). For example, his Table 6-6 indicates only a question mark for total occupational risk for the solar component of the hybrid (solar and conventional back-up) plants he considered. The total deaths per 100 megawatt-years for the solar component are listed in the same table as negligible, in contrast to his other tables such as 1-1 and 6-8 which indicate otherwise. The risk due to operation and maintenance of solar systems is apparently not included, although they are mentioned in related reports by JPL. The risk for the coal back-up is taken as the geometric mean of the maximum and minimum values, a procedure which I believe is generally not used. I also had difficulty in determining precisely how the occupational risk was calculated. For this and a variety of other reasons, I referred to Caputo's report only seven times in a total of about 180 references, although I also referred to related JPL reports.

My recollection of the phone call to me mentioned by Caputo is not, I am afraid, the same as his. Caputo was invited by the AECB to write me directly, outlining specific areas where my report could be improved. To date, I still await the letter.

Yours sincerely,

Dr. Herbert Inhaber
Scientific Adviser
Technology Impact Division

HI/bd

Atomic Energy Control Board
Assessment and Research Directorate
P.O. Box 1046
Ottawa, Canada
K1P 5S9

cc: Dr. Chauncey Starr

(1) R. Caputo, An Initial Comparative Assessment of Orbital and Terrestrial Central Power Systems, Report 900-780, Jet Propulsion Laboratory, Pasadena, California, March 1977

p.s. I understand that your general policy is to allow authors to respond to technical comments made in your letters section. I trust that this response meets your criteria.

June 4, 1979

Dr. P.H. Abelson
Editor, Science
1515 Massachusetts Ave. NW
Washington DC 20005

Dear Dr. Abelson:

The following is in response to your letter of April 25, in which you indicated that you would not be publishing my reponse to letters commenting on my Feb. 23 article in Science. Let me first state that what follows is in no way meant to challenge your ruling as editor. Some of my colleagues have suggested that I place my position on record, and I am taking this opportunity to do so.

My reading of Science leads me to believe that when an author's paper is strongly criticized in the Letters section, the author generally has the chance to respond. I would not claim that this is an unchallenged right, but merely one which is frequently exercised.

Your disinclination to allow the same procedure in my case is apparently based on the fact that I had published two articles (one in New Scientist and another in Energy) dealing with the topic of my Science article before the last-named publication appeared. You are, of course, well within your rights to set any conditions that you, your editorial board and referees wish to in connection with Science articles.

Allow me, however, a brief comment. I was unaware that it was Science's policy to restrict or prevent an author's response on the basis of what he or she had published previously on the same topic. I can understand papers being rejected or modified on the basis of previous publication (apparently this was a policy of an editor of the New England Journal of Medicine, but I frankly do not comprehend the connection between an author's response to letters and his or her previous publication. Are they not two separate matters? I might add in partial defense, in case one is needed, that I believe (a) New Scientist is usually regarded as a news journal rather than a journal of record, and (b) my article in Energy dealt solely with one energy system (hydroelectricity) rather than the 11 I covered in the Science article. My publication in your journal is the only one I have written on the overall risk of energy in any journal to date.

Your letter goes on to note that I "have responded extensively to (my) detractors". You may be referring to an exchange of letters I had with Prof. Holdren in Nuclear News. Let me state first that I have published no articles in Nuclear News, and do not propose to do so. A report which I wrote was commented on briefly in that publication some months ago, and Prof. Holdren decided to write that journal on the subject of that report. I was invited by its editors to respond, and did so. Again, I do not see any connection between my response to Nuclear News and your ruling that I cannot respond in Science. I do not believe that there is great overlapping of the two audiences. Perhaps I am missing something here.

In all of this, I have no intention of silencing any critics. While I clearly do not agree with some of their comments, science can only progress by careful analysis and re-analysis of results. I only ask for myself what they have been granted. Some of my colleagues with whom I have discussed this matter have been equally mystified by the ruling in your Apr. 25 letter. As a result, I would respectfully ask you to reconsider, and allow my letter of May 3 in response to Lemberg and Caputo to be published.

Yours sincerely,

Dr. H. Inhaber
Sessional Lecturer

Department of Physics
Herzberg Laboratories
Carleton University
Ottawa, Canada K1S 5B6

6 July 1979

Dr. H. Inhaber
Department of Physics
Herzberg Laboratories
Carleton University
Ottawa, Canada K1S 5B6

Dear Dr. Inhaber:

I have your letter of 4 June requesting a reversal of our decision not to print your letter of 3 May.

As you have suggested, we do not follow rigid rules on letters. I reiterate that your views have enjoyed extensive space and coverage--more so than that of your critics. A continuation of this controversy in Science is not justified.

Sincerely,

Philip H. Abelson
Editor

PHA/rls

SCIENCE
American Association for the
Advancement of Science
1515 Massachusetts Avenue, NW
Washington, DC 20005

Science Vol. 204, pp. 564-566, 11 May 1979

Energy: Calculating the Risks (II)

Herbert Inhaber, in his article "Risk with energy from conventional and nonconventional sources" (23 Feb., p. 718), concludes that the health hazards of deriving energy from wood, wind, and sunlight are comparable to those of using coal or oil and much greater than those of using nuclear power. The article, however, displays none of the calculations on which this surprising conclusion supposedly rests, but simply describes the author's approach and summarizes the results. For all the details, readers are referred to Inhaber's report for Canada's Atomic Energy Control Board [(*1*), hereinafter referred to as AECB 1119]. We have examined AECB 1119 in some detail, having been motivated to do so in part because one of us was named in the article and in the report as the "well-known nuclear critic" whose data Inhaber says he has used to preclude accusations of pronuclear bias. A report we coauthored (*2*) is indeed the source of a number of AECB 1119's citations, but Inhaber has both misrepresented and misused our results.

We are not the only ones thus abused. Comparison of AECB 1119 with its references reveals instance after instance where Inhaber misread his sources or propagated errors. As we shall show here, in fact, Inhaber's report is a morass of mistakes, including double counting, highly selective use (and misuse) of data, untenable assumptions, inconsistencies in the treatment of different technologies, and conceptual confusions. Several statements in the *Science* article about how the numbers in the underlying report were derived, moreover, are misleading or wrong. When the effect of the major errors and inconsistencies in AECB 1119 are removed, the *Science* article's conclusions change drastically: the difference between coal's health hazards and those of nuclear power shrinks, and the calculated hazards of the renewables fall to near or below those of nuclear.

These are serious charges. We document them here at such length as *Science*'s space limitations permit, and in greater detail elsewhere (*3*). We begin by comparing some of the article's assertions with what one finds in the underlying report.

Inhaber makes many statements in the article conveying the impression that he has treated conventional and nonconventional energy technologies on the same footing. But examination of AECB 1119 shows that the implied systematic approach and consistency are absent. Indeed, with all Inhaber's emphasis on the occupational risks of constructing energy facilities, he clearly has not included the occupational risks of building coal, oil, or nuclear power plants in the risk figures for these technologies. The numbers tabulated in AECB 1119 for occupational deaths and injuries in the coal, oil and nuclear fuel cycles and summarized in figure 5 of the

Science article come for the most part directly from Smith *et al.* (*2*) and are for operation and maintenance only; they include no contribution from materials acquisition, component manufacture, or plant construction. Inhaber lists materials requirements and partial labor requirements (onsite construction but not materials acquisition or component manufacture) for the conventional technologies in the article and in AECB 1119, but he does not apply the methodology he has described to translate this information into occupational risks for inclusion in his totals. If he had used this methodology for nuclear power in the same fashion he did for the renewables, the lower bound of nuclear's occupational risk as presented in the article's figure 5 would have been about 1.7 times higher and the upper bound about 1.15 times higher.

Inhaber claims at several points in the article that he has bent over backward to avoid any bias toward nuclear power. Concerning public risk from reactor accidents, for example, he writes: "To avoid any bias in favor of nuclear power, I used the highest values of public risk from reactors taken from a wide number of sources (in some of these, Rasmussen's values were used)." Indeed, the *Science* article and AECB 1119 lead the unwary reader to believe that the references from which "the highest values of public risk from reactors" were taken include not only the Rasmussen report but also the Ford/MITRE study (*4*) and Smith *et al.* (*2*). In fact, however, Inhaber's "upper limit" figure for reactor risks is about three times smaller than the upper limit given in the Rasmussen report, more than 40 times smaller than the upper limit given in Smith *et al.* (notwithstanding Inhaber's taking credit, based on his citation of this reference, for using the values of a "well-known nuclear critic"), and more than 200 times smaller than the upper limit implied in the Ford/MITRE study. Had Inhaber actually used, say, the Smith *et al.* upper limit, the upper limit on public man-days lost per megawatt-year of nuclear-generated electricity would have been about 60 rather than the 1.5 shown in figure 6 of the *Science* article, and the upper limit on nuclear's total man-days lost per megawatt-year in figure 7 would be about 70 rather than the 10 shown.

Inhaber's declaration that "present-day technology, models, and systems with their corresponding risk, are used" is also deceptive. His occupational and public risks from the coal fuel cycle, for example, are based in part on practices that are either illegal in present U.S. operations (coal-dust levels in mines) or in new plants (SO_2 emissions). Present dust standards imply occupational deaths from black-lung disease as much as 60 times lower than the figure used by Inhaber. Correction of the black-lung figures would lower Inhaber's upper limit of the occupational man-days lost per megawatt-year of coal-generated electricity by a factor of 1.4.

The sulfur dioxide emissions Inhaber says he considered (*1*, ed. 3, p. A-1) fall within the New Source Performance Standards in force at the time the report was written; but the upper limit of the number of public deaths from sulfur dioxide-related disease (table A-2, all editions), which is used in the risk calculations, corresponds not to these emissions but to emissions five times higher, exceeding the New Source Performance Standards by a factor of 3.3 (*5*). The net inflation of public man-days lost from the coal fuel cycle is a multiplicative factor of 1.3 compared to what would be obtained by consistent use of the New Source Performance Standards.

Inhaber's exaggerations of the risks of coal use also inflate substantially the apparent risks of the renewables, since most of the upper-limit risk of the latter comes from coal "back-up" in all cases where back-up is assumed to be required. (The *Science* article's figure 4 indicates that 62 percent of the upper-limit risk of wind, 74 percent of the upper-limit risk of photovoltaics, and 85 percent of the upper-limit risk of solar-thermal-electric systems come from the assumed coal back-up.) Since Inhaber's entire treatment of storage and back-up for renewables is intricately fallacious, however, one cannot get reasonable figures for this part of the risk simply by removing the inflation from coal's effects. Inhaber has used the same (wrong) ratio of back-up to renewable energy and the same quantity and type of storage, for example, for all three kinds of systems, notwithstanding their entirely different characteristics and the different roles they would play in utility grids.

Inhaber's common ratio of back-up to renewable energy is derived for the case of the solar-thermal-electric system (*2*, p. E-7, all editions), but he has it wrong. He assumes (and tallies up the materials requirements for) an energy-storage capability of 16.5 hours of operation at 70 percent of rated capacity, which his references indicate would permit an annual load factor of about 85 percent. Yet to the risk computed for each 1000 megawatt-electric years delivered by this system, he adds the risk for 19 percent as much energy—190 megawatt-electric years—from coal as "back-up." Here Inhaber appears to have misunderstood his source on back-up requirements (*6*). The solar plant described needs no *net* back-up energy at all to be the energy-producing equivalent of a conventional base-load plant *with the same annual load factor*, although it needs some back-up *capacity* if the reliability characteristics of the grid are not to be altered by the solar plants. Removing the risk of the superfluous back-up energy would reduce Inhaber's upper limit estimate of the risk of solar-thermal-electric systems more than sixfold.

The manner in which Inhaber has sampled the literature for the data he uses in his report is also remarkable. In AECB 1119's treatment of photovoltaics, for example, he starts with materials requirements from an unpublished Jet Propulsion Laboratory (JPL) interoffice memorandum dated May 1976, ignoring in so doing the somewhat lower numbers published in the final report of the same project (*7*), which report he also cites. Then Inhaber asserts (*1*, F-1, all editions), ostensibly on the basis of a report in a 1971 conference proceedings, that the land requirement for photovoltaics is 34,500 square meters per megawatt-year of electrical output rather than the 3800 square meters per megawatt-year used by JPL, and that it follows that the JPL materials requirements per megawatt-year must be multiplied by a correction factor of 2.27. Inhaber's basis for making the inflation factor 2.27 instead of 34,500/3800 = 9.08 is his supposition that the

34,500 square meters per megawatt-year "refers to peak power" and should therefore be divided by 4 to correspond to the average output. This is an astonishing bit of reasoning. First, peak power is measured in megawatts, not megawatt-years. Second, it takes about four times *less* area to make a peak megawatt than to make an average megawatt, not four times more. Third, it is an elementary calculation to verify that the JPL land requirement was correct in the first place (*8*), so no "correction" to the JPL materials requirements on this basis is warranted at all. Removal of this error alone reduces the nonback-up part of Inhaber's upper-limit estimate of risk from photovoltaics by a factor of about 1.7.

As our final detailed example, we consider Inhaber's treatment of methanol from biomass. He assumes that the methanol is made from wood obtained in conventional logging operations (in the treatment of which he makes many errors we will not detail here) and that the product is used to drive automobiles at 12 percent efficiency (mechanical work at the wheels divided by chemical energy in the fuel). Inhaber contends it is fair to consider a megawatt-year of electricity produced at a power plant to be equivalent to a megawatt-year if mechanical energy delivered to the wheels of automobiles because the electricity "could have been used to drive autos and buses." The absurdities in this contention are too many to explore thoroughly here. We note only that (i) losses between the power plant and the wheels of electric autos (transmission and distribution, battery charging and discharging, and losses in the controller and in the electric motors themselves), completely ignored in Inhaber's comparison, are typically around 50 percent; (ii) if electric vehicles really made more sense, one could easily burn the wood directly to make electricity without suffering the significant conversion loss in going from wood to methanol.

Inhaber's actual numerical procedure to calculate occupational risks of building methanol plants is to take numbers for oil refineries from Comar and Sagan (*9*) and multiply them by "correction factors" of 3.0 × 2.0 × 1.5 = 9.0. The 3.0 is the ratio of the efficiency of electricity generation with oil (0.36) to the assumed fuel-to-work efficiency of methanol in automobiles (0.12); multiplication by this factor is completely incorrect, as noted above. The factor of 2.0 Inhaber explains as being due to the fact that methanol contains only half as much energy per gallon as does gasoline; hence, he contends, it requires twice as much materials and labor to build a methanol plant as to build an oil refinery. This, too, is wrong. If volumetric energy density of the product governed the size and complexity of the facility, coal-gasification plants would be impossible. The fact is that methanol-from-biomass plants require fewer and less complicated operations than oil refineries and would probably require less construction material and labor, not more. The factor of 1.5 comes from Inhaber's assumption that methanol plants last only 20 years, while oil refineries last 30. His reference on this point gives 20 years as the owner's depreciation period for accounting purposes, having nothing to do with physical lifetime. Thus the whole factor of 9 is an arbitrary and unwarranted inflation of the materials and labor requirements of methanol; what crowns the performance is that the values from Comar and Sagan that Inhaber multiplies by the factor of 9 are not for construction at all, but for operation and maintenance. Removal of the first unwarranted "correction" factor would reduce the total risk due to methanol as shown in the article's figure 7 by a factor of 3.0 since this factor pervades every methanol calculation in AECB 1119. Removal of the other inflation factors and errors in the methanol calculations [see (*3*)] would reduce the various components of Inhaber's methanol risk by additional factors of 1.5 to 10.

We could go on and on, but we believe we have presented enough detail to give the reader the flavor of what is in and behind Inhaber's article in *Science*. Correcting just his largest errors completely transforms his results, raising the upper limit of nuclear risks to public and occupational health into the lower part of the uncertainty range for coal and oil, and dropping the health risks of the nonconventional sources into the middle of the uncertainty range for nuclear. Even correction of *all* of Inhaber's errors would not produce the "right" answers about relative risks of conventional and nonconventional energy technologies, of course, because many needed data are as yet nonexistent, and because important categories of harm are left out of his approach altogether. But by propagating an analysis riddled with distortions, errors, and inconsistencies, Inhaber has muddied rather than illuminated even the circumscribed part of the risk problem he tackled.

John P. Holdren

Energy and Resources Group
University of California, Berkeley 94720

Kirk R. Smith

Resource Systems Institute
East-West Center
Honolulu, Hawaii 96822

Gregory Morris

Energy and Resources Group
University of California, Berkeley

References and Notes

1. H. Inhaber, *Risk of Energy Production* (Report AECB 1119, Atomic Energy Control Board, Ottawa, Ontario, March 1978): *ibid.*, ed. 2, May 1978; *ibid.*, ed. 3, November 1978. The *Science* article does not specify to which of the three editions it refers, and some of its numbers differ from those in all three. Our comments on AECB 1119 here refer to the third edition unless otherwise specified.
2. K. R. Smith, J. Weyant, J. P. Holdren, *Evaluation of Conventional Power Systems* (Report ERG 75-5, Energy and Resources Group, University of California, Berkeley, July 1975). Inhaber's first reference in his AECB report contains 30 citations to this report, 13 direct ones, plus 17 more where Inhaber took the data from our report but mentioned also the original source we had cited.
3. J. P. Holdren, K. Anderson, P. Gleick, I. Mintzer, G. Morris, K. R. Smith, *Risk of Renewable Energy Sources: A Critique of the Inhaber Report* (Report ERG 79-3, Energy and Resources Group, University of California, Berkeley, April 1979).
4. Nuclear Energy Policy Study Group, *Nuclear Power: Issues and Choices* (Ballinger, Cambridge, Mass, 1977). The authors state on p. 179 that "the expected number of cancers could be several times higher, depending on the assumed dose-response model used in deriving the risk estimates," than the values given in the Rasmussen report. On the same page, they note that "the WASH-1400 probability estimate could be low, under extremely pessimistic assumptions, by a factor of as much as 500." The implied upper limit on the product of probability and consequences is a factor of 1500 to 2500 larger than the WASH-1400 "best estimate." Inhaber's "upper limit" is only 6.7 times the WASH-1400 "best estimate."
5. To derive this result we used the upper limit of the National Academy of Sciences' dose-response relation referenced by Inhaber, for the most unfavorable location that the Academy considered (a plant sited 60 kilometers upwind from New York City) and worked backward from Inhaber's figure for public deaths to determine the emissions needed to produce these. See National Academy of Sciences, *Air Quality and Stationary Source Emission Control* (Government Printing Office, Washington, D.C., 1975), chap. 13.
6. R. Manvi, *Performance and Economics of Terrestrial Solar Electric Central Power Plants* (JPL Internal Report 900-781, Jet Propulsion Laboratory, Pasadena, Calif., October 1976). We have consulted the head of the JPL solar project of which this work was a part, and he confirms our analysis of the point and of Inhaber's error (R. Caputo, private communication, March 1979).

7. R. Caputo, *An Initial Comparative Assessment of Orbital and Terrestrial Central Power Systems* (Final Report, Report 900-780, Jet Propulsion Laboratory, Pasadena, Calif., March 1977). Inhaber propagated a number of errors from the 1976 JPL internal memorandum, despite early warnings from Caputo that this material was unreliable (R. Caputo, personal communication); in fact the memorandum appears to have been Inhaber's main source for his methodology and for much of his data relating materials requirements to occupational injuries and diseases.

8. Average insolation on a horizontal surface in the United States is about 180 watts per square meter (averaged over seasons and night and day). Assuming the collectors cover half the land area charged to the plant and that the efficiency of the cells in converting sunlight to electricity is 10 percent, and using the same 30-year lifetime assumed by Inhaber, yields $180\ W/m^2 \times 0.10 \times 0.50 \times 30$ years = 270 watt-year/m^2, which gives 3700 square meters per megawatt-year.

9. C. L. Comar and L. A. Sagan, *Annu. Rev. Energy* **1**. 581 (1976).

June 11, 1979

Philip Abelson, Editor
Science
1515 Massachusetts Ave NW
Washington DC 20005

Dear Dr. Abelson:

The letter of Holdren, Smith and Morris [Science, 204 564 (May 11, 1979)] makes a great many statements. I find that most of them are without merit. Of those with merit, most do not change the results of my article substantially, and are well within the range of differing system models. That is, there is room for honest difference of opinion on certain of the matters raised, because much of the basic data is still not well known.

Holdren et al also make a number of charges about my motivation, as well as personal accusations. I will not dignify these charges with an answer.

Holdren's statements are with little or no exception, made from one viewpoint: reducing the risk from all energy systems except nuclear power. He is entitled to his attitude, but it seem unlikely that I would have uniformly and substantially overestimated the risk of ten systems and underestimated one. A good case can be made that I underestimated the risk, not of nuclear power, but of some of the other systems. This case is made in some appendices of the report (7) on which my article was based. Holdren ignores this case. I shall repeat some of the reasoning in the following reponses.

Holdren's statements comprise a mixture of both major and minor charges. In order to simplify the reader's task, I will treat each statement as it arises. I will take the liberty of rephrasing the statements to shorten them.

p. 564, column 2. Occupational risk of building coal, oil or nuclear plants was not included. The basis for this claim is the statement that "occupational deaths and injuries" in these fuel cycles come "for the most part directly" from Holdren's work (1). In the first place, this is not the complete truth. The data I used in Appendices A through D of my report (7) for the occupational risk of electricity production were taken only in part from Reference 1. For example, the number of occupational man-days lost for coal-fired electricity production as shown in my Table A-2 is larger than Holdren's (in his corresponding Table A-3). Other examples could be given where my risk data are much larger than that in Holdren's own work. The implication of the above paraphrased statement is that Holdren's estimates are the only valid ones, which I think has yet to be proved.

In the second place, even if Holdren were right, we would have a most curious situation. Tables A-3, D-3 and E-3 in Holdren's work (1), dealing with the risk of coal, oil and nuclear, respectively, are flatly labelled "Health Effects". No indication is given that any significant quantities are excluded. The same situation prevails in a later article which Holdren co-authored (2). Table 4 of this article shows "occupational accidental deaths and injuries" for four conventional energy systems. Values generally agree with Holdren's previous work (1), indicating that the latter was not a typographical error. Again, no statement is made that "substantial" occupational risk was excluded.

The curious situation mentioned above arises because Holdren, in his Tables A-1, D-1 and E-1 (1) shows the labor and material requirements for three conventional energy systems. Now he says: By the way, the risk attributable to this labor and material simply weren't counted in the total. I admit that it is difficult to tell from Holdren's labelling of his tables whether or not the risk was included. We then either have an ex post facto changing of the facts to suit claims made in the letter being analyzed, or an omission on Holdren's part of "substantial" occupational risk from his Tables A-3, D-3 and E-3. I suggest that neither position is tenable.

p. 564, column 2, paragraph 2, line 20. The article does not include materials acquisition risk for conventional systems. This is again only part of the truth. Risk attributable to producing the coal, oil and uranium was definitely included. Otherwise how could Holdren claim (column 3) that this risk was overestimated?

The only questions which then arise are (a) was the risk attributable to non-fuel materials used in building a conventional plant such as turbines, concrete, etc. excluded, and (b) how big is this risk compared to that of mining the materials, transporting them, operating and maintaining the plant, and all other activities not directly related to building the plant?

The immediately preceding response indicates that it is indeed difficult to tell from Holdren's data whether or not non-fuel materials acquisition was included. Let us suppose, for the sake of argument, that it was not. I suggest that in the case of coal, that the occupational risk attributable to mining the fuel for 30 years, upgrading it, transporting it and operating the coal plant will be far greater than the risk in building the turbines and supplying the concrete for the generating plant. The same argument can be made for other conventional systems.

How large is the risk which may have been left out? Using Holdren's Table A-1 (Ref. 1, p. 113), one notes a total of 4.3 tonnes of metals and about 500 man-hours of labor are required to build the coal-fired generating plant. Assuming that the metals are steel and the labor falls into the category of "miscellaneous contracting", it can be shown that the risk value in dispute is about 1.0 man-day lost per unit energy. However, the total occupational risk I estimated for this system was 18 - 73 man-days lost per unit energy. I do not regard the quantity in dispute as significant in comparison. If I have inadvertently left out this component of risk because of confusion over the labelling of Holdren's tables, I apologize; however, the final risk results do not change greatly.

In terms of nuclear power, the number of man-days lost per unit energy for non-fuel materials is about 1.4. The values used in my article for occupational risk were 1.7 - 8.7, so the addition of 1.4 would increase the lower and upper values by factors of about 1.7 and 1.15, as Holdren states. However, the relative rank of nuclear power would not change.

It is of interest to contrast this with the values Holdren himself uses (Ref. 1, Table E-3). Values of 1.9 - 3.9 occupational man-days lost per unit energy are recorded there. If I had taken Holdren's values directly and added the 1.4 man-days lost which are in dispute, the resultant maximum would have been about 40% lower than the one which I used in my article. I find it curious that Holdren does not refer to his own work in this discussion.

p. 564, column 3. The upper limit of nuclear public risk should be 40 - 200 times higher. I did not state in my article that I used the highest values for nuclear public risk which have been claimed anywhere. The report on which the article is based (7) notes that the highest public risk values for nuclear power were taken from two literature reviews: one done by Holdren (1) and another by two highly-respected researchers, Comar and Sagan (8). The values used by Holdren (Ref. 1, summarizing Table E-3) are 17 - 42 times lower than the latter's values. Again, I find it curious that Holdren did not refer to his own work.

Most values in the literature are lower than those I used. To take only two examples out of many possibilities, a Finnish study (9) has values for maximum public deaths from reactors as about 0.27 of that which I used. A Norwegian study (10) has maximum deaths as 0.06 of those which I used. Many other examples could be cited.

There are, of course, some estimates of nuclear public risk in the literature higher than that I used, for example, the Ford/MITRE study to which Holdren refers. This was cited neither in my article nor the report (7) which I wrote, not because the Ford/MITRE estimate is invalid, but because it is an example of what could be called a "worst case" estimate.

With respect to the Ford/MITRE study, the upper limit they used was qualified by the statement that it dealt with the most pessimistic circumstances. I am surprised that this qualification has been relegated to a footnote in Holdren's letter. The Ford/MITRE value and other "worst cases" were not included in my article for at least two reasons. First, there is no real evidence for some of these claims in terms of nuclear power. Secondly, worst cases can easily be hypothesized for most, if not all, energy systems. For example, one could consider the public risk of hydroelectricity only for Italian dams in 1963, where and when the largest disaster associated with this energy form took place. These worst case assumptions would indicate that most energy systems had high and relatively uniform levels of risk, which would be little proof of anything in particular.

Let us now consider whether the values for nuclear public risk were, as is claimed, 200 times too low. I used an upper value of 23×10^{-5} deaths per megawatt-year. Multiplying by 200 yields 4.6×10^{-2} deaths. About 60,000 megawatt-years of nuclear energy were produced in the Western world in 1978, according to Nucleonics Week. This would imply about 2800 deaths directly related to nuclear accidents in 1978, a value which I do not find reasonable. If others find this reasonable, they are entitled to use it.

It can be shown (11) that if nuclear power is treated on a historical basis, as is done for hydroelectricity and other systems, the public risk per unit energy, including that produced by the accident at Three Mile Island, is much lower than the estimate I used. As a result, I believe that the values used in the article are fairly conservative, erring in the direction of overestimating risk.

p. 564, column 3. Occupational risk from the coal cycle is lower than quoted in the article. Occupational risk for conventional technologies was taken from either Comar and Sagan (8), a comprehensive review article, or from Holdren (1). In some cases Comar and Sagan's values were higher than Holdren's and in other cases lower. For example,

Holdren estimated a maximum value of 2.2×10^{-3} accidental deaths per unit energy; Comar and Sagan had a value of 1.5×10^{-3}, which I used. I felt that both review articles presented a reasonable picture of occupational risk in conventional systems in the mid 1970's.

With respect to what this risk might be in the future, I did not consider this question. If it were to be considered, identical criteria would have to be applied to all energy systems, not just coal. In this context, I find it remarkable that Holdren did not note that occupational risk in uranium mines is probably falling at the same rate as in coal mines.

It should not be assumed that occupational risk will fall solely because of the governmental regulations that Holdren mentions. If it were that simple, then all occupational risk could be abolished overnight by government decree. For example, consider the period between 1971 and 1976 in the United States, a time of unprecedented laws and regulations on both the federal and state levels dealing with occupational health. Yet of 23 industries reporting injury frequency rates in that country between those dates, 18 showed an increase, not a decrease (3). This does not, of course, mean that we should abandon efforts at government regulation; merely that in considering occupational risk we should evaluate what is, not what we would like it to be.

It should also be noted that occupational risk can vary strongly from one country to another. For example, the death rate in Norwegian coal mines is about 50% higher than the value I assumed (12).

p. 564, column 3. Sulfur dioxide emissions from coal are too high, so the public risk from this source is overestimated. I used Holdren's values (Ref. 1, Table A-3) for the number of disabilities (between 1 and 216 per unit energy) due to air pollution produced from coal. These disabilities formed the majority of the total man-days lost due to air pollution as calculated in the article.

Holdren does not dispute this. The area of difference lies in the deaths (as opposed to disabilities) from air pollution. I took the values of deaths from Hamilton (13) and Sagan (14). While they are higher than Holdren's, I believe that a reasonable case can be made for their validity. In any case, even if Holdren's values were used for death, the total number of man-days lost due to air pollution would drop by only 30% for the maximum value. The difference for the minimum value is 14%.

In terms of my article, using Holdren's values for public risk due to coal would decrease the height of the appropriate bar of Fig. 7, showing the total risk, by about 1.5 mm; that of the minimum by a non-visible amount. Neither change would alter the relative rank of coal.

Holdren notes [footnote (f) to his Table A-3 (1)] that his values for public risk from coal may be low. He states that other pollutants such as oxides of nitrogen, ozone and carbon monoxide may also have health effects, but were not included. I then conclude that the range between Holdren's and my estimates is well within the differences among risk analysts.

Holdren also assumes pollution control technology which, I believe, is superior to that which is in present average use. These assumptions, in effect, give the benefit of the doubt to coal and oil systems. For example, he assumes [footnote (g) to his Table A-2] that 80 - 90% of sulfur is removed by scrubbers; 99 - 99.5% (by weight) of particulates are removed; and 40% of the total sulfur is removed by cleaning. Similar stringent assumptions are made for the removal of oxides of nitrogen.

I find it strange that Holdren does not quote actual values for emissions as opposed to those specified by governmental regulations. As mentioned above, there is often a substantial difference between the two levels. For example, it can be deduced that the regulations which Holdren quotes imply 11 - 38 metric tons of sulfur oxides emitted per unit energy. While data is somewhat difficult to find, there is evidence from Finland, the Soviet Union and Norway (9) (15) (16) that "emissions from modern European plants put into operation during the last 5 - 6 years" are substantially higher than those Holdren assumed. It can therefore be deduced that Holdren's estimates for emissions correspond to some time in the future. I made it clear that I was considering, as much as possible, present-day conditions.

p. 566, column 1. Most of the risk of "renewables" comes from using coal as a back-up. The assumption that coal would be used as a back-up is traceable to Jet Propulsion Laboratory studies [references 6 and 21 in my report (7)]. Their reasoning was that oil and natural gas would not be available for much longer, hydroelectricity was not widely available in the United States, leaving only coal as a future back-up. I think there is some merit in this argument.

Only three of the six non-conventional (or "renewable" systems) were assumed to have back-up, so Holdren's statement cannot be valid for the other three. Because assuming coal as back-up raised the overall risk of non-conventional systems which required it, I re-did the calculations assuming a low-risk back-up such as natural gas or nuclear power. This was almost equivalent to assuming no back-up at all. I realize that some advocates of solar energy might be uncomfortable with nuclear power as a back-up, but my calculations were hypothetical.

These results were shown as jagged lines in Figures 5 - 7 in my article. I am surprised that Holdren does not make reference to these results, implying that only coal is assumed as a back-up. While the absolute values of total risk, as shown in Figure 7, do change when low-risk back-up is substituted for coal, the relative rankings of the three systems assumed to require back-up do not change substantially. I then conclude that the rankings are not highly sensitive to the type of back-up chosen.

p. 566, column 1. <u>Back-up for solar-thermal electric plants requires capacity, but no energy</u>. I find myself mystified by this contention. This implies that the back-up source is never used. This surely is a remarkable state of affairs.

Manvi [reference 65 of my report (7), p. 7], states: "Back-up is required because of the unreliability of solar insolation. In these studies back-up is generally achieved with fossil energy". Manvi (p. 11) goes on: "Also this solar plant in order to substitute for a best baseload plant...must receive about 19% of its rated energy from a non-solar source in a 'hybrid' operation". Surely there can be no misunderstanding that the word "energy" is used here.

Both the amount of back-up capacity and energy required will depend on quantities such load factors, the degree of penetration into the electrical grid, and other variables. Manvi notes in his Figure I-6 that the higher the degree of penetration, the greater the back-up capacity required. This implies that if centralized solar systems are used more in the future, they will need more back-up capacity and likely more conventional energy.

Estimates of the amount of back-up energy required can vary. Reference 6 of my report (7), prepared by Richard Caputo (quoted by Holdren as saying that no back-up energy would be required) notes on p. 1-3 that 10% coal energy would be needed for solar thermal and solar photovoltaic systems. In spite of this variation, it is clear that some back-up energy will be needed for certain non-conventional systems, if they are to be used as baseload requirements, as Caputo assumes.

In any case, the question of how much back-up will be required is moot, since Figures 5 - 7 of the article showed results with essentially no back-up (i.e., low-risk back-up).

p. 566, column 1, para. 1. <u>74% of the upper-limit risk of photovoltaics and 85% of the upper-limit risk of solar thermal electric is due to back-up</u>. The Jet Propulsion Laboratory study by Caputo which Holdren quotes deduces comparable values. On p. 6-37 of the JPL study [Ref. 6 of my report (7)] it is noted that 90% of the deaths attributable to the photovoltaic system is due to back-up; 78% of the deaths attributable to the solar thermal electric system is due to back-up.

p. 566, column 2, para. 1. The upper-limit estimate of non-back-up risk from photovoltaics should be reduced by a factor of 1.7. There are considerable differences in the literature as to the amount of materials required per unit energy output for photovoltaics. For example, the JPL study noted above assumed aluminum support structures for the photovoltaic elements. Another study by Sears (5), assuming steel support structures, requires much greater amounts of materials than the JPL study. All other factors being equal, the risk attributable to a ton of steel is probably higher than that for a ton of aluminum. As a result, the risk calculated on the basis of the Sears model (which I did not use) would be much higher than that computed using the JPL specifications, whether or not the disputed factor of 2.24 which Holdren mentions is used.

There is some reason to believe that I underestimated the risk of photovoltaics, rather than overestimated it. To take a few examples at random, I assumed zero risk from manufacturing the photovoltaic materials themselves, as opposed to the support and auxiliary structures. I assumed a 30-year plant lifetime. Present arrays are guaranteed for one year (4). Other assumptions relatively favorable to photovoltaics were also made.

All this being said, there is some evidence that the factor of 2.24, reducing the non-back-up risk by a factor of 1.7, may be incorrect. This will be investigated in detail before any further editions of the report (7) are issued. However, I find that making this change does not alter the relative rank of solar photovoltaic significantly.

p. 566, column 2. Methanol is considered to be used at 12% efficiency in automobiles, but other energy systems are used to produce electricity at higher efficiency. Since the main use proposed for methanol is in vehicles, it seemed logical to assume that it would be used that way. However, it was pointed out after the first two editions of the report (7) were printed that this assumption, while reasonable, might be unfair to methanol in terms of risk calculation because the efficiency of turning a fuel into electricity (usually around 35%) was greater than the 12% efficiency of turning fuel into mechanical energy in a vehicle.

In the interest of giving non-conventional energy systems the benefit of the doubt, the efficiency was assumed to be about 0.33. Due to a typographical error in the text of the report, this assumption was not made as clear as it could have been, although the calculations show the correct assumption being used. I regret that Prof. Holdren was unable to phone or write me to clarify this point, as a number of others did.

I then was assuming electricity from biomass (or methanol) and comparing it to electricity from petroleum or coal. This is precisely what Holdren suggests.

p. 566, column 3. The assumption of multiplying occupational risk of oil refineries by a factor of 2 to take account of the fact that methanol has an energy-to-weight ratio about half that of oil is incorrect. No experimental evidence is given that the assumption is incorrect. Instead, statements about coal gasification are made which are not relevant to the subject at hand. While oil refineries handle a fluid, methanol refineries would be handling large masses of logs or wood chips, an inherently more risky operation. To give the benefit of the doubt to methanol, I did not even count in the risk from handling wood products. The factor of two that I chose, based on energy-to-weight ratios, is probably a conservative one in terms of risk. If there is evidence to the contrary, I would appreciate seeing it.

P. 566, column 3. The lifetime of the methanol plant should not be 20 years. The reference (17) which I used for this assumption is, as far as I know, one of the most comprehensive studies of methanol ever produced in Canada. The value of 20 years for plant lifetime was taken from this source; if there is evidence to the contrary, I would be pleased to see it. It is inadequate to say merely that an assumption is incorrect without facts.

p. 566, column 3. The risk calculated from Comar and Sagan is for operation and maintenance of the methanol factory, not material acquisition and construction as its title suggests in the report (7). Holdren *et al* are correct here; the title is a mislabelling. It will be corrected in any future editions. This then implies that I have underestimated the risk of methanol, since I have not counted in the risk of acquiring the non-wood materials used in the factory. It is difficult to judge how large this undercounting is.

In terms of operation and maintenance of methanol refineries, there are estimates of refinery risk in the literature (8) substantially higher than those which I used. It is regrettable that Holdren has not called attention to this, especially when it is clearly noted in the report (7).

p. 566, column 3. Making the above changes completely tranforms the results, raising the upper limits of nuclear to the lower part of coal and oil, and dropping the risks of non-conventional sources substantially. As noted above, there are strong reasons to reject the claim with respect to raising nuclear risk. The claim is greatly different from that stated in Holdren's own work (1). Since, as far as I know, Holdren has not repudiated or corrected this reference, there seems to be a contradiction here.

The other claims made do not substantially affect the rankings of systems, as noted above. Absolute values of risk of many systems will be subject to some debate until further data is gathered.

There is one point in this comment which deserves discussion. The implication is that the upper risk limits of one system should be compared with the lower limits of another. This is a remarkable contention, one that is at variance with the practise of most risk analysts, as well as the practise of Holdren himself (1). Surely there must be a typographical error here. With assumptions of this type, one could "prove" anything one wished about risk.

p. 568, footnote 7. A JPL internal memorandum (18) is the source of the report's (7) entire methodology. I am confident that Dr. McReynolds, the author of the memorandum, is a brilliant and accomplished scientist. I am equally confident that he would not claim that he originated the idea that a ton of steel is associated with a certain number of deaths and accidents. In fact, the idea is ancient; neither McReynolds nor I originated it. My reading of history leads me to believe that there are few really new ideas.

The memorandum in question was a useful compilation of data, almost all of which could have been obtained elsewhere. It also explained some of the methodology in the Caputo and Herrera studies [refs. 6 and 21 of my report (7)].

This concludes the discussion of the major technical issues which Holdren raised. In terms of risk from conventional energy systems, Holdren's report (1) and I are in agreement with Comar and Sagan (8), Gotchy (19) and other risk analysts who have concluded that the total risk from coal and oil is much higher than from natural gas or nuclear power. Those of my values which differ from Holdren's were generally taken from Comar and Sagan. Since Holdren is an editor of Annual Reviews of Energy, where Comar and Sagan's article appeared, I am surprised that he did not bring any errors in their work to their attention.

In terms of non-conventional energy systems, as noted above, there is room for disagreement on the exact data and assumptions. I find that Holdren has concentrated solely on potential overestimation of risk, as opposed to the potential underestimation I detailed in the report (7). He is certainly entitled to his beliefs, but one may wonder if such an approach is the correct way to arrive at scientific truth. In summary, then, Holdren's letter contains little that affects the results in a substantive way.

I would not claim that no errors were made in the report (7) on which the article was based. In a lengthy study of this type, using varied sources of data, some of which are less than superb in quality, mistakes were bound to occur. I have tried to correct them as quickly and diligently as possible. But they must be put in context. None that have been pointed out to me have strongly affected the results and conclusions.

In terms of these mistakes, I would quote the words of Walter Patterson, one of the anti-nuclear leaders of Friends of the Earth in Britain (6): "Of course we shift our ground; as we learn, our views evolve and we hope the same holds for you".

Yours sincerely,

Dr. Herbert Inhaber

Department of Physics
Herzberg Laboratories
Carleton University
Ottawa, Canada KIS 5B6

References to preceding letter

REFERENCES

1. X.R. Smith, J. Weyant and J.P. Holdren, "Evaluation of Conventional Power Plants", Energy and Resources Program, University of California, Berkeley, CA, July 1975, Report ERG 75-5.

2. R.J. Budnitz and J.P. Holdren, "Social and Environmental Costs of Energy Systems", Annual Review of Energy, 1 553 - 580 (1976).

3. "Accident Facts - 1977 Edition", National Safety Council, Chicago, illinois, 1977, p. 27.

4. Office of Technology Assessment, "Application of Solar Technology to Today's Energy Needs", United States Congress, Washington DC, June 1978, p. 394.

5. D.R. Sears *et al*, "Environmental Impact Statement for a Hypothetical Photovoltaic Solar-Electric Plant", Lockheed Missiles Co., Huntsville, Alabama, July 1976.

6. As quoted in "Nuclear or Not", edited by G. Foley and A. van Buren, Heinemann, London 1978, p. 134.

7. H. Inhaber, "Risk of Energy Production", Atomic Energy Control Board, Ottawa, Canada, 1978, Report AECB 1119.

8. C.L. Comar and L.A. Sagan, "Health Effects of Energy Production and Conversion", *Annual Review of Energy*, *1* 581 - 599 (1976).

9. J. Miettinen, I. Savolainen, P. Silvennainen, E. Tornio and S. Vuoti, "Risk-Benefit Evaluation of Nuclear Power Plant Siting", *Journal of Nuclear Energy*, *3* 489 - 500 (1976).

10. Norges Offentlige Utredninger, "Nuclear Power and Safety", State Printing Office, Oslo, Norway, June 1978, Report NOU 1978: 35C, p. 127.

11. Ref. 7, *op. cit.*, fourth edition, in preparation.

12. Ref. 10, *op cit.*, p. 284.

13. L.D. Hamilton, ed., "The Health and Environmental Effects of Electricity Generation - A Preliminary Report", Brookhaven National Laboratory, Upton, NY 1974.

14. L.A. Sagan, "Health Costs Associated with the Mining, Transport and Combustion of Coal in the Steam-Electric Industry", Nature, 250 107 - 111 (1974).

15. E.I. Vorob'ev, L.A. Ilin, V.A. Kniznikov, D.I. Gusev and R.M. Barkhudarov, "Atomic Energy and the Environment", Atomnaya Energiya, 43 (5) 374-84 (1977).

16. Ref. 10, op. cit., p. 253.

17. Intergroup Consulting Economists, "Economic Pre-Feasibility Study: Large Scale Methanol Production from Surplus Canadian Forest Biomass", Fisheries and Environment Canada, Ottawa, Sept. 1976.

18. Interoffice Memorandum to T.D. English from S.R. McReynolds, May 26, 1976, Jet Propulsion Laboratory, Pasadena, California.

19. R.L. Gotchy, "Health Effects Attributable to Coal and Nuclear Fuel Cycle Alternatives (draft)", U.S. Nuclear Regulatory Commission, Washington, DC, Sept. 1977, Report NUREG - 0332.

Science Vol. 204, p. 1154, 15 June 1979

Risk Accounting

It seems to me that there is cause for concern that the recent flurry of letters (4 May, p. 454; 11 May, p. 564) attacking the recent article by Inhaber (23 Feb., p. 718) will cause readers of *Science* to lose sight of the fundamental methodological change in risk accounting proposed by Inhaber, namely, that one should charge for those risks incurred in the acquisition of materials of construction and the construction of the facility. In fact, this aspect of risk accounting has not generally been included previously. Had it been, we would not have read a continuing series of scientific and public pronouncements in the past that solar energy is benign, either ecologically or from the point of view of health and safety. No energy source is. Just think about the pollution from copper mines and smelters when you consider a technology that needs copper.

I myself have had questions about some of Inhaber's data and results, as have had others. More detailed and accurate studies are needed to confirm or negate his general results than have been afforded by the letters of criticism. Whatever the eventual results, he has made an important contribution to our thinking.

I also question two of Lemberg's criticisms (4 May) of Inhaber's method. I find no basis for charging only incremental risks connected with material acquisition, as proposed by Lemberg. If a coal miner is killed in a coal mine accident, we don't say, "But he might have been killed constructing some skyscraper" when we count the risks from coal-generated electricity.

Second, society cannot exist only with energy systems subject to daily or other frequent loss of all or most generating capacity. There must be storage or back-up systems, and a proper risk accounting has to include a charge for this aspect of an unsteady supply source. Lemberg might have suggested that Inhaber should have used the risk corresponding to the average societal energy mix rather than that due to coal, which Inhaber estimated to have the highest risk. Or Lemberg should have proposed less risky back-up.

In his risk accounting, Inhaber did not allow for one potentially important aspect. If society spends $1 billion a year more to make electricity, it does not have that billion dollars to improve the health and safety from whatever risks provide the greatest risk reduction potential per dollar expended. If one could "save" a life (defer a premature death) by expending $200,000, a billion dollars per year saved by use of a cheaper energy source could enable the "saving" of 5000 lives a year. This contribution to risk accounting could be dominant if there is a major disparity in costs of energy among various sources.

David Okrent

School of Engineering and Applied Science, University of California, Los Angeles 90024

Anne and Paul Ehrlich and *Mother Earth News*

Anne and Paul Ehrlich at Stanford University wrote two articles on AECB 1119 that were published in *Mother Earth News* in the July/August and September/October, 1979 issues. The Ehrlichs refused to permit these to be reproduced here. The articles are far too long to be paraphrased; readers are referred to the original journal. The editor of *Mother Earth News* declined to publish my two replies; they are presented here.

Response—Part 1

The following is a response to the article by Anne and Paul Ehrlich in *Mother Earth News*, entitled "The 'Inhaber Report,' Part 1."

Let me say from the start that there were indeed errors in various editions of the report (AECB 1119). As they were pointed out, they were corrected in subsequent editions. Some numbers have gone up and others have decreased. I find upon making these changes that the relative risk rankings, if not the absolute values, of the eleven energy systems which I considered remained about the same.

The authors make a great many statements about my motivations, reasoning ability, etc. I will not dignify these with a reply, since quite frankly I believe they are beneath the dignity of the authors themselves.

In order to make the readers' task as simple as possible, I will try to rephrase briefly the authors' comments, page by page, along with my reply. I will deal only with substantive issues, leaving the rhetoric to others.

p. 116. *Windpower material requirements were a factor of 1000 (or 20 or 50) times too high.* It is difficult to judge from the authors' comments exactly what factor they believe to be correct. They introduce a statement about 150,000 1979 Pontiacs, irrelevant since there is no inherent way to gauge the weight of cars which would provide enough steel to build a given number of windmills.

There was an error of about a factor of 20 in the wind materials requirements. The calculations were based on a U.S. government report which, to put it mildly, was confusing to read. When I phoned the Department of Energy to have it clarified, nobody at the other end could explain it to me. All this may indicate is that there are problems in certain parts of DOE, but I have found that many of my colleagues have had difficulty in interpreting this table (originally taken from the Project Independence report of 1974). I might add that there was at least one large typographical error in this table, uncorrected after many printings.

It is invalid for the authors to state that there was an error of 1000 in the intermediate calculations. This is simply not the case.

Readers may get the impression that if materials requirements are in error by a factor x, then the overall risk is in error by the same factor. However, there are contributions to risk from other sources, such as storage, maintenance, emissions from materials production, etc. These add together, rather than multiply. As a result, the overall risk of wind changed by much less than a factor of 20 when the error was removed.

The correct materials requirements apply to large windmills, which, if they are ever built in quantity, will likely be owned by public utilities. I gather from the authors' other writings that they are primarily interested in small, decentralized windmills, as opposed to large ones. However, Simmons, an authority on windpower, notes that small windmills weigh, per unit output, up to ten times as much as large ones. As a result, if I had considered small wind turbines my results would have been off by a factor of 20/10 = 2, not small, but also not overwhelming.

What did I do when the error was pointed out to me? (I might point out here that Prof. Holdren, whose comments the authors quote at length, had a copy of the report for about 10 months before responding to it.) I immediately corrected the galley proofs for a forthcoming article in *Science*. The authors state (p. 117) that I have ignored constructive criticism. This instance, as well as many others, shows that just the opposite is true.

p. 116. *Air pollution and coal mine dust levels well above the limits set by law for new facilities were assumed.* I made it clear in my list of assumptions that I was considering present-day conditions, whether or not any laws or regulations were broken. In the case of air pollution, even the EPA notes that present-day emissions are far higher than the hoped-for levels of 1990. As long as each energy system is treated the same, my assumption is a reasonable one. A case can be made that it is the only possible assumption, since nobody can predict the future.

The authors' statement is unfortunately incorrect as well. In the first three editions of the report, air pollution levels were assumed that were much lower than present-day conditions. To put it quantitatively, I assumed 11–38 metric tons of sulfur oxides per megawatt-year; values for the U.S. in the mid-1970's were about 130. In effect, I assumed super-clean coal, rather than "dirty" coal.

The authors display what could be called a touching faith in laws and regulations. I have no doubt that the accident at Three Mile Island violated a number of regulations. Using the reasoning of the authors, it simply didn't take place. One should leave the laws and regulations to lawyers, and the evaluation of actual conditions to scientists.

p. 116. *Construction risks for solar systems were included, but not those of conventional systems.* This is correct, due to misinterpretation of a document by Smith, Weyant, and Holdren. They had gone to the trouble of listing material and labor requirements for conventional systems (taken from a previous publication, WASH-1224), but had not incorporated the associated risk in their tables for occupational risk. I assumed that they had done so, and was surprised when told that they had not.

Readers might get the impression that other writers on risk always included dangers attributable to building conventional energy systems, and I was the first (or one of the first) to neglect them. This is not the case. As noted above, the Holdren document did not include it, nor did other documents by Comar and Sagan, Gotchy, WASH-1224, etc.

All this said, what was the magnitude of the error? The risk attributable to building conventional energy facilities was about 1 man-day lost per megawatt-year. For coal and oil, this addition was negligible, since the maximum risk of these systems was stated to be of the order of 1000 man-days lost. The change is more significant for natural gas and nuclear, where the rest of the risk is 1.1–5.7 and 2.1–11 man-days lost, respectively. While the lower ends of these two systems show a large increase when the addition is made, their relative ranking remains the same, because there are such wide differences between systems. I find that the conclusions remain unaltered when this error is corrected.

p. 116. *The risk for reinforcing rods was counted twice.* Since there is no documentation for this statement, I cannot comment on it.

p. 116. *Unjustified differences in facilities' lifetimes were assumed.* The authors may be referring to the fact that lifetimes of 20 years were assumed for wind, methanol, and solar space heating systems. Other facilities had 30-year lifetimes assigned. The shorter lifetimes were assumed on the basis of the scientific literature, and in some cases could be regarded as overly generous. For example, many of the windmills that have been built to date have operated much less than 20 years. The Meinels, two of the founders of modern solar energy thinking, note that it will be difficult to build solar systems that will last longer than ten years. For each case where I assumed a lifetime less than 30 years, there is detailed reference to either experimental results or other scientific data.

p. 116. *Renewable energy technologies are built and maintained largely by "roofing and sheet metal" workers, which inflates the risk greatly.* This statement is far from being true. For example, no roofing and sheet metal workers are assumed in methanol, hydroelectricity, and ocean thermal. For wind and solar space heating, only half the construction times are assigned to this trade.

Even if another trade, such as "miscellaneous construction", were assumed to replace all roofing work, the risk per unit time would drop by only about 25%—hardly a "great" deflation. Since, as mentioned above, there are contributions to risk from many sources, this possible change will alter the total risk of most systems by much less than 25%.

As in many other instances of the authors' statements, no proof is supplied. If they have evidence that roofing should or should not be assumed where I have done so, let them supply experimental data. It is insufficient to make these statements without some facts.

p. 117. *It was assumed that materials were transported the same average distance that coal is. This produced a 25-fold excess risk in one group of risks.* As a first approximation, I made that assumption, since I could find little evidence on how far and by which mode (rail, truck, or barge) particular materials were transported. If materials are moved by truck, rather than by rail, as I assumed, the risk per ton-mile will probably be greater. In turn, this would probably increase the risk of non-conventional systems, since they tend to require more materials than conventional systems. A case can then be made that my assumption underestimated the risk of non-conventionals, rather than overestimating them.

The assumption is probably not valid for materials like sand and aggregate. In any future work, I will assume that these materials are moved only short distances, with negligible risk. It is not clear how the authors derived their value of a "25-fold" excess risk. The only energy system where assuming that rock and earth are moved with negligible risk lowers the total risk by more than a few percent is hydroelectricity. Even there, the decrease was about 10–20%. The relative rank of hydroelectricity did not change after this alteration. Unfortunately, readers might get the impression from the authors' statement that the total risk for this source was strongly overestimated, rather than only one component.

p. 117. *"Dirty" coal was assumed as back-up. No back-up is required.* The point about "dirty" coal was answered above. To summarize, the coal I assumed was much cleaner than is presently the case.

Whether or not back-up is required at all is a matter of contention. A Jet Propulsion Laboratory study, some of whose data I used, assumes the use of back-up energy (in distinction to capacity) on about a dozen pages and tables (all references available on request). The amount seems to vary; sometimes it is 10%, other times between 0% and 20%. Reckard, in a recent study for Brookhaven National Laboratory, assumes 30% for both solar thermal electric and solar photovoltaic. Iannucci assumes 16% for one system in a paper presented to the 1978 International Solar Energy Society Congress. Many other scientific references could be supplied. Readers might get the impression that I arbitrarily decided that back-up would be required, but my assumption was based on previous studies. As noted above, an implicit assumption in the report was that baseload reliability was required. Since the studies on risk of which I am aware assume a baseload mode, I broke no new ground.

The solar space heating system I considered also required back-up energy constituting 30% of total output. To give this system the benefit of the doubt, I assumed zero risk from the back-up required.

p. 117. *Non-conventional systems will be used as "fuel-savers," not as baseload, so no back-up is required.* Opinions also differ on this point. At least one solar proponent (F. Bove, "Here Comes the Sun: The Government Discovers Solar Energy," *Science for the People*, vol. 11, p. 7, 1979) feels that baseload systems are desirable: "(solar systems) are most economical when they are allowed to operate as much as possible (providing 'base-load' power) because they are capital intensive (requiring a large initial investment) but have a relatively low operating cost." In other words, if a "fuel-saving" mode were assumed (roughly similar to what utilities call a "peaking" mode), the energy produced per unit capacity would be comparatively low. Since my study discussed risk per unit energy production, not risk by itself, in some instances the assumption of a fuel-saving mode could produce higher risk per unit energy than assuming a baseload mode. The Jet Propulsion Laboratory study mentioned above clearly assumed baseload for both solar thermal electric and solar photovoltaic, since they required enough storage and back-up to produce baseload reliability.

p. 117. *The report was reviewed negatively before publication by Tupper and Lemberg, and their comments were ignored.* The Lemberg comments were received after the first edition went to press. More substantively, neither set of comments were ignored.

Dr. Tupper's main point was that occupational risk can vary from one company to another within an industry. This is true, and note was taken of it (see footnote 126 of the report). However, the occupational statistics published by bodies such as the U.S. Bureau of Labor Statistics do not show the risk company by company. While it might be useful to show a range of risk for all industries, this data does not generally exist. In Canada, even national occupational statistics on risk, let alone on a company basis, are not generally available.

Lemberg prefaced his comments by noting that the report was diligently done and an effort had been made to avoid "pro-nuclear" bias. He suggested that I "challenge my assumptions". This was done. For example, in my *Science* article and the third edition of the report, risks both with and without back-up were shown.

However, there were philosophical points in Lemberg's comments which were not acceptable. One of his major suggestions was that the risk attributable to producing materials should not be included in the total. This would then imply that the risk of mining coal or uranium would be excluded for coal and nuclear, respectively. In turn, this would lower the risk of nuclear by up to 60%. This is an assumption which has not been made by scientists studying the entire fuel cycle, to my knowledge. The Ehrlichs would never endorse such a suggestion themselves, since they argued that I had left out the risk attributable to the materials for building conventional electrical plants.

It is, I suggest, unfair to say that Dr. X has criticized a report without giving the basis of that criticism, especially when one would disagree with the criticism oneself. This tactic may work on readers who are unfamiliar with the literature on risk, but will not work with those who spend much time on the subject.

p. 117. *Individual scientists have repeatedly asked the Board to publicly and prominently repudiate the report.* Since no names are mentioned, nor reasons given, it is difficult to assess this comment. The authors may be referring, among others, to letters from Amory Lovins received in 1978. In them, he made no specific comments on the methodology or data in the report. As a result, his call for a retraction of the report was not accepted.

p. 117. *The report has been progressively revised to eliminate the most obvious mistakes.* Here we have a case of "damned if you do, damned if you don't." If I correct the report, this shows how mistaken it was. If I don't correct it, critics are being ignored. I suggest that the authors cannot have it both ways.

The report has been revised as new data has come to light. In this, I am doing no more (or less) than others who work in a relatively new field. The report clearly points out that much data relating to risk is either vague or subject to controversy, and any report based on this data cannot hope to be perfect.

p. 117. *Inhaber denied writing what was clearly in print.* There is no basis for this statement. Since it is not specified further, I cannot comment on it.

p. 117. *The report confused thermal energy with electrical energy in the section on methanol.* There is no basis for this statement. In the first two editions of the report, it was assumed that the prime use of methanol would be in transportation. Most studies on the subject have made this assumption. The relative efficiency of using methanol for transportation is about 0.12; the relative efficiency of burning it for electricity is about 0.33. There was no confusion of thermal energy with electricity, since thermal energy was not considered as an end-use for methanol.

It was later pointed out that to make a fair comparison, one should consider electricity produced from methanol, since the other energy systems considered (with the exception of solar space heating) assumed electricity as an end product. This assumption lowered the risk from methanol, since the assumed efficiency increased, as noted above. This gives methanol the benefit of the doubt, since most who have advocated have specified transportation as its end use.

p. 117. *Risk of methanol was at least 10 times too high.* As mentioned immediately above, one could assume electricity instead of transportation as the end use, decreasing this aspect of risk by about a factor of 0.33 /0.12 = 2.8.

However, there were a number of areas where the risk was underestimated, rather than overestimated. For example, air pollution risk from the steel required in the methanol plant and for the equipment for cutting trees was greatly underestimated. At least two scientific papers suggest that the trees themselves will produce sulfur oxides, the basis for my air pollution and risk estimates. Producing methanol chemically will require substantial amounts of electricity and oil. The risk attributable to supplying this electricity and oil has to be added to that of other sources, if one considers the entire energy cycle.

When these recalculations are done, both raising and lowering values, the overall risk of methanol does drop by between 35 and 60%. Its rank in terms of risk would then be about fourth highest instead of its previous sixth, since the risk of other systems has changed as well.

p. 117. *Every competent scientist who has examined the report in detail agrees it should never have been issued.* I will not list here the approximately forty universities, government agencies, national academies, etc. to which I have been invited to speak, the scientific journals to which I have been asked to submit papers, the commendatory letters from scientists working in the field of risk. All are available to anyone who asks, but cannot be demonstrated on this page.

Rather, I would call attention to the work of Dr. Lemberg, whom the authors quote. While Lemberg did make criticisms—and I outlined above where I agree or disagree with them—he did not call for retraction of the report. If he had, the authors would surely have pointed this out. Lemberg is the only one, to my knowledge, who has been paid for reading the report, so I presume he had some incentive to go over it in detail.

The overwhelming bulk of comments have been favorable, not critical, but we are not weighing statements in the balance here. The authors specify that "every" scientist has called for its retraction. In their own work they supply an example of one who did not.

On the other hand, the authors qualify their statement with the words "examined the report in detail". They may be implying that only they (and perhaps Prof. Holdren) are qualified to do so. This assertion, quite frankly, would hold no weight in the scientific community.

p. 117. *If criticism is restrained, it is ignored; if it is frank, it is still ignored.* This is a rephrase of a previous comment, and is still invalid. Appropriate criticisms have been taken into account in subsequent editions, as even the authors have admitted. To take another example, staff of the International Atomic Energy Agency sent an extensive list of comments after the first edition was published. Almost all were incorporated in later editions.

Both "frank" and "restrained" criticisms have been incorporated in the report, when they dealt with particular aspects of methodology and data. General mud-slinging and name-calling cannot be considered part of a scientific discussion.

p. 117. *The approach in the report was taken from published and unpublished Jet Propulsion Laboratory studies.* This statement unfortunately implies some sort of plagiarism, and there is no basis for this claim. First, plagiarists rarely cite the work from which they lift their ideas, and I cited the JPL work a number of times. Second, the JPL approach is itself hardly unique or original. The idea that a ton of steel or aluminum has associated with it a certain number of deaths or accidents is an old one, going back many years. Even the JPL studies did not claim any particular originality for this—neither did I. The idea that the entire fuel or energy cycle should be considered is again an old concept, and neither JPL nor I can claim originality here. One of the first to discuss this was Dr. Hamilton of Brookhaven, but I am confident the idea goes back before him.

Summary

Most of the statements made by the authors are invalid. Of those which have some degree of validity, the alteration in the report which would be required would not change the conclusions significantly. Finally, the tone adopted by the authors is most unfortunate, unless they had something other than a scientific comment in mind.

Response—Part 2

The following is a reply to the article by Anne and Paul Ehrlich entitled "The 'Inhaber Report', Part 2," as published in *Mother Earth News.*

As in my reply to the first part of this series, I will rephrase and shorten the authors' comments. Many of their statements are either complete or partial repetitions of criticisms made in the first part, and these will be noted as appropriate. Again, I will deal only with substantive issues.

p. 116. *The report purports to show that solar energy sources are nearly as dangerous as conventional sources like coal.* No such statement was made in the report. The maximum risk of coal was estimated as around 2000 man-days lost per megawatt-year. The risk of solar space heating was estimated to be around 110 (in the same units); solar thermal electric, around 600; solar photovoltaic, around 700. The authors may regard these numbers as "nearly" all the same; I do not.

p. 116. *Indirect damages to health, such as climate disruption and nuclear proliferation, were excluded. They put more lives in danger than the direct risks.* It was clearly stated in the report that these considerations would be excluded. I do not know what more can be done.

Whether or not they put more lives in danger is a matter of opinion. Certainly there is not general agreement in the scientific community on their extent, or whether they will affect human health at all. If the authors have their own quantitative estimate of this indirect damage, they owe it to all of us to publish it along with the assumptions on which it is based. Until they do, the discussion is somewhat sterile.

p. 116. *No risk estimates were made of such effects as oxides of nitrogen, trace metals, etc., in air, water pollution effects, etc.* The answer to this is almost a replica of the one immediately preceding. It was clearly stated in the report that those aspects would not be considered.

If the authors have their own estimates, I and others would be pleased to see them. Of the papers on risk of which I am aware, none combines these admittedly important factors with other conventional factors like direct deaths and disease. Even a paper by Holdren and Budnitz (the authors have relied heavily on analysis by Holdren) does not do so. Readers may get the impression that I am the first to write on the subject without including numerical estimates of these effects. This is incorrect.

As mentioned in the preceding response, there is dispute in the scientific community over the size of health effects due to these pollutants. Neither I nor the Ehrlichs are in a position to resolve these disputes.

p. 116. *There are repeated assertions that the analysis is a comprehensive treatment of the accident and disease risks of energy sources.* The word "comprehensive" is not used in the report, to my knowledge, to describe the analysis. Naturally, an attempt was made to make the study as comprehensive as possible, under the circumstances. However, as noted in the preceding two responses, it was made clear that areas possibly relating to human health were not included. Even though no report on risk can be perfectly comprehensive, the study has been occasionally described by colleagues as being more comprehensive than some related studies. This question I leave for others to judge.

p. 116. *The risk of constructing conventionally fired electricity plants was not included.* This statement was made in Part 1, and was answered in my response to that part. Briefly, the authors are correct. However, the correction is very small, and makes no change in the relative ranking. Needless to say, the error will be corrected in any future work.

p. 116. *Risks for present-day conventional systems were compared to present-day non-conventional systems, not future ones, which might be better.* This is correct. It was clearly stated in the report that only present-day systems were being evaluated. This is not because of discrimination against non-conventional systems, but because I found that in studying the literature that if one compared a system which might exist in the year 2020 or 2050 to one which actually existed in the 1970's, one could "prove" anything one wished. If one can "prove" anything, one is proving nothing.

For example, one might assume at some unspecified future time that the collection efficiency of solar collectors will be much higher than it is now, or that the load factor (ratio of actual energy produced to that which could be produced under optimum conditions) of wind plants will increase greatly. In turn, the materials requirement and associated risk per unit energy would decrease. However, this "proof" of low risk of non-conventional systems would be only hypothetical.

p. 116. *The report chooses non-conventional models which are most demanding of materials and labor.* The opposite is true. Models were often chosen which gave the benefit of the doubt, in terms of materials and labor, to these systems. A number of examples could be given. In wind, the lifetime assumed was much greater than has been achieved experimentally. In terms of storage, the storage time assumed was much less than that specified by the U.S. National Academy of Sciences or in a paper presented at a recent International Solar Energy Society Congress. In terms of solar space heating, the construction times were much less than can be deduced from the work of the Meinels, two of the founders of modern thinking on solar. In terms of methanol, a risk per unit time spent logging was assumed that was lower than recent Canadian experience.

This list could be extended. I would not claim that the benefit of the doubt was given to non-conventional energy systems in each and every instance; that would be unreasonable. However, I have given this benefit in enough cases to show that the authors' statement is incorrect.

p. 116. *It is assumed that one BTU or joule is the same as any other.* This statement has no basis in fact. As an example, consider the sections dealing with coal and oil. I noted the efficiencies assumed, around 35–40%, and based the risk calculations on the final output of electricity. If I had assumed that 1 BTU of heat equals 1055 joules of electricity in every instance, I would have been assuming an efficiency of 1.0, which is not achievable. I clearly did not make this assumption.

There was one place in the text where the wording was somewhat obsucre, but the proof is in the calculations. They show I did not make this assumption.

p. 116. *Assuming that one BTU is the same as any other produced an overestimate of methanol risk by a factor of 3–6.* As noted above, the assumption was not made. This statement differs from the one made in Part 1, where the authors stated that the methanol risk had been overestimated by a factor of 10. In consequence, I do not know which factor the authors believe is right. As noted in my response to Part 1, a good case can be made for using methanol for transportation instead of electricity. In that instance, the risk would be relatively higher.

p. 117. *Back-up is not needed for solar, since the load factor is the same as for conventional systems.* This is a partial restatement of Part 1. As noted in my previous response, a number of reports and papers have indicated that back-up will be required for some, but not all, non-conventional systems. The back-up is needed to generate a higher load factor than that of comparable conventional systems, in the words of a Jet Propulsion Laboratory study, "to make the ground solar plant as reliable as conventional plants not subject to the sporadic unavailability of sunlight". In other words, it is insufficient to merely supply energy, i.e., to have an appropriate load factor. The energy must be supplied with a reasonable reliability if baseload conditions are to be met. Back-up was not assumed arbitrarily, but to make all energy systems as comparable as possible.

p. 117. *Back-up risk constituted 65–85% of the upper limits of risk for wind, solar photovoltaic, and solar thermal electricity.* This is correct. Two observations may be made. First, it is unfair to discuss only the upper limits, especially when the authors imply they are presenting a balanced view. For example, back-up risk ranged from 4–13% of the lower risk limits of these three systems, but the authors do not mention this. Second, the implication of this statement is that back-up is only something that can be dispensed with if it raises the risk too high. However, as noted above, it is essential to baseload reliability. Even if back-up were not included, the relative ranks of the systems remain fairly similar.

p. 117. *Occupational and public risk are added together.* Both are shown separately. For the reader who might wish to combine them, this is done in a few graphs, but only after each was shown separately in tables. Readers might get the impression that the two types of risk were lumped together so they could not be considered individually. This is incorrect.

p. 117. *The same number of days lost was assigned to deaths from respiratory disease as from cancer.* This is correct. The crude assumption was made because there seemed to be no adequate estimates of the different days lost for various diseases. This assumption was also made by Holdren, on whose analysis the authors rely. In a report by Smith, Weyant, and Holdren, the same number of days lost are assigned to all causes of death. If the authors can say that the scientific community has agreed on values for different diseases, I would be pleased to use these values. However, I find that the overall risk of most systems is fairly insensitive to the precise value used, since much of the risk comes from accidental deaths, not disease.

p. 117. *For coal, conditions and practices were assumed that are illegal.* This was answered in Part 1. I tried to consider actual conditions, not the state of the law.

p. 117. *Wind material requirements were overestimated by a factor of 50.* This was answered in Part 1.

p. 117. *For ocean thermal, the generating capability was understated by a factor of more than two, and the construction labor overstated by 50–100 times.* There is some confusion in the literature between gross and net power. The two are not always the same, and the differences are not always specified. This led to an error of some material requirements by a factor of about 1.7. Changes in risk due to material acquisition because of this change would lower the total risk by about 5%. Because of the relative novelty of ocean thermal systems, the construction times are not known precisely. I believe that the values chosen are approximately correct.

p. 117. *The report claims to have used the highest available nuclear accident risk values, but hasn't. The upper limit used in the report is much lower than that given in three of the four references on reactor safety cited.* No such claim was made. Such a claim would have been unreasonable, since somebody, somewhere may have asserted an unreasonably high value for this quantity, without any justification. Neither I nor anybody else would be under an obligation to accept it.

There were about seven references dealing with nuclear public risk, so I do not know how the authors narrow it down to four. I have checked the references, and I find that they either do not discuss nuclear public risk quanititatively, use a value lower than that in the report, or sometimes use a "worst case" value. As noted above, "worst cases" are not used in the report, because by using them, one could again "prove" anything one wished. The values for nuclear public risk were taken from a paper by Comar and Sagan, in a journal one of whose editors was Holdren. The values are much higher than those quoted in Norwegian, Finnish, and Soviet studies, a report by Science Applications Inc., etc.

p. 117. *It is implied that waste management is considered, but it isn't.* The data on waste management were taken from the report by Smith, Weyant, and Holdren, mentioned above. The title of the data was as they had specified, and the footnote was similar. Since that time, I have come across data in a Norwegian publication from which one could estimate waste management risk. I find that the values used in the report are pessimistic, i.e., much higher than one can deduce from the Norwegian and other studies.

p. 117. *Risks for building a methanol factory are based on operation and maintenance, rather than construction.* This is correct. The error will be removed in any future work.

p. 117. *Materials and labor data for construction of natural gas-burning facilities are not included in risk. Data are taken from fuel-oil and coal gasification facilities.* The first part of this statement was answered in Part 1. The change is more significant for natural gas than for other conventional systems, but its relative rank remains unchanged.

Natural gas occupational (not public) risk was taken from fuel-oil and coal gasification statistics because the paper by Smith, Weyant, and Holdren (noted above) did not include values for natural gas. If the authors have data which disagree with mine, I would be pleased to see it.

p. 117. *The storage system for solar thermal electricity stored heat, not electricity. It should then not have been used for solar photovoltaic and wind, which produce electricity.* This storage system was not assumed for photovoltaics and wind. It was clearly stated in the report that this would not be done. The heat storage system was probably a "lower limit" in terms of risk, as compared to storage systems such as steel or lead–acid batteries. In order to give the benefit of the doubt to photovoltaics and wind, I assumed that the risk attributable to the solar thermal storage system would be applied to these two systems–not the storage itself. There apparently was a misreading on the part of the authors.

p. 117. *In translating southwestern U.S. solar conditions to those of Canada, the report should not have multiplied all sources of risk by 1.32.* The authors are correct in two ways. First, the factor should not have been 1.32, but rather 1.67. That is, the risk for solar thermal electric and photovoltaic was underestimated. Second, the back-up should not have been multiplied by this factor. However, there is evidence suggesting that other components of risk should be multiplied by this factor. For example, Iannucci uses a storage time of 12–75 hours for Madison, Wisconsin, where weather patterns are roughly similar to that of Canada. The maximum value is greater than the 16.5 hours used in the report.

When the two changes are made, i.e., multiplying the risk by 1.67 instead of 1.32, and multiplying back-up by 1.0 instead of 1.32, the total risk for solar thermal electric rises by 9–50%. Similar results are obtained for solar photovoltaic. Correcting these two errors then produces a higher risk, not a lower.

p. 117. *The material requirements for solar photovoltaic were overestimated by misunderstanding the difference between peak and average power.* There was no misunderstanding of peak and average power. A mistake was made because one document did not specify clearly what fraction of the land in a photovoltaic station would actually be covered by the collectors. The ratio of collector area to land area was used in one calculation. Changes in risk from material acquisition due to this alteration would lower the total risk of this source by about 0.6–3%.

p. 117. *The highest material requirements of non-conventional systems were combined with the lowest performance figures available.* This statement was made on p. 116, and is answered above. To give further examples why this statement is not correct, a solar collector evaluated by the Ontario Ministry of Energy had a collector efficiency lower than that assumed in the report, as did the first report on the Lorriman house in Toronto. A number of solar collectors have a higher ratio of material weight to area than I assumed (see the compilation by Rogers). The Provident house in Toronto has a larger storage tank than I assumed. Whether or not these examples are perfectly representative is a matter of some debate. However, the authors state that I have used some of the most unfavorable data available. These examples demonstrate otherwise.

p. 117. *Risk estimates are inflated by assuming most of the work done on non-conventional systems is by roofers, a dangerous trade.* This statement was made in Part 1, and answered previously.

p. 117. *The wind load factor of 0.34 from Project Independence was ignored, and a lower value substituted.* The value of 0.34 was theoretical. I checked the literature to see if it had been achieved in practise, and found that except for one obscure report about a Soviet windmill in the 1930's, there was little or nothing to indicate it was being achieved regularly at present. On the other hand, there was evidence from Simmons and other compilations that a lower value was implied by today's technology. As a result, I did use a lower value.

The issue raised by the authors is broader than the question of load factors. In their work as biologists, they undoubtedly come across scientific papers which disagree—sometimes violently—with each other. One can, of course, say that one paper is correct and all the others wrong, but this is rarely done. The same situation applied in risk studies. There is often contradictory evidence, and one attempts to choose data on as consistent a basis as possible, given these contradictions. Especially in the case of non-conventional energy systems, the data are not always as clear and consistent as one would like.

p. 117. *It should not be assumed that facilities producing methanol have a lifetime of only 20 years, as opposed to the lifetime of 30 years for most other systems.* This is a partial repetition of a statement made in Part 1. The lifetime assumed was based on reports of the Alberta Energy Resources Conservation Board and a report to the U.S. Presidential Energy Resources Council, and used in a report to the Canadian Department of Fisheries and the Environment.

p. 117. *Assuming a methanol plant will be twice as big as an oil refinery per unit energy because the heat energy per unit weight of methanol compared to gasoline is one-half is unwarranted.* Data on methanol plants are scarce. The approximation made in the report was a reasonable one, given the lack of other data. If the authors have information to the contrary, I would be pleased to see it.

p. 117. *Materials used in non-conventional energy systems are assumed to be transported by rail the average distance that coal is.* The statement was made in Part 1, and answered previously. Briefly, it is not clear that this assumption leads to an over- or an underestimate of transportation risk of non-conventional systems.

p. 117. *A generous estimate was made of the time professional roofers would take to maintain solar rooftop collectors.* A case can be made that I was overoptimistic, i.e., that I underestimated the time. For example, using data from a recent New England Electric experiment using 100 solar collectors could lead to an estimate of 30 man-years per megawatt-year, or 17 times the value estimated in the report. If one uses the economic data in a book by Field and the New England Electric capital costs, one can estimate labor of 5.5–25 times that in the report. I then conclude that the report's value of about 9 professional hours per installation is not "generous."

p. 117. *The ratio of risk per hour for "amateur" as opposed to professional roofers was assumed to be too high.* It is likely that amateur roofers will have a greater risk per unit time than professional, since falls in and around the home are one of the greatest sources of risk. Amateur as opposed to professional roofers were assumed since most advocates of solar heating have stressed its decentralized, do-it-yourself nature.

The exact factor to be used is a matter of some debate. In a recent book, Schurr and colleagues implicitly used a factor of 5–10, greater than my assumption. If the authors have data showing what the factor should be, I would be pleased to see it.

p. 117. *None of the report's detailed calculations–and few of its assumptions–were included in the articles submitted for publication in journals.* The authors refer to a paper which was published in *Energy*. They will find a considerable number of calculations there. The other two papers, in *Science* and *New Scientist*, list many of the assumptions which were made.

Of course, one cannot condense a report of about 150 pages into the 8 or 10 pages (in the case of *New Scientist*, 3 or 4) one is allowed in a scientific journal. Journal editors have been generous to me, but there is a limit to their generosity.

The appendices attached to the report show the calculations and assumptions in considerable detail. If these appendices had not been attached to the report, then the authors would clearly have a case that there had been concealment.

In Part 1, the authors state that the Board had been vigorous in its distribution of the report. In Part 2, they imply there has been an attempt to conceal the data in the report. I suggest that one cannot have it both ways.

Summary

Some of the statements made are partial or complete repetitions of Part 1, and were dealt with previously. Most of the new points made here are invalid, for the reasons noted above, Those which are valid do not strongly affect the overall conclusions of the report.

Quebec Science

G. Provost wrote a critique of the report in the Canadian journal *Quebec Science*. The author's response follows. Provost's article and the (abbreviated) response were printed in French; the two are given in translation here.

Quebec Science, Vol. 16 (June 1978), pp. 15-17

A Facelift (Cover-Up) for Nuclear Energy Risks
A critique of a report by the Atomic Energy Control Board on the risks of various forms of energy production by Gilles Provost.

Per unit of electricity produced, the danger to the life and health of Canadians from nuclear energy would be 1/200 of the danger from oil or coal, 1/5 of the danger from hydroelectric power, so dear to Quebecker's hearts, and 1/60 as harmful as solar energy. Moreover, electric heat from nuclear sources would be about 1/10 as risky as an equal amount of heat from solar energy (for example, using rooftop solar collectors).

At any rate, those are the general conclusions (which are nothing if not grandiose) to be culled from a report recently released by the Atomic Energy Control Board, an organization much given to official silence, whose task it is to defend the public interest against the nuclear industry and to ensure that security standards are established and enforced.

This study is extremely important not only because of its unusual nature but also because of the methods of comparison it uses: for the first time, at least in the Canadian context, an attempt has been made to compare nuclear energy with the "soft" or new technologies, and also for the first time a broader perspective, going beyond possible catastrophes, allows us to take into account all the dangers of each energy system, from the mining of material needed for the plant to the final disposal of wastes, including the various forms of toxic pollution, work accidents due to construction or handling of combustible materials and, of course, the risk of a major catastrophe. It is a highly technical study, packed with figures, which cannot be assessed without a large amount of data.

On the whole, the lofty goals of the study have not been met, and the conclusions cited above remain highly debatable, for two main reasons: first, a study of the Report reveals several unrealistic hypotheses and contradictions or careless errors which strain its credibility; the second, more basic reason is even more of a paradox. This original study, designed to widen the traditional debate, limits too strictly its perspective on the number of accidents attributed to each energy source. The major bias inherent in this method is that it greatly favors those technologies which, because they require a minimum of staff, provide a very few jobs. It is obvious that when a society such as ours, preoccupied above all by the unemployment problem, chooses an energy system, we must take into account its effects on employment and the economy as much as the quantity of electricity for a given price. Second, it is clear that in a centralized and automatic system requiring only a small plant, the risk of accidents and death is much lower than in another system requiring huge plants and a large workforce. Pushed to the extreme, if no workers are required, there can be no accidents in the workplace.

Some Very Different Conclusions

In fact, several days of calculation and analysis in collaboration with Pierre Sormany, a colleague at *Quebec Science*, showed me that it is possible to come to somewhat different conclusions by correcting small errors, by "adjusting" the most unlikely hypotheses and especially by expressing the results in such a way that the number of jobs provided by each energy system is taken into account. In this way, for each job created per unit of electricity produced, nuclear energy causes 8 times as many lost days (among workers and the public) than wind energy, 3 or 4 times as many as solar electric energy, twice as many as natural gas, and about as many as hydroelectricity or proposed plants powered by wood alcohol. In this light, only coal and oil plants are more dangerous than nuclear plants, and the difference is only a factor of three.

The incredible differences indicated in the Control Board's Report melted away like snow in the sun, and the case for nuclear energy seems much less attractive. Note as well that the figures for direct heating by solar energy is not included because this alternative, so different from the others, involves too many unknown factors. Depending on the mood of the moment, and with a slight modification of certain basic data about which there is no valid information, the conclusions can be entirely contradictory. By softening the unnecessarily severe hypotheses about solar heating it can be shown for example that it is 6 times less dangerous than nuclear energy for each job created per unit of energy produced. Other too favorable hypotheses, if also amended, would show the nuclear option to be much more dangerous. Finally, it is a definite mistake to compare an energy system so decentralized (down to the individual rooms of a dwelling) with other systems while including only the large production plants and neglecting consideration of the distribution system needed to bring the electricity to each urban center, to each household and even to each outlet in a house. In conclusion, it is impossible to make a comparison of these two systems based on the data from the study or other data presently available, and the Energy Board's conclusions in this area must be completely rejected.

Between 15 and 2,000 Days Lost!

The quality of the assessment of the dangers of each system varies from chapter to chapter. In certain cases, like the use of wood alcohol produced by an intensive exploitation of forests, the analysis is invalidated by gross errors, and we had to redo it in its entirety. As for hydroelectric power, it was treated only in an appendix, using a different methodology; and the conclusions contradict the findings of other chapters. Nevertheless, by adding supplementary data available from Hydro Quebec, one can come to some apparently valid conclusions. Finally, in most of the other cases, the "improvements" one can make on the Board's official calculations make little difference. Even when the conclusions are substantially affected, moreover, the basic results remain the same and one reduces the difference between the various systems rather than overturning their relative positions.

The most "solid" chapters in this study concern traditional energy forms, especially coal, oil or gas thermal plants as well as nuclear plants. Obviously these are the most current technologies, for which most data are available, and subjectivity therefore plays a less important part.

Even so, there are mjaor difficulties. In the case of coal, for example, the greatest danger seems to come from toxic emissions of sulphur dioxide. Obviously the effect of this pollutant varies according to the anti-pollution devices used, the kind of coal, the location of the factory (in the country or in the middle of an already polluted or heavily populated area), etc. Finally, the Board's study estimates the number of days lost to the public for each megawatt of coal electricity from 15 to . . . 2,000. The enormity of the "imprecision" can be appreciated by noting that the total number of days lost for all risks in the coal system, including mining, transportation, etc. runs between 38 and . . . 2,005. To give another point of comparison, the total risk involved in the nuclear system varies between 2 and 10 days lost per megawatt-year of electricity. Estimates of the danger of coal plants will vary a great deal according to one's optimism or pessimism. The Report's use of the highest possible values has repercussions on every other energy form since every steel mill requires coal and therefore produces toxic gases.

When We Forget Technological Progress

This uncertainty about the real dangers of coal electricity produces another major imbalance because Mr. Herbert Inhaber, author of the Report, assumes that most of the new technologies (sun, wind, etc.) cannot function at all times, and therefore a back-up system is needed for days without wind or sun. He chooses a coal back-up system, always assuming the highest possible danger levels, with the strange results that 95% of the dangers ascribed to a solar electric system come, in fact, from the coal-powered back-up system. As well, 2/3 of the dangers ascribed to a wind-powered system come from the coal back-up system.

That example of the risks of coal plants shows up another basic weakness in this kind of study: a modern technology with the best security and anti-pollution devices is compared to often antiquated industries. For example, it is unlikely that coal mine statistics reasonably reflect the true dangers of an ultra-modern mine opened today. As well, the thermal coal plants, built for the most part when no one really bothered about pollution certainly cause much more pollution than would a modern plant. Furthermore, the average level of risk in steelmaking is certainly higher than in a modern steel mill.

Evidently, this method of comparison especially favors nuclear energy over the "new" systems which require large quantities of raw materials or a large workforce. In the light of these considerations it is much more likely that the number of days lost per megawatt-year is between 33 and 105 for coal (rather than 38 to 2,005) and from 7 to 40 for oil (rather than 11 to 1,938 as calculated in the Board's report). The maximum figures are still higher than for nuclear energy, but nevertheless the gap is much reduced. Incidentally, some reassessment of nuclear risks led us, Pierre Sormany and me, to raise the maximum number of lost days per megawatt-year of nuclear energy to 15 (rather than 10). This does not make much difference in the overall picture.

Hydroelectric Risks Overestimated

Moving to hydroelectric power, largely neglected in the Report, we note that the risk of fatal accidents estimated for the construction of dams and the transport of material represents a total of 200 to 500 deaths for a major complex like Hydro Quebec's Manic-Outardes project which will produce 165,000 megawatt-years during its life of 50 years (assuming that the plants produce at an average of 60% of their capacity). Such a conclusion obviously makes the people at Hydro Quebec wild, since according to them the figures are at least 10 times too high. This changes the conclusions entirely because, according to the Report, construction accidents alone account for 2/3 of the risks of a hydroelectric system. It is equally curious that, according to the Report, per unit of energy produced, maintenance of a hydroelectric plant seems to be 10 to 15 times more risky than maintaining gas, oil, or nuclear plants, even though the latter are much more complicated and more susceptible to breakdown. Moreover, the Report itself cites an American study which says that injuries at hydroelectric plants in the U.S.A. are half as frequent and 1/10 as serious (in terms of days lost) as the North American average of all plants.

Finally, the other large risk factor in hydroelectric plants is the possibility of a catastrophic dam burst. There again, the comparison is made on the basis of the world average, forgetting that dams are generally situated much closer to the cities and designed according to standards different from those in effect at present. In addition, there may be reason to suppose that dams in poorer countries are of lower quality than those in Quebec, for example, just as nuclear plants in poor countries are clearly less secure than in industrialized nations. Therefore, the risk of hydroelectricity seems overestimated once again. If the figures are corrected, hydroelectric power appears to represent a danger of the same magnitude as nuclear energy and not 5 to 10 times higher as claimed.

The same kind of critique can be made for most of the other energy systems studied, especially for the methanol system where the Report's author includes the operational risks twice, without noticing it, and he compares electrical plants with methanol used as automobile fuel for no apparent reason. He also assumes, again arbitrarily, that a methanol factory would be just as dangerous as a petroleum refinery, but that its life would be 10 years shorter (20 years rather than 30).

In conclusion, it is evident that this Report does not end the debate it opened. It is very fortunate that an attempt has been made to compare more closely the advantages and disadvantages of various means of energy production, but the methods of comparison must be considerably more refined if the conclusions are to be of any value whatsoever.

There Was No Cover-Up
by Herbert Inhaber

The question of how much risk to human health that different energy systems produce is far from being resolved. As Dr. A.T. Prince, President of the Atomic Energy Control Board, said when issuing report AECB 1119, *Risk of Energy Production* (to be referred to as "the report") "We hope it will generate a useful debate on the relative risk estimated for the various systems, as well as on how to reduce risk in *all* energy systems." The debate has been started by the most interesting discussion by Gilles Provost, *Maquiller le risque nucléaire,* (to be referred to as "the article"), which appeared in *Québec Science* in June.

I would like to thank Mr. Provost (and his co-worker, Pierre Sormany) for taking the time to go through many of the calculations and tables of the report. Because of its length of about 150 pages, many people have not had the time to do a detailed analysis.

It would be surprising if in the course of this analysis, Mr. Provost agreed with all of the report. The available data are not known well enough to excpect perfect agreement. M. Provost did point out areas of divergent opinion. They may be summarized as follows: (1) the equation of risk; (2) job creation; (3) centralization; (4) hydroelectricity; (5) time-frame of data; and (6) other minor issues. In the spirit of continuing a constructive debate on risk of energy systems, I will comment briefly on each.

(1) The report defines risk as the number of man-days lost per unit energy output. The article defines risk as the number of man-days lost per unit time required for material acquisition, construction, operation and maintenance. There is, of course, no way to prove that one definition is better than another, and M. Provost is certainly entitled to his opinion.

However, changing the formula (nowhere defined in the article) does not imply a cover-up, as stated in the article's title. There was then no cover-up. In fact, scientists who have written on relative risk of energy systems have used the definition employed in the report. I may be guilty of not using as much imagination as M. Provost, but I have used a standard practice.

Using a different denominator in the risk equation naturally produces different results. In addition, the time noted in the denominator of the article's equation is incomplete. For example, no time is allowed for construction of energy storage and back-up, transportation of materials, and other categories. If this were factored in, the results of the article would change substantially.

As well, the equation used in the article neglects completely risk to the public from the operation of energy systems. This is borne out by the low implied values for man-days lost using coal. For this system, most of the risk is due to air pollution, which is a public as opposed to an occupational risk. Considering only occupational risk is permissible, although it should have been made clear to the readers. However, much of the concern about nuclear power centers on potential risk to the public, and I would have thought that the article would discuss it.

(2) Job creation. A main goal of the article is to show that some energy systems create more jobs (per unit energy output) than others. There is no question that this is true. However, the object of the report was not to discuss jobs, but risk. The subject of job creation and energy deserves a report of its own.

As mentioned above, the discussion of job creation in the article is flawed because some aspects were left out. In spite of this, Provost has performed a service by calling attention to the subject.

Individually, most staff members of the Control Board are concerned about the high unemployment rate. However, the Board does not exist to create jobs, but to create safety. Job-creation is a task for federal and provincial departments of finance and labor.

On a philosophical basis, the energy system which creates the most jobs per unit output is not necessarily the best one for Canada. Suppose that the one million unemployed in this country were paid to generate electric power by running on treadmills. At a stroke our unemployment problem would be wiped out. But how much energy would be produced? On average, one person can produce about 50 watts of power. Assuming that each of these people worked about 2000 hours a year, and that there were no power losses in the system, a brief calculation shows that they would produce about 11 megawatts of power. By contrast, a nuclear reactor like Pickering produces about 2000 megawatts. It would take the entire population of the United States to produce the power of one nuclear reactor. While all of us want full employment and adequate energy, we should not choose the most inefficient way of producing this energy.

(3) Centralization. The article implies that decentralized energy systems are inherently better than centralized ones. This contention has been made by commentators like Amory Loving. The object of the report was not to evaluate energy systems in terms of centralization, but in terms of risk.

The systems with the highest risk per unit energy were coal- and oil-fired electricity plants, both highly centralized. They were then followed by decentralized systems like wind and solar space heating. Lowest in risk were centralized systems like natural gas-fired electricity and nuclear. The conclusion to be drawn from this is that we cannot say that centralization inherently produces greater or less risk.

(4) Hydroelectricity. Risk from hydroelectricity is mentioned prominently in the article. However, it is not, as claimed, treated in a different way from the other systems. The calculations are presented in full, but due to lack of time some results were not shown graphically. This problem has been corrected in the second edition, which is now available. I apologize for any inconvenience this may have caused.

The article states that Hydro-Quebec officials feel that the risk values for hydroelectricity (presumably using the report's equation for risk) are too high. If these officials have data that will be useful in assessing risk from this energy source, I and others will be pleased to read it in the scientific literature. I can say quite frankly that my research assistant and I had great difficulty in assembling information on hydroelectricity risk in spite of its wide use in Canada. We used some data from Ontario Hydro and less from Hydro-Quebec. The reason for this choice was a recent series of articles in *Le Devoir* which stated that Hydro-Quebec officials had not thought much about the overall risk of hydroelectricity. The articles were written by Gilles Provost.

There are a number of other points about hydroelectricity which deserve clarification. The article states that (a) most hydro dams are built closer to cities than those in Quebec; (b) older dams have standards lower than those being built; and (c) dams in underdeveloped coutries are of lower quality than those of Quebec. All of these points deal with the risk to the public of dam failures. I calculated that this portion of hydroelectricity risk is only about 20% of the total, so we should remember we are talking about a relatively small proportion. With respect to (a), I have seen no data discussing this, and would appreciate receiving the calculations used to prove this. We do know, as stated in the *Le Devoir* articles mentioned above, that the collapse of certain Quebec dams could cause substantial loss of life.

With respect to (b), I can only supply an example. The dam accident which caused the greatest loss of life occurred not in the 1920's, but in 1963. As far as I know, nobody has done a statistical study on failure rates of dams built long ago as compared to new ones.

Finally, point (c). Dams in underdeveloped countries are often designed by Western engineers on contract. As a result, they are probably not substantially different from those of Quebec. A table in the report shows that of the seven worst dam disasters, six occurred in industrialized countries.

To summarize the comments on hydroelectricity, we must be sure we are including all the associated risk, not just part.

(5) Time-frame of data. The article implies that a comparison is being

made between modern industries and those out of date. Data used in the report in all cases were either from the present or as recent as possible. However, no attempt was made to extrapolate into the future, and this decision may be the basis for the comments in the article. For example, it is possible that with improved air-pollution devices, the risk from burning coal or oil may decrease in the future. Similar comments can be made about any of the 11 energy systems considered. However, nobody knows exactly how risk will change in coming years. It would be unrealistic to expect only one or a few systems to have their future risk values change.

(6) Other minor issues. There were other issues raised in the article which deserve comment. *Quebec Science* is not a journal like *Nature* or *Science*, in which all the equations and assumptions can be specified, so perhaps the points in the article could have been clarified if more space was available.

There were some calculational errors in the section on methanol, as the article indicates. Others have also noticed this. The error is corrected in the second edition. The relative ranking of methanol risk in comparison to others does not change significantly. The article takes issue with a number of other assumptions dealing with methanol. A lifetime for a methanol factory lower than that for most other energy facilities was assumed because a recent Canadian study indicated it was the best value. The article states that the risk of methanol factor operation was counted twice. As far as I can tell, this is not the case. Perhaps there has been a misunderstanding of the methodology.

Considering that the word "nuclear" appears in the title, there is comparatively little discussion of nuclear power in the article. It does state, however, that reactors in underdeveloped countries are less safe than those of industrialized countries. The issue bears little relationship to the report, which considered risk primarily in the North American context. However, there is little evidence to support the contention. I have seen none. If M. Provost has calculations on the subject, I would be glad to see them.

Finally, the article raises the risk of nuclear power from 10 man-days per megawatt-year as noted in the report to 15. It is unfortunate that no explanation is given for this alteration. The report explains its major assumptions; the readers of *Quebec Science* deserve no less.

To summarize, the article has explored the report in detail. The main point of difference is that the article uses a different equation to calculate risk. The lack of explanations in the article may leave some readers confused. I have tried to clear up this confusion.

Environment

J.H. Herbert *et al.* wrote an article in *Environment* commenting on some of the more technical aspects of AECB 1119. While no public response to this article was ever made, the author wrote a memo to file which follows.

Environment, July-August 1979, p. 28–33
Published by Heldref Publications
H. H. Herbert, C. Swanson, and P. Reddy,
"Energy Production and the Inhaber Report"

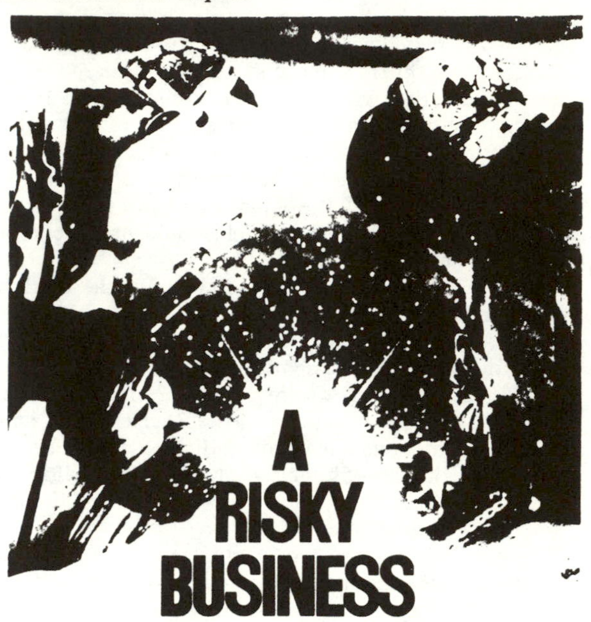

BY JOHN H. HERBERT, CHRISTINA SWANSON, and PATRICK REDDY

Energy Production and the Inhaber Report

JOHN H. HERBERT is an economist/econometrician, formerly with OSHA, he is presently associated with PRC Energy Analysis Company in McLean, Virginia. **CHRISTINA SWANSON**, an economist, and **PATRICK REDDY**, a meteorologist, are both associates at PRC.

THE PRODUCTION OF ENERGY entails human as well as economic costs. A number of recent studies have attempted to identify and quantify the risks involved in different types of energy production. Probably the best publicized of these analyses is that of Herbert Inhaber. A summary of the study, *Risk of Energy Production*, which was prepared originally for the Atomic Energy Control Board of Canada, appeared earlier this year in *Science.*[1] The *Science* article compared public and occupational risks associated with energy production from different sources. It indicated that the risks involved in producing nuclear energy were much lower than those associated with any other source of energy except natural gas (see Figure 1).

Risk, and especially risk of energy production, is currently eliciting considerable public concern. Because of this concern and because Inhaber's report is being referenced as a summary of the risk of energy production, it is essential that there be widespread critical evaluation and discussion of this report and of the risks of energy production.[2] The purpose of this essay is to identify some basic inadequacies in the study's approach and use of data, especially as regards nonconventional (i.e., solar) technologies, and to suggest some alternative approaches and data sources.

Three points are critical for the development of any risk analysis of energy technologies: first, the importance of the *technological assumptions* concerning applications of energy technologies made before a risk analysis is begun; second, the need for a qualitative and, if possible, a quantitative assessment of the types of risk associated with each energy technology—the *risk assessment* methodology; and finally, the usefulness of having a systematic source and framework for calculating different amounts of risk associated with each energy technology—the *sources of risk*.

Technological Assumptions

In *Risk of Energy Production*, Herbert Inhaber made numerous assumptions concerning the application of nonconventional energy technologies and the comparability of conventional and nonconventional technologies. Such assumptions need to be addressed before any risk analysis is begun. Many of these assumptions, however, were not explicitly stated or clearly delineated within the text. Readers who fail to identify all of the study's major assumptions may conclude that the risk calculations it contains apply to a broad class of potential real-world situations, when, in actuality, the calculations are appropriate only to a limited range of energy technology scenarios.

One of the most limiting assumptions implicit in the study is that all of the energy technologies are producers of, or substitutes for, base-load electrical energy of utility grid quality. Because of this assumption, generic solar technologies were not compared with conventional energy technologies. Rather, the comparison was limited to versions of each technology that would produce base-load electrical energy (or what was defined as its equivalent in the report) suitable for use in a utility grid. This limitation requires that hypothetical solar thermal, photovoltaic, and wind energy systems considered in the study be equipped with both storage and backup components, which would significantly increase the materials and labor requirements of these systems. The additional materials and labor requirements of storage and backup systems correspond to increased occupational safety and health risks per unit of energy produced. Furthermore, the systems are dependent specifically upon conventional coal-fired backup units,[3] which, in Inhaber's assessment, are associated with the highest level of risk of all the technologies, thus leading to greatly increased calculated risks for the solar technologies. Thus, storage and backup requirements substantially increase the overall risk of the evaluated solar and wind systems.

The proportion of total risk that derives from backup and storage requirements is very high. For solar thermal electric 89 percent of the risk measure is the result of backup and storage. For photovoltaics and wind energy conversion systems the proportions are 78 percent and 66 percent, respectively.

At the present time, many energy analysts expect future solar electric and wind-power electric applications to provide peak, intermediate, or base-load power, or some variable mix of these power load categories. If this is true, it is unlikely that, in actual applications, solar electric and wind-power electric will use as much backup or storage as specified in Inhaber's assessment. Many systems, particularly wind-power electric, may do without storage and backup altogether, whereas others will require either backup or storage, but not both. For example, peak-load and intermediate-load facilities could consist of solar/natural gas hybrids. In this case, the need for storage would be eliminated. Furthermore, because of lower air

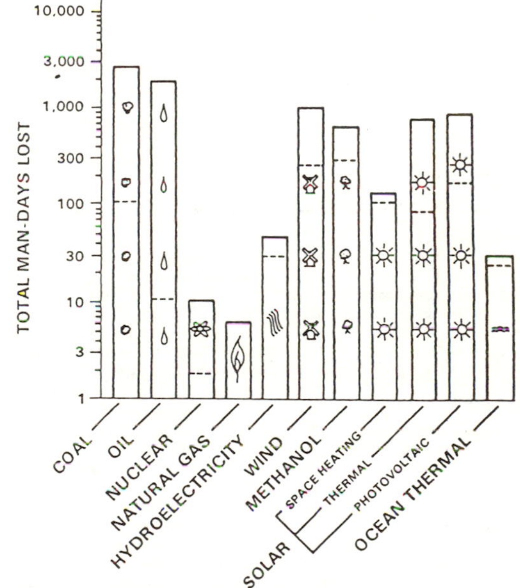

FIGURE 1. Inhaber's ranking of energy sources in terms of their riskiness as measured by total man-days lost per megawatt year of net output over the lifetime of the energy system. Each bar represents a range of man-days lost per unit of energy as a result of the implementation of a particular energy system. The dashed line within each bar represents the lower end of the range of man-days lost, while the top of each bar is the upper end of the range. Thus, for example, the total man-days lost per unit of energy produced by a nuclear plant ranges from approximately 2 to 10. (Source: Inhaber, *Risk of Energy Production*)

INCIDENCE RATES

According to Inhaber's evaluation, the risk calculations were obtained by the following procedure: an incidence rate X_i (average number of man-days lost per day of work in industry i) was multiplied by Y_i (the number of man-days required within industry i to produce a certain amount of material T_i).

The procedure is symbolically represented below:

$LWD_i = Y_i \cdot X_i$ (Lost Work Days for industry i)
$R_i = LWD_i \div T_i$ (Risk per unit of production for industry i)
$\Sigma R_i = \Sigma Y_i \cdot X_i \div T_i$ (Total risk per unit of production)

where $i = 1, \ldots, k$ industries, which supply the workers and materials required to put in place and maintain an energy facility of a specified size over a certain period of time.

LWD_i is equal to the lost workdays that result in industry i from building, operating, and maintaining a particular type of energy facility. If lost workdays are calculated for all the industries that supply goods and services for building and maintaining an energy facility over time, the sum of the calculations (ΣLWD_i) is the total number of lost workdays associated with that particular energy facility. Incidence rates and their sum over all industries are significant because they identify the *risk per day* of work activity associated with one industry, and with all industries that supply materials and services to an energy facility.

In practice, however, the risk calculation procedure outlined above is not applied by Inhaber to the *conventional technologies*. Instead, statistics for occupational and public accidents/diseases and for man-days lost per unit energy were extracted directly from *Evaluation of Conventional Power Systems*, which is itself a summary of such data taken from numerous other sources. Because of this, it is difficult to ascertain how individual risk calculations were derived. The manner in which the risk calculations are reported further obscures important information. For the conventional technologies, Inhaber presents calculations of occupational and public man-days lost in relation to gathering and handling fuels, transportation, electricity production, and (for nuclear systems) waste management. A detailed analysis of the risks accruing to each industry associated with particular types of power plants is not provided.

emissions and reduced materials and labor requirements, the risk to human health would be much lower than for base-load systems, which require both storage and backup, or systems with coal-fired backup.

Assessing the Risks

The manner in which the risk calculations are reported in Inhaber's report obscures important information (see box). *Incidence* rates by industry per day or year of work are not reported although these statistics are of great interest. For example, lost workdays per year of work incidence rates can be used to compare the individual riskiness of each of the various work activities required to support an energy technology. Furthermore, such statistics indicate the risk associated with any activity per day of activity, whether it be the illness risk of jogging in the rain or the accident risk of working in a mine. From an individual's viewpoint, therefore, they are among the most meaningful statistics in risk analysis. It could be contended that the mode of presentation used by Inhaber renders the risk calculations useful primarily from a management point of view—assuming that management's sole objective is to minimize the total number of lost workdays per unit of energy output. Information provided in this manner is significant to risk analysis but it is limited in its potential application.

Severity is not considered in Inhaber's risk analysis even though individuals and safety officers in industries assign greater importance to avoiding severe accidents than minor accidents. Risk analysts[4] have also recognized the analytic importance for risk analyses of severity measures that assign increasing weight to each sequential lost workday. The discussion of severity and, if possible, the calculation of severity measures for different energy technologies stress the fact that the social cost of one individual being disabled for a month is greater than that of thirty separate individuals being disabled for one day each (see box on page 31).

The importance of the distinction between *catastrophic and non-catastrophic events* is also not taken into account by Inhaber although many risk analysts note that avoidance of catastrophic events is assigned a much greater importance by individuals and societies than avoidance of non-catastrophic events, even though the non-catastrophic events may affect the same number of people over time as the catastrophic event does in a single instance.[6] Essentially, if people consider catastrophic events to be a greater evil than non-catastrophic ones that have the same long-term average impact, they are willing to make greater efforts (or to pay more) to avoid catastrophic events. Although it is mentioned that certain energy technologies are associated with a greater likelihood of catastrophic events, the importance of this distinction for risk analysis is not developed within *Risk of Energy Production*.

Figure 2 illustrates the difference between Inhaber's risk measure and other possible measures of risk. Inhaber's risk measure, which does not take into account either severity or catastrophe, is indicated by C_1. C_2 takes account of severity and indicates that lost workdays become increasingly costly as their sequential number increases. Catastrophes are associated with a family of cost curves that are dependent upon the nature of the catastrophe. This is indicated by the cost functions C_3 and

C_4, in which catastrophic events are assigned a greater cost per event by individuals and society than non-catastrophic events; for example, the consequences of an LNG explosion or a plane crash might best be represented by one of these curves. For any specified number of lost workdays, an examination of Figure 2 suggests how great a miscalculation can result from estimating the cost of accident or illness events exclusively on the basis of the procedure outlined in Inhaber's study.

SEVERITY

Risk analyst Chauncey Starr has proposed the calculation of a simple risk number based on severity,[5] which is similar to the following total social cost formula:

where V_i	=	Social cost per industry *i*,
i	=	1, . . ., *k* industries required to build and maintain a particular energy facility,
N_t	=	number of individuals that experience *t* consecutive disability days,
C_1	=	personal cost of one disability day for a non-catastrophic event,
r	=	social interest rate,
$C_1\,(l{+}r)^t$	=	average cost of a particular number of lost workdays, and
t	=	number of consecutive lost workdays being considered (*LWD* in previous equations).

Severity measures, such as this one, emphasize that the social cost of one individual being disabled for a month is greater than that of thirty separate individuals being disabled for one day each.

Sources of Risk

Seven general sources of risk in energy production are delineated in Inhaber's risk analysis. These sources represent the distinct industrial activities that comprise a fuel cycle and are associated with the material and manpower required to put an energy technology in place and to maintain it in operation. These requirements are, in turn, associated with incidence rates to obtain risk estimates which are expressed as total lost workdays per source for a specific fuel cycle. (Eight, rather than seven, sources of risk are listed below, because one of the categories in the analysis—raw materials and fuel production—has been subdivided.):

- *Raw material* required to build a component for a plant;
- *Fabricating* the component;
- *Building* the plant;
- *Operating* and *maintaining* the plant;
- *Transporting* material used for component replacements, material component fabrication, and plant construction;
- *Fuel production* for use by the plant;
- *Waste disposal*; and
- *Public health.*

One difficulty encountered in evaluating the sources of risk is that the data elements required for the quantitative analysis are not listed in *Risk of Energy Production*, nor are they collated by a common coding scheme such as Standard Industrial Classification (SIC). Data elements include: the industries involved in the fuel cycle; the number of workers employed in these industries; the average number of

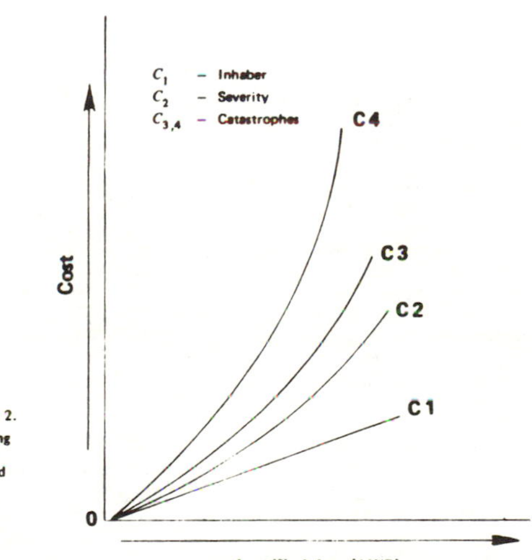

FIGURE 2. Estimating the costs associated with different types of work

accidents, illnesses, and lost workdays incurred by these workers; and the amount and type of environmental effects produced by these industries.

A more systematic approach would employ input-output tables as a basis for analysis. Such tables would identify, by SIC code, the amount of products and services required from supporting industries to produce and sell one dollar's worth of an industry's output (for example, $1.00 of electric energy from a coal-burning electrical utility or $1.00 of space heating energy from solar collectors requires a certain amount of purchases from supporting industries). Dollar amounts could also be expressed in terms of physical units such as tons of materials. Occupational safety and health statistics and statistics on environmental health effects for these industries when they are operating at a particular level of activity can also be obtained. Collation of this data in terms of SIC codes encourages a systematic and comprehensive treatment of occupational and public risks.

A fuel cycle—in which the *principal* industrial activities required to build, operate, and maintain an energy facility over time are identified—is neither the only nor necessarily the best framework for comparing risks associated with different energy technologies. Another possible approach would be to examine in greater detail the total number of *primary and supporting* industries required to build, operate, and maintain a particular type of energy facility over time. Still another approach would be to look at the change in risk that results from a relative change in the amount of energy supplied by one energy source when an alternative source is substituted (for example, if space heating and water heating energy requirements supplied by a coal-burning electric utility are provided instead by solar collectors). Such comparisons are possible if input-output tables are used as a basis for analyses.

A fuel cycle input-output analysis results in an examination of impacts on only a few industries. A *comprehensive* input-output analysis, however, would result in the calculation of impacts on *all* the industries affected, for example, a decreased demand for space heating and water heating supplied by coal-fired electric utilities that would be offset by increased demand for solar collectors.

An advantage of using input-output tables as a framework for comparison is that, given alternative energy technologies supplying a certain amount of energy demand, the numbers of employees, identified by SIC, required from supporting industries can be easily determined.[7] Moreover, once this determination is made, the expected number of occupational accidents and illnesses incurred by these employees can be computed by using the Bureau of Labor Statistics' (BLS) Occupational Safety and Health Statistics Series.[8] Furthermore, the amounts and type of emissions produced, identified by SIC code, associated with the supporting industries also can be obtained from data bases such as EPA "SEAS MODEL" data base;[9] using this data base, public health impacts can be computed from site-specific morbidity and mortality models. Thus a systematic and comprehensive analysis of the occupational and public safety and health impacts can be achieved by employing input-output tables as a basis for comparing the risks of the various energy technologies.

Table 1

THE AVERAGE AND TOTAL INCREMENTAL IMPACTS OF SUBSTITUTING SPACE HEATING DEMANDS SUPPLIED FROM AN ELECTRIC UTILITY WITH A SOLAR SOURCE

SIC	Industry	Jobs	AV-LWD/ Emp./Yr	Total LWD
3351	Copper Rolling & Drawing	1643	1.133	1861
3331	Primary Copper	294	1.689	493
10*	Copper Ore Mining	602	.692	417
491	Electric Utilities	-6362	.616	-3919
3353	Aluminum Rolling & Drawing	794	.734	583
381	Eng. & Sci. Instruments	1836	.221	406
331	Blast Furnace & Basic Steel Prod.	3834	.773	2964
3334	Primary Aluminum	235	1.045	245
10*	Iron Ore Mining	97	.692	67
3339	Other Non-Ferrous Metals	174	2.023	352
243	Plywood & Other Wood Prod.	1262	1.221	1541
124	Other Fed. Enterprises	-635	.360	-229
125	State & Local Gov. Enterprises	-2884	.420	-1211
356	General Industrial Machinery	1269	.730	926
282	Plastic Products	1856	.385	715
322	Glass, Flat	680	1.023	696
32	Cement, Clay & Concrete Prod.	1094	1.085	1187
12/11	Coal Mining	-590	(1.015** 2.113***)	(-599** -1247***)
14	Other Non-Ferrous Mining	78	.456	36
30	Plastic Materials & Synth. Rubber	247	.345	85
32	Misc. Stone & Clay Prod.	290	1.085	315
14	Stone & Clay Mining & Quarrying	144	.456	66
285	Paint	107	.570	61
242	Logmills, Sawmills & Planing Mills	398	1.710	681
Totals		*4800*		*(6459**, 5811***)*
Average			*(1.32)** (1.21)****	

* Information was only available on metal mining.
** Bituminous Coal
*** Anthracite Coal

Sources: U.S. Department of Labor, **Chartbook on Occupational Injuries and Illnesses in 1976**, Bureau of Labor Statistics Report 535, pp. 19-29, 1978; U.S. Department of Labor, **Occupational Safety and Health Statistics for the Federal Government in 1975**, Occupational Safety and Health Administration, Occupational Safety and Health Report 2066; H. Craig Peterson, **Output and Employment Impacts of Solar Heating**, Proceedings of the American Section of the International Solar Energy Society, pp. 450-455, 1978.

A comprehensive input-output analysis of the employment impacts of a transition from a conventional electric utility to solar collectors supplying 2.29×10^{10} kwh of energy by 1985 for space and water heating purposes was recently completed by H. Craig Peterson.[10] Peterson's employment impact results can be used to examine the occupational safety and health impacts of this transition. This is accomplished by multiplying the changes in employment computed by Peterson by average lost-workdays-per-employee incidence rates from BLS sources.[11] Furthermore, if the change in total lost workdays (TLWD) is divided by the change in total employment (TEMP), the average lost workdays per employee incidence rate associated with the transition can be computed as follows:

$$AVLWD = \frac{TLWD}{TEMP}$$

Using Peterson's employment results, our calculations indicate that the AVLWD incidence rate associated with the transition is between 1.21 and 1.32 average lost workdays per year (see Table 1). This result might be considered high since the rate associated with all manufacturing industries is approximately 0.79. The high number is partially explained by the high lost-workday impacts associated with increased activity in such industries as copper and nonferrous metal fabrication and in logmills, sawmills, and planing mills which results from the transition. It is also explained by the fact that building an industry is usually much more hazardous than operating and maintaining an industry. Accordingly, if the average lost-workday impact were averaged over the life of the system so as to include operating and maintaining the solar system, the calculated number should be lower. Finally, the public health risk associated with a solar energy system should be much less than that associated with alternative systems of energy production, primarily because few if any harmful environmental effects would result from using solar energy as a fuel.

The type of framework described above can be employed to calculate the impact of a transition from a conventional electric utility to an alternative energy source such as methane, solar, peat, or wood for space heating and water heating energy demands. The numbers calculated apply at that point in time at which a certain amount of capacity from an alternative source is made available. It does not compare the impacts of building, operating, and maintaining alternative systems over the lives of the systems. This would be a useful additional method of comparing energy technologies.[12]

In Focus

The analysis presented here strongly suggests that the results of *Risk of Energy Production* are based on inappropriate assumptions and inadequate methodology. In addition, much of the data used are not well documented, or are outdated,[13] or are not comprehensive enough.

Inhaber's assumptions appear to us to be inappropriate in that his analysis requires that all of the technologies examined produce base-load electrical energy or its equivalent. This requirement places solar and wind energy systems at a disadvantage because of the high storage and backup capacity that would be needed for base-load applications.

Inhaber's methodology is inadequate in three ways. First, Inhaber does not account for qualitative distinctions in the nature of accidents, their varying impacts on society, and their differing degrees of acceptability within society. Secondly, the usefulness of the risk measure of man-hours lost per unit of energy is limited. Activities related to one system may have lower risks per man-hour than activities for another; yet, if the former system requires more man-hours of these activities, then the associated risk per energy unit produced will be higher. Finally, a systematic source and framework to serve as a basis for calculating the differences in risk associated with each energy technology is not provided.

These problems suggest that *Risk of Energy Production* fails to offer a definitive comparison of the risks of various energy technologies. However, the objectives of the analysis are worthy of continued research, and *Risk of Energy Production* does serve to bring into focus many of the issues that need to be resolved to develop useful comparisons that are based on clearly interpretable analyses.

NOTES

1. Herbert Inhaber, **Risk of Energy Production** (AECB1119), Atomic Energy Control Board, Ottawa, Canada, 1978; "Is Solar Power More Dangerous than Nuclear?" **New Scientist**, May 18, 1978; "Risk with Energy from Conventional and Nonconventional Sources," **Science** 203 (February 23, 1979).

2. For three letters highly critical of Inhaber's data sources and the manner in which he manipulated his data, see **Science** 204 (May 4, 1979): 454, and **Science** 204 (May 11, 1979): 564.

3. R. S. Kirk Smith et al., **Evaluation of Conventional Power Systems**, California Institute of Technology, July 1975.

4. Chauncey Starr, Richard Rudman, and Chris Whipple, "Philosophical Basis for Risk Analysis," **Annual Review of Energy**, 1976, pp. 629-662.

5. Starr et al., note 4 above.

6. See, for example, Paul Slovic, Baruch Fischhoff, and Sarah Lichtenstein, "Rating the Risks," **Environment**, April 1979, p. 14.

7. U.S. Department of Labor, **The Structure of the U.S. Economy in 1980 and 1985**, Bureau of Labor Statistics, Bulletin 1831, 1975.

8. U.S. Department of Labor, **Occupational Injuries and Illnesses in the United States, by Industry**, Bureau of Labor Statistics, Bulletin 1932, 1974, pp. 78-79.

9. Peter House, **Trading off Environment, Economics, and Energy**, Lexington Books, Lexington, Mass., 1977, pp. 82-110.

10. H. Craig Peterson, "Output and Employment Impacts of Solar Heating," **Proceedings of the American Section of the International Solar Energy Society**, 1978, pp. 450-455.

11. Note 8 above.

12. This example is not comprehensive or definitive in that the severity and the catastrophic nature of the accident events are not considered and the statistics are provisional. It represents just one stage of a risk analysis.

13. P. J. Reddy, J. H. Herbert, A. Ryan, and C. A. Swanson, **A Review of the AECB Report "Risk of Energy Production,"** PRC Energy Analysis Company, March 1979, unpublished final draft, upon which part of the present discussion is based. See also note 2 above.

Memo to file, from H. Inhaber, Sept. 17, 1979, on above article.

The following are notes dealing with this article in case a written reply is made. For simplicity, the comments in the article will be paraphrased and in some cases combined.

p. 27, column 1. *Many of the assumptions of AECB 1119 are not stated clearly, in particular that noting that the energy mode is baseload.* Most assumptions were stated in detail in the main text. However, that relating to baseload was not clearly specified, although it was implied in a number of places. For example, on p. 20 of the third edition, it is noted that the reliability aimed for is that of conventional electrical distribution systems, i.e., one hour of loss in a year. There will be a clearer discussion of baseload and other modes in the fourth edition.

p. 29, column 1. *Storage and back-up would significantly increase the labor and materials requirements.* I do not believe this is the case. In fact, in terms of storage I did not specify the labor and materials requirements at all, but only the risk associated with storage. I did the same for energy back-up. In the table and graph summarizing labor and materials requirements, those attributable to storage and back-up are not included. I admit that this was not clearly spelled out, and I will rectify this in the future. The risk of storage and back-up was included in appropriate graphs, but the labor and materials requirements were not. The latter probably constitute a small proportion of (the) total.

p. 29, column 2. *The proportions of risk attributable to back-up for the three systems assumed to require it are high.* This is correct. The proportions in a similar study (1) are also large. However, even if one assumes no back-up, the relative rankings of the systems remain about the same. In this context, it is unfortunate that the graphs showing what the results would be if no back-up were assumed were not shown on p. 29. These graphs were included in the third edition, and in the *Science* article, to which the authors refer. The inclusion of these values would have made the relative rankings with and without back-up much clearer.

p. 29, column 3. *Solar electric and windpower could be used without storage and back-up.* Correct. However, my assumptions were made clear, and at least one other study (1) assumed storage and back-up. If no storage and back-up are used, the question arises as to the usefulness of the energy supplied by certain systems. For example, consider a windmill producing energy at 2 a.m., without storage. At this time, it is likely that the grid is operating solely on baseload, which is not usually switched on and off quickly. As a result, the wind energy probably could not be used. If this situation arises frequently enough, the load factor, or ratio of energy used to that which could be produced, could go down substantially. If the load factor decreases, the materials and labor risk per unit energy output will increase. This could be offset by decreases in risk attributable to storage and back-up, but it is not clear how much change there would be in the total risk.

There do not seem to be many studies which calculate what the load factor would be for solar electric or wind systems without storage and back-up. If there were, appropriate calculations could be done. It is of interest to note that a solar electric system being built in California has a storage system (although no back-up). It has an estimated load factor of 0.4, lower than the value of 0.7 assumed in AECB 1119.

p. 30, box. *Risk calculation procedure used for nonconventional systems was not shown for conventional systems.* Correct. In order to save space, readers were referred to Smith et al. (2) who in turn took most of their results for conventional systems from WASH-1224 (3). I felt there was not too much point in repeating in detail material which had already been published.

p. 30, column 1. *Incidence rates of injury or disease per year of work are not reported.* These incidence rates are shown in great detail in the appendices dealing with nonconventional energy systems, although they are not for conventional systems. For example, in Table E-2, 27.7 hours of iron ore mining is stated to correspond to 0.012 man-days lost due to accident, 1×10^{-6} deaths, etc. I do not know why these calculations seem to have been overlooked.

p. 30, column 3. *Severity is not considered.* Since this is not defined, it is difficult to tell whether or not this was considered in AECB 1119. If by severity is meant noting that different accidents can produce different man-days lost, then the subject was included in the calculations. The man-days lost per unit time include both severe and less severe accidents, and so are an average over severity.

Severity could also be defined, for example, by assuming, that an accident which causes 30 man-days lost should be counted in a "risk number" as more than 30 times the value of an accident which causes 1 man-day lost. This was not considered in AECB 1119 for two reasons. First, data are generally not available in the form that would be required for this calculation. One would need to know that x man-years in a given trade produced y injuries of duration 1 day, z injuries of duration 2 days, and so forth. The U.S. Bureau of Labor statistics generally show the average number of man-days lost per unit time worked, and not a breakdown by length of time per injury or disease. If the latter data were available, then a "severity" measure could, in principle, be calculated.

Second, there is no general statement on how the calculation should be done. The measure proposed by Starr (box, p. 31; the actual equation seems to be missing) was put forth as a suggestion, not as a universally agreed on measure. If there had been substantial agreement among risk analysts on the equation to be used, it would have been incorporated in AECB 1119.

p. 30, column 3. *The distinction between catastrophic and noncatastrophic events is not given sufficient importance.* The relationship of catastrophic to noncatastrophic risk was discussed early in the report, on p. 4. However, the report was not concerned with the psychological aspects of risk. These have been discussed in the work of Slovic and many others, and there was little I could contribute to the discussion. AECB 1119 was concerned with the actual risk, not what people may or may not perceive. If it could be shown that on average people overestimate catastrophic risk by *x%* compared to noncatastrophic risk, this could be taken into account in calculations. This, however, would be a task for another report.

p. 30, column 3. *Risk should be calculated as in Figure 2.* It was somewhat difficult to interpret Figure 2, so what follows is subject to revision. The notion of "cost" to society is a useful one, but one which is unfortunately not defined. Until it is, Figure 2 can only be viewed as a general diagram, not a method of objectively assessing risk. One is left with the questions: is "cost" an economic, health, psychological, or social cost, or some combination of all of them? If it is a combination, who is to do the combining, and on what basis?

A problem with curve C_4 of Figure 2 is that it seems to go to infinity. This implies that for some energy system, a small increase in lost workdays produces an infinite "cost" to society. This does not seem to make sense. If a coal-miner breaks a leg, does this mean that society is wiped out? This question has to be answered before schemes like Figure 2 are adopted in risk analysis.

p. 31, column 1. *Data elements for quantitative analysis are not listed in AECB 1119, nor are they shown in a scheme like the Standard Industrial Classification (SIC).* The data elements, such as the industries in the fuel cycle, and the accidents, illnesses, and deaths per unit time in each industry were shown clearly in each appendix. The number of workers in each industry was shown in Appendix M of the third edition, although not in the first two editions. However, this last piece of information is not necessary for the calculations.

Data on risk per unit time worked were not shown by Standard Industrial Classification, but by industry, such as steel making, iron ore mining, etc. It was expected that the industry description would be adequate for most readers, but those readers who wish SIC numbers can easily supply them.

p. 32, column 1. *Input-output (I–O) tables should be used as a basis for analysis.* At present, I–O tables exist only in terms of money. While one can conceive of I–O tables in terms of risk, where, for example, the risk attributable to supplying copper for the electrical motor industry would be tabulated, at present these tables do not exist. Risk I–O tables will not necessarily have the same numerical values as economic tables. For example, a dollar spent in coal-mining will produce greater risk to human health than one spent on office work. It would be desirable to have risk I–O tables for risk studies, and the authors of this article are to be encouraged in their pursuit. These tables clearly could not be used in AECB 1119 because of their nonexistence.

p. 33, column 1. *Table 1 is an example of the use of an I–O table.* Table 1 is not an example of an I–O table, but lists working days lost for certain industries. It is somewhat difficult to interpret since the amounts of copper, aluminum, plywood, etc. for the solar collector are not specified, nor the energy per unit area collection, nor risk associated with construction, operation, etc. It is presumed that these are specified in the original paper. When this information is supplied, the reader may be able to interpret the change in risk in going from coal-fired electricity to solar.

A further problem with this table is that deaths are not mentioned. These constitute most of the total man-days lost if one assumes that a death is about 6000 man-days, an assumption that is made in many risk analyses. For example, hard coal mining produces about 2 man-days lost due to accidents per employee-year of 2000 hours. However, there are about 0.0012 deaths per employee-year. If each death corresponds to 6000 man-days lost, deaths are the equivalent of 7.2 man-days lost.

In addition, as noted in footnote 12 of the article, Table 1 says nothing about severity or catastrophic accidents. This indicates that these concepts are far from universally applied.

p. 33, column 1. *The risk per worker should be calculated, not the risk per unit energy.* This concept is consistent with the suggestion of a severity measure, but inconsistent with the suggestion of a measure of catastrophe. The latter is concerned with public risk whereas the risk per worker deals solely with occupational risk. If risk per worker is to be used as the sole measure of risk, those who are interested in public risk of energy will have cause for concern.

Almost all the risk analyses of which I am aware–WASH-1224, Comar and Sagan, Gotchy, Smith *et al.*–used risk per unit energy as the main measure, not risk per worker. Just because a method has been used almost universally does not mean it should always be so. However, there are strong reasons for doing this. First, as mentioned above, public risk is unaccounted for in the "risk per worker" concept. Second, consider two cases: (a) a type of coal is discovered which is exactly the same as regular coal except its heat production when burned is twice that of ordinary coal; or (b) a solar collector is invented, exactly the same as other collectors except that it gathers twice as much energy per area as the others. According to the "risk per worker" concept, in these two cases nothing has changed. According to the "risk per unit energy" concept, the occupational risk has dropped by half. Readers can judge for themselves which concept makes more sense.

p. 33, column 2. *Much of the data in AECB 1119 is not well documented, or is outdated.* In terms of documentation, there were almost 200 references by the third edition. There will be about 400 in the proposed fourth edition. This does not imply, of course, that all numbers in AECB 1119 are explained perfectly. With respect to data being outdated, this is a charge which cannot be defended against unless specific examples are cited. However, efforts were made to use as up-to-date data as possible, although some older data may have been used inadvertently.

p. 33, column 2. *AECB does not offer a definitive comparison of risks.* No such claim was made in the text. It was noted that it was a crude attempt, and that there were bound to be changes. However, these changes should be workable and bear some relationship to the work on risk which has gone before.

HI/bd
Attachment

References

1. R. Caputo, "An Initial Comparative Assessment of Orbital and Terrestrial Central Power Systems," Jet Propulsion Laboratory, Pasadena, California, March 1977.
2. K. Smith *et al.*, "Evaluation of Conventional Power Plants," University of California, Berkeley, July 1975, Report ERG 75-5.
3. "Comparative Risk-Cost-Benefit Study of Alternative Sources of Electrical Energy," Atomic Energy Commission, Washington, December 1974, Report WASH-1224.

Other Comments

This section contains a wide variety of sources, some commenting directly on AECB 1119, and some indirectly. Lord Rothschild of Britain used some of the values of the report in the widely-broadcast Dimbleby lecture. For his trouble, he was criticized in the British journal *The Economist.* A response by the author follows.

The *New Scientist* took note of the controversy, and published a brief report discussing Rothschild, the Holdren letter in *Science*, and P. Ehrlich's comments in a speech. A letter from the author follows.

M. Flood, of Friends of the Earth, wrote an article for the British *Royal Society of Health Journal*, part of which was concerned with AECB 1119. Only that part is reprinted here. A response from the author follows.

S. Weaver, a member of the editorial page staff of *The Wall Street Journal,* wrote a perceptive article for the *AEI Journal on Government and Society.* This was based in part on her experiences in writing an article for *The Wall Street Journal*; the article follows.

W. Greider wrote a story for the Washington *Post* which summarized the conclusions in a novel way. P.M.S. Jones of Britain analyzed the results of AECB 1119 in a British context.

Finally, P.H. Gleick stated in a letter to *The New York Times* in late 1979 that AECB 1119 had been withdrawn. This statement was contradicted by H.J. Spence, spokesman for the Atomic Energy Control Board.

THE ECONOMIST DECEMBER 2, 1978

Nuclear safety

Figures are risky

There is nothing quite like quoting other people's numbers to get one into trouble. This is what Lord Rothschild, founder head of Mr Heath's Downing Street think-tank, did last week in his BBC Richard Dimbleby lecture on risk, which defended nuclear power as safer than most alternative energy sources, even including windmills.

The letters column of The Times erupted with abuse. One professor claimed that the crux of the debate is about values, not facts, giving pro-nuclear readers the excuse to point out that zero growth is not everybody's favourite value. But the hardest knock at Lord Rothschild came from one of the country's chief advocates of windmills, Mr Peter Musgrove, of Reading University. Mr Musgrove said his lordship had got his figures out by a factor of 100 on the amount of materials needed to make a windmill. Others say a factor of 50.

The safety of windmills is closely related to the cost of mining, processing, and constructing the materials needed to build them—steel, cement and so on. Lord Rothschild had cited a report by Mr Herbert Inhaber of the Atomic Energy Control Board of Canada. This voluminous report had the figures worked out in impressive detail. But Mr Inhaber relied for source material on an American government report published in March, 1977. Windmill technology has moved on since then, and monster 200-foot diameter windmills now being built in the United States need only a fraction of the materials of their predecessors. Lord Rothschild did say in his lecture that some people thought Mr Inhaber had stuck his neck out too far, but windlovers did not notice that.

The fact is that the more you look into safety statistics the more complicated they become. On past form, deaths from nuclear power are almost nil. But to put figures on the risk of a Baader Meinhof suicide attack, as Lord Rothschild did, is . . . well, risky.

It is not only terrorist risks that are hard to pin down. Take coal. Assessment of the effects of pollutants on health is at the frontiers of science, and therefore an issue on which few people agree. Mr Inhaber's own upper and lower estimates of coal risks differ by a factor of 30. Or windmills. Put a few thousand of them in the North and Irish Seas—the most likely choice in Britain—instead of on land, and the construction and maintenance risks would be anyone's guess. Up again, by a factor of what?

A tilt at windmills

THE ECONOMIST JANUARY 6, 1979

Tilting at nuclear power?

SIR—Lord Rothschild was accused (December 2nd) of using incorrect data when comparing nuclear risk to that of windmills. Lord Rothschild had noted in his recent BBC Richard Dimbleby lecture that, when the entire energy cycle is considered, nuclear power apparently has a much lower risk to human health than windmills. This statement was based on a report I had written ("Risk of energy production", AECB-1119).

Professor Peter Musgrove of the University of Reading said that the values for windmill materials' requirements, on which part of my risk calculations were based, were incorrect. I have carefully examined Lord Rothschild's face, at least figuratively, and can find no trace of egg.

You do not have the space to allow a full technical discussion but the upshot of the matter is that Professor Musgrove is talking about windmills which have not been built and may never be built. One of your countrymen, C. P. Snow, used the phrase: "Jam tomorrow, jam yesterday, but never jam today." In the context of windmills, the phrase can be modified if we take "jam" to mean windmills that require few materials per unit of energy output: "Jam tomorrow, but no jam today or yesterday." In other words, I based my report on what the facts are today, not what they might be tomorrow. A similar report in the year 2000 would require recalculation of the risks for all energy systems, not just windpower.

In any case, even if Professor Musgrove were correct in his claim, which I do not admit, there is substantial contribution to risk from construction, energy storage, energy back-up and transportation materials. That is, the risk of windmills does not vary directly with the amount of materials needed to build them. Assuming Professor Musgrove is right, the risk from these other aspects of the wind energy system is still about 40 times that of nuclear power. I then conclude that Lord Rothschild need not spend any sleepless nights.

As one who believes that windpower can eventually make a contribution to our energy needs, I feel that exaggerated claims of its benefits will ultimately hinder, rather than help, its progress. I am afraid that the statements made about Lord Rothschild's talk fall into this category.

HERBERT INHABER
Ottawa Atomic Energy Control Board

New Scientist 17 May 1979

The risks of researching energy safety

A vitriolic attack on Herbert Inhaber, the man who said that unconventional power sources could be more dangerous than nuclear energy, is under way in the US. Last Wednesday, Paul Ehrlich said that the wide dissemination of Inhaber's study (for example, in *New Scientist*, vol 78, p 444) was a "scandal". And on Friday, John Holdren, from the Energy and Resources Group at the University of California, described Inhaber's report as "a morass of mistakes" in a letter to *Science* (11 May, p 564). Holdren has previously written to the Atomic Energy Control Board of Canada, which published Inhaber's study, calling the report, "the most incompetent technical document ever circulated by grown ups" and demanding its withdrawal.

The antagonism to Inhaber's work owes much to the fact that supporters of nuclear power hold it up as evidence that the anti-nuclear movement is misleading people about the hazards of alternative energy systems.

Holdren's attack in *Science* barely manages to restrain itself to the usual scientific niceties. Holdren says that Inhaber's report is riddled with mistakes such as "double counting, highly selective use (and misuse) of data, untenable assumptions, inconsistencies in the treatment of different technologies, and conceptual confusions". He describes several of these errors in detail.

For example, Inhaber compared in great detail the risks of various ways of making energy—including, for instance, the risks of fatal accidents in building windmills, solar power plants and so on. But, says Holdren, he did not include similar risks during the building of nuclear and coal-fired power stations. Inhaber relied on published studies for estimating the risk to the public from a reactor accident. He said in his report that he deliberately used the most pessimistic figures. But, according to Holdren, he did not. Holdren says that Inhaber's "upper limit" figure for reactor risks is three times below that in the Rasmussen report, commissioned by the US Atomic Energy Commission, and is 200 times smaller than the figure implied in a Ford/MITRE study of reactor risks.

Holdren claims that removing the most important errors from Inhaber's study, "transforms his results, raising the upper limits of nuclear risk to public and occupational health into the lower part of the uncertainty range for coal and oil, and dropping the health risks of the nonconventional sources into the middle of the uncertainty range for nuclear".

Ehrlich directed his attack more to those who accepted Inhaber's report too uncritically. He said at a meeting at the University of Northern Colorado last Wednesday: "The burning question now is to discover why the reviewing processes of distinguished journals like *Science* [which published a paper by Inhaber in February] utterly broke down, whether newspapers like the *Wall Street Journal* and well-known scientists like Baron Rothschild [who used Inhaber's analysis in his BBC Dimbleby Lecture last year] that were fooled will print prominent retractions, and whether the Atomic Energy Control Board of Canada will at long last state publicly that the Inhaber Report is hopelessly flawed." He went on to claim that the Inhaber report, and the Three Mile Island near-miss, demonstrate "one of the key problems of the nuclear establishment—the stupidity and dishonesty that permeate it". □

New Scientist
14 June 1979

Not Guilty

I found it interesting to read the recent discussion (This Week, 17 May, p 526) on a report I had written on the risks of energy production (described in the pages of your journal, vol 78, p 444). Professors Holdren and Ehrlich in the United States have been most prominent in commenting on it. I regret to say that I have not heard personally from either one of them; if I had, I might have been able to correct some of their impressions.

Allow me, for lack of space to discuss them all, to evaluate one of Holdren's more serious charges: I failed to use the most pessimistic figures for public risk from nuclear reactors, as I had claimed. I never made that claim. What I did say was that I used the most pessimistic figures from two extensive literature compilations. One was a lengthy report of which Holdren was himself the senior author. The second was a paper by two highly respected American risk analysts, Drs. Comar and Sagan, which appeared in *Annual Reviews of Energy*.

There are two curious points about these two data compilations. First, Holdren's values for estimated public risk were substantially lower than those of Comar and Sagan. I chose to use the latter's figures. As far as I know, Holdren has not repudiated, retracted or corrected his report, so I presume that he still stands behind its values. Secondly, Holdren is an editor of the journal in which the Comar and Sagan article appeared. I would not hold editors responsible for everything that appears in their journals, but I am mystified why, if Comar and Sagan were in error, Holdren did not bring it to their attention.

Holdren may wish to use the "worst possible" values for the risk of certain energy systems, but this method is not employed by serious risk analysts. If it were, one could "prove" anything one wished. For example in terms of hydroelectricity, one could consider only the risk of Italian dams in 1963 (the year of the Vajont disaster); in terms of coal occupational risk, one could consider solely a shift in which a cave-in occurred; in terms of wind, one could consider a system which had a lifetime of only weeks instead of years. Studying only one the "worst possible" case would "show" only that all forms of energy, conventional or non-conventional, coal, nuclear, wind, oil or solar, should be avoided like the plague. Perhaps this is what Holdren has in mind?

Despite Holdren's tone in his letters to *Science* and elsewhere, I have carefully studied each of his criticisms. I find those which are valid do not substantially alter my report's conclusions. I hope that a discussion of the relative risk of energy can proceed in the future on a more constructive basis.

Herbert Inhaber *Quebec*

Royal Society of Health Journal
Vol. 99 No. 6 December 1979

The Health Hazards of a Renewable Energy Strategy for Britain

Michael Flood, B.SC., Ph.D.
Energy Consultant to Friends of the Earth

. . . Although much has been written over the last few years about the dangers of pursuing the 'conventional' energy path–the 'hard' path in Lovins' terminology–comparatively little has been published about the possible risks associated with low energy scenarios. A number of individuals have attempted to ridicule the whole concept of industrial society being dependent on renewable energy systems but few of their studies are worthy of detailed comment. (*c*) However, one major study, published in March of last year has attracted a good deal of attention. Its startling conclusions are frequently quoted by ardent proponents of nuclear energy.

Risk of Energy Production was researched and written about by Dr. Herbert Inhaber, a science adviser to the Atomic Energy Control Board of Canada. It is an attempt to compare and contrast the risk potential of ten different energy technologies including four 'conventional' ones–coal, oil, natural gas, and nuclear, and six 'unconventional' ones–solar thermal electricity, solar photovoltaic, solar space heating, wind, ocean thermal, and methanol (from biomass). (*d*)

From his analysis, Dr. Inhaber concludes that: "Contrary to the intuition of many people, the risk to human health (and its resulting consequences) per unit energy from unconventional energy sources such as solar and wind are apparently higher than those of conventional sources such as electricity produced from natural gas and nuclear power."

Clearly, a few remarks about the report and its methodology are in order since Dr. Inhaber implies that the 'soft' path may well be more dangerous than the conventional one.

The following is far from being a comprehensive list of criticisms of the report:

1. Dr. Inhaber is concerned to establish the risk per unit of energy produced and not the more relevant consideration, the risk per unit of energy produced *and* consumed. Significant conversion and transmission losses are not included in the analysis.
2. Dr. Inhaber's comparisons are asymmetric; he compares heating by electricity and heating by solar panels, yet neglects the risk incurred in producing, fabricating, and installing transmission lines.
3. Dr. Inhaber ignores important technologies like passive solar heating which, under his own classification, would not be considered as an

energy 'producer'. Passive solar heating is relatively low risk since it requires merely a different design of building–although there may be slightly higher material requirements than in traditional structures.

4. Although acknowledging that 'many quantities (used in the calculations) are only estimates' he omits to show the uncertainty in his figures. (*e*)
5. Some of his figures have already been challenged by authorities on specific technologies. Dr. Peter Musgrove, an acknowledged expert on wind power, claims that some of the quantities used in calculating the risks associated with windmills are 100 times too high, and that Dr. Inhaber's conclusions for wind are therefore 'quite wrong'.
6. About *two-thirds* of the risk associated with wind is derived, not from the technology itself but from the pollution associated with a dirty coal-fired back-up system! (*f*) It is not reasonable to regard wind as a source of firm power like electricity from the grid. For many applications such security of supply would in any case not be necessary (*g*)
7. He greatly favors those technologies which, because they require a minimum of staff, provide few jobs. Understandably there are fewer accidents in the workplace. (The criticism has been forcefully developed by Gilles Provost.)

Dr. Inhaber has clearly failed to grasp the main concepts underlying low energy futures. His methodology may be of academic interest, but it has no relevance for policy making.

References (partial)

(c) Using data culled from a 1958 history book, Sir Fred Hoyle has calculated that the number of large windmills that would be necessary to meet Britain's future energy requirements would be around 20 million. 'The area occupied,' he studiously calculates, 'would cover more than half the area of England. When in full operation such an ensemble of mills would make an appalling roar, and the number of serious accidents they would cause would probably run into hundreds of thousands each year. To arrive at this bizarre conclusion he arbitrarily assumes a per capita energy consumption figure of *three* times the present figure and that *all* of this demand will be met by an inefficient design of wind generator. More interesting examples are to be found in Petr Beckmann's book: 'The Health Hazards of Not Going Nuclear' (The Golem Press, Boulder, Colorado, 1976).

(d) Ocean thermal and solar thermal systems are means of generating electricity. The ocean thermal concept is designed to use the thermal gradients between the warm surface and cold lower levels of the ocean; the solar electric system–Dr. Inhaber assesses the 'power tower' variety–uses tracking mirrors to focus the sun's rays on a receiver (in this case at the top of a tower) and the heat is used rather in the conventional way to drive a turbo-generator.

(e) One critic has noted that 'If Dr. Inhaber's histograms showed error bands, or confidence bands, all the newer energy technologies would be seen to have wider bands on them than the better developed technologies. The results would show general implications but would not give anything as apparently explicit as Dr. Inhaber's figures.'

(f) Dr. Inhaber's choice of coal as a back-up is a little puzzling especially in Canada where it would most likely be low-risk hydro.

(g) It is interesting to note that according to Professor Bent Sørenson a mere 10 hours' storage make a typical Danish wind machine into as reliable a source of firm power as a typical light water reactor.

Editor, Royal Society of Health Journal
13 Grosvenor Place
London SW1X 7EN, United Kingdom

Dear Sir:

My attention has been drawn to an article by M. Flood entitled "The Health Hazards of a Renewable Energy Strategy for Britain", in your December issue. I regret to say that this article contains a number of mis-statements about my work (1). In the following, I will condense, as necessary, Dr. Flood's statements, and attach my response.

(1) Ref. 1 does not include "significant conversion and transmission losses" in calculation of risk. Transmission losses are small. The McGraw-Hill Encyclopedia of Energy (2) can be used to show that the average loss due to transmission and distribution of electricity in the United States in 1975 was about 11%. Canadian conditions are similar (3). I suspect that the proportion for Britain, with its generally smaller distances between energy production and consumption centres, would be even smaller.

Because there are so many other sources of uncertainty in the data, this 11% does not significantly change the overall results. For example, if one considers the bar graphs showing total risk per unit energy in my *New Scientist* article (1), reducing the values by 11% would decrease their height by about 0.5 mm, a negligible amount.

With respect to conversion losses, these were included for systems which produced electricity. For example, I noted that the assumed coal conversion efficiency was 37%, oil 37 - 39%, etc. The statement that conversion losses were omitted is simply incorrect.

(2) The risk in producing, fabricating and installing transmission lines is neglected. In the first two editions of AECB 1119 (1), this was indeed correct. The third edition made a rough calculation of the risk attributable to this source. It was found to be about 2% of the total risk of hydroelectricity, and much less than that for other energy systems. Similar results have subsequently been found by Donakowski of the U.S. Institute of Gas Technology. I then conclude that neglect of this source of risk is not of great consequence.

(3) Uncertainty in the figures is not shown. Fig. 1 in the New Scientist article (1) states: "The maxima are at the top of the bars, the minima are the horizontal dotted lines". I do not know how this can be made clearer.

(4) Dr. Peter Musgrove has challenged some of the quantities used in windpower. There were indeed errors in some of the numbers for materials requirements in this section. This has been admitted elsewhere, and corrections made to subsequent versions of the results. However, it is incorrect to state, as Dr. Musgrove is quoted as saying, that the overall risk of windpower was thus rendered 'quite wrong'. Total risk of this system has many contributions: materials acquisition, construction, operation and maintenance, back-up energy, storage, transportation, etc. When the error in materials requirements is corrected, the overall ranking of wind in terms of risk remains about the same, because of the contribution of other sources.

(5) About two-thirds of the risk associated with wind is produced from a "dirty" coal-fired back-up system. The coal system assumed was not "dirty". On the contrary, it was much cleaner than the average of present-day systems in

Canada, producing 11% - 37% of today's sulphur oxides, 4% - 11% of the particulate matter, etc. I suspect that the ratio of pollutants for the system I considered to that of British coal-fired plants would be similar.

The proportion of total wind risk associated with energy back-up ranged from 4% - 64%, so it is misleading to use only the upper value. The implication of the underlined statement is that the risk associated with the energy required when the wind doesn't blow can somehow be dismissed. If windpower is to be used in anything except negligible amounts, it will require back-up, as D. Miller of your South of Scotland Electricity Board has stated (4). The risk associated with this back-up must then be incorporated with all other sources of risk.

(6) Ref. 1 "greatly favours" those technologies which provide few jobs. No statement favoring one technology over another was made in Ref. 1. On the contrary, it was clearly stated that society would need a variety of energy systems in the future. The underlined comment is untrue.

Based on the above-noted mis-statements made in only a few lines of the article, your readers have cause for wondering how much validity exists in the rest of the article.

Yours sincerely,

Dr. Herbert Inhaber

79 Boucher St.
Hull, Quebec
Canada J8Y 6G7
Feb. 27, 1980

REFERENCES

1. H. Inhaber, "Is Solar Power More Dangerous than Nuclear", New Scientist, May 18, 1978, pp. 444 - 446; H. Inhaber, "Risk of Energy Production", Atomic Energy Control Board, Ottawa, Canada, 3 editions (March - November 1978), Report AECB 1119.

2. D. Lapedes, editor, "McGraw-Hill Encyclopedia of Energy", New York, 1976, p. 219.

3. W.L. Wardrop & Associates Ltd., "Net Utilization Efficiency of Oil, Gas and Electricity as Energy Supply Systems for Buildings", Canadian Electrical Association, Montreal, Canada, August 1979.

4. D.J. Miller, "Alternative Energy Sources", in G. Foley and A. van Buren, Nuclear or Not? Choices for our Energy Future", Heinemann, London, 1978, pp. 101 - 106.

AEI Journal/July-August 1979

Inhaber and the Limits of Cost-Benefit Analysis
Suzanne Weaver

In January there was an exchange in the *Washington Post* between Mark Green of Congress Watch and Peter Schuck of the American Enterprise Institute on how much cost-benefit analysis could be reasonably used for the purpose of social regulation. I read through that exchange and, unsurprisingly for a representative of the *Wall Street Journal*, came to a somewhat different conclusion from Mr. Green's. Yet as I read his article, I was struck most of all by the extent to which I agreed with his analysis.

His argument, to quote from the *Post* article (January 21, 1979, section C1), is that "given the state of economic art, mathematical cost-benefit analyses are about as neutral as voter literacy tests in the Old South." That is not exactly a dispassionate way of putting it, but I think he is on to something. It may well be that by thinking explicitly about costs and benefits, even apart from the final verdict one reaches in a given case, one injects something into the debate that is not politically neutral and is, in some fundamental way, hostile to a large part of the current movement for social regulation.

This possibility came to me quite powerfully out of a piece of work that I did recently, part of whose results appeared in the *Journal* as a feature piece. The subject was the report produced a couple years ago by Dr. Herbert Inhaber, a physicist working for the Atomic Energy Control Board of Canada. Inhaber made a first cut at going through the existing literature on the risks associated with various energy sources, in an attempt to figure out how the sources ranked relative to one another.

The *Journal*'s involvement with the Inhaber report began some months ago when an article in the paper by another author mentioned the document in passing. Following that mention, a strange thing happened. Our features editor, Tom Bray, began to get detailed and passionate mail on the subject, telling what an egregiously bad piece of work the Inhaber report was. Some of this mail directed our attention to studies that, we were told, destroyed poor Inhaber quite completely. Tom Bray–being,

---

Suzanne Weaver is an editorial page writer for the Wall Street Journal *and the author of* Decision to Prosecute, *a book on antitrust. These remarks were delivered at the AEI-*National Journal *Conference on Regulatory Reform, May 21, 1979.*

among other things, a first-rate journalist—thought that if Inhaber could make so many people so mad, he must have struck a nerve somewhere. I was asked to see what all the noise was about.

It turned out that Dr. Inhaber had backed into these attacks by attempting a variety of cost–benefit analyses. What he set out to do was to make a pretty straightforward calculation of various kinds of energy risks. He took all the literature on the various energy sources—conventional systems like coal, oil, and nuclear power, as well as the newer and more decentralized technologies like wind and solar energy—and he added up the various risks to life and limb from each of them, all the way from the mining of the materials necessary to constructing each system, through the generation of power (including the back-up facilities needed by decentralized systems), to final waste deposit. The resulting report was fairly widely distributed, in no small part because Inhaber was the first person to do the unattractive but useful work of going through all the existing sources in the field and making the rather tedious calculations necessary to extract some kind of comparable data from them. But Inhaber's report also got attention because it reached a rather startling conclusion. He said that the energy systems some people have touted as "clean," like solar energy and wind, could actually be riskier to society than some conventional systems, including nuclear power.

In retrospect, it is no great mystery why Inhaber's method of risk analysis would come up with such a conclusion. There are two kinds of structural reasons that contribute to the final verdict.

First, when one deals with a power source like nuclear energy, the maximum risks may be huge, but they are also remote. So any risk assessment method that counts both these factors will judge nuclear energy as relatively safe. Second, some of the new decentralized technologies are at the moment relatively inefficient and unreliable compared to the older systems. The low efficiency means that relatively large collectors are needed. Large collectors take substantial amounts of material to build. And these materials pose risk to life and limb as they are mined, manufactured, and transported. Further, the low reliability of the new systems makes it necessary to take account of their back-up systems, and add the attendant risks of those into the final risk total.

One can begin to see some of the structural problems with a method like the one Inhaber used. Moreover, there are limitations from obvious and ordinary failings in the data—from the gaps, the ambiguities, and the errors in some of the studies, and the lies as well. Also, it is hardly likely that any researcher doing the first comprehensive collection job in this area can avoid contributing some plain ordinary mistakes of his own.

Besides those kinds of problems, there are middle-level conceptual problems that risk assessment of this sort is only beginning to deal with. For example, is it helpful to count up all the risks involved in huge solar collectors, if in fact the solar collectors will not be built until some way is found to bring the size down and the cost into line?

There are also the larger conceptual dilemmas. For instance, are there some possibilities connected with some kinds of energy production that are so horrible that it is quite reasonable to refuse to risk them now no matter how remote the chances of their occurrence? Or are there some kinds of dangers—for instance, the risk of producing a deformed child—that should be counted as being worse than illness or death or shortened lifespan?

As can be readily imagined, people wrote to Inhaber and the Atomic Energy Control Board with criticisms like this and, as a result, his report has been undergoing continual updating and correction. But none of these limitations explains the phenomenon that I was asked to investigate. The letters the *Wall Street Journal* was getting about the Inhaber report were not filled simply with criticism and suggestions for amendment. Instead, admitting to having no interest in amending the Inhaber report, they thought it would be better if the document were obliterated altogether. They said the report deserved obliteration because it was so badly done, but made no attempt to improve upon its analysis, or come up with a fundamentally different conclusion.

The chief opponent of the Inhaber report has been John Holdren, a Berkeley physicist active in the movement against nuclear power and in promoting research into nonconventional energy sources. In their letters, it is true, Dr. Holdren and his allies have included specific criticisms of the report. They have argued that Inhaber overestimated material requirements for windmills, used conservative assumptions for some nonconventional energy sources that he did not apply to other systems, and misquoted sources on the various kinds of conventional energy waste disposal risks. But in addition to the specific criticisms, these various letters and publications have also assailed Inhaber with a most extraordinary kind of general invective. They have charged not only that he is in error, but also that he is deliberately lying. They have called his document "by far the most incompetent technical document . . . ever known to have been distributed by grownups." They have called it "garbage" and Inhaber a "buffoon." They have claimed flatly that any expert who defends the report either has not read it or is not an expert (which, in fact, is not true).

They have attempted, when talking to me by telephone, to make contact with my dim journalistic mind by explaining that they have used all these strong words because their situation is similar to that of the journalists who attacked Watergate. We had used very strong language, I was told, to bring home the horror of the Watergate offense against society and, for the same reason, they were using strong language to attack Inhaber.

Even for scientific controversies this is a very strange and, of course, grossly indecent way to conduct a debate. And for a while, I confess I was mystified by the spectacle. But now I realized that Inhaber's critics were precisely right to be so upset. The offense of a document like this is not that it opts for one energy system over another. Instead, what is really going on is that it launches, in large part unintentionally, a much more dangerous kind of attack. A document like this operates on the assumption that the risks of the various energy sources may be different in degree, but are not in critical respects different in kind; if they were different in kind, one could not presume to compare them on any scale whatsoever.

To put it another way, the method of inquiry in the Inhaber report asserts by implication that energy sources are all somehow ethically equivalent and that they can, in the large, be judged according to the same standards. The enterprise suggests that a person who argues for one energy source over another cannot legitimately be judged a friend or enemy of the people simply by the choice he finally makes.

This kind of comparison takes a subject that has been spoken of in moral absolutes and presents it as something uncertain, ambiguous, subject to doubt and, perhaps, to compromise. Making an assertion like this—saying that an area of inquiry is subject to doubt—may seem a minimal thing. After all, it certainly does not guarantee what one's verdict will be in a particular case. One might follow a method like Inhaber's and then decide to ban nuclear power, or a suspected carcinogen, altogether. But Inhaber's kind of discourse is, in spite of such uncertainties, a significant threat to the strategy by which today's new regulation has been making its gains.

The argument surrounding much of the new regulatory movement is that technological capitalism is poisoning people. The corollary of the argument is that the poisons produced by this capitalism must be removed from the environment, that this is the only ethically responsible course for public policy to take, that those who raise cost considerations should not be trusted because their arguments are only smokescreens thrown up by the forces of greed.

For instance, Mark Green, in that *Washington Post* article, says that cost-benefit analysis might have prevented the Salk polio vaccine from coming on the market. He says cost-benefit analysis would have killed the idea of abolishing slavery. He says it would never have supported the child labor laws. He tells us that we cannot put a price on the child who can be saved from disfigurement from flammable sleepwear, or a price on the worker who is saved from asbestos-induced cancer. And the implication, which in its own way pollutes—indeed, poisons—the debate on regulation, is that those who talk in terms of cost-benefit analysis do not care about the child, do not care about the death.

I agree that these worries about children and lives are worries of the most profound kind. But in a way I think the people who have been attacking cost-benefit analysis in this way have not expanded their field of moral worry far enough. For instance, on the matter of toxic chemicals, we are increasingly able to discover more not only about the toxic character of the workplace, which is serious enough, but also about the toxic character of natural processes. To take one example—and this is not a smokescreen, but a set of issues that at some point we must face in terms of public policy—by now we are aware of huge numbers of carcinogens occurring naturally in the food supply, in foods ranging from fish to green vegetables. Does our growing knowledge of these things come without some obligation to act? And if we do have to act, how can we do so without some form of cost-benefit analysis, imperfect though it may be?

Mr. Green is right in calling attention to the fact that cost-benefit analysis is the enemy of the unbridled agenda of social regulation that he has been promoting. But I think we will soon reach the time when regulatory trade-offs in this area cannot be avoided. It might be good for public debate on the issue if more people began to point this out.

The Wall Street Journal
April 24, 1979
Reprinted by permission of
The Wall Street Journal

THE PASSIONATE RISK DEBATE

by Suzanne Weaver

In the wake of the accident at Three Mile Island, it has become clear that the country's future energy choices are going to be heavily influenced and perhaps even dominated by issues of safety and risk. A controversy now brewing among "risk assessment" specialists suggests that unless the climate of discussion changes, the debates on these issues are going to be bitter, confusing to the public and less useful than they could be in helping us make policy.

Recently this page printed excerpts from a speech that Baron Nathaniel Rothschild originally gave in Britain on the subject of risk. It is impossible, Lord Rothschild argued, to create a risk-free society. For instance, some people say that nuclear power is unacceptably risky. But a recent study by a Dr. Herbert Inhaber, Lord Rothschild continued, has concluded that some of the alternatives suggested–solar or wind power, for instance–may be even riskier than atomic energy.

We soon began to get letters responding to the Rothschild remarks. Were we aware, one of them asked, that the Inhaber report was flawed by "fallacious reasoning" and "serious misstatement of fact"? Did we know, said another, that its distribution had caused "a minor scandal in the energy field"?

The intensity in these letters raised questions of its own. Who was this Inhaber? And why were people so mad at him?

It turns out that Herbert Inhaber is a physicist, author of books on the physics of the environment, and an Associate Scientific Adviser to Canada's Atomic Energy Control Board. In 1978 he published a report, drawn from existing literature, comparing relative risks among both conventional energy sources, such as coal, oil, and nuclear power, and nonconventional sources, like wind and solar power. His was the first such survey that set out to cover the new decentralized technologies as well as the more established ones and to count risks for each energy source all the way through its fuel cycle, from mining and construction through generation and back-up systems to disposition of wastes.

Dr. Inhaber wrote his report in the summer of 1977. It was sent around to colleagues at the AECB for criticism and released in the spring of 1978.

"This field of study is young," the Board's official preface said of the report. Nevertheless, "we hope it will generate a useful debate on the relative risk estimated for the various systems."

A Hot Item

The report has become, as scientific papers go, a hot item. Several hundred copies were printed originally, but by now, 2,200 have been distributed. Journals like Nuclear News, New Scientist, and Energy have summarized it, and it was the basis of a recent article by Dr. Inhaber in the prestigious magazine Science.

All the interest is hardly mysterious. For one thing, as the first such comprehensive survey it was bound to become a starting point and a standard reference for future work in the field.

But just as important, Dr. Inhaber had reached a rather startling conclusion: All things considered, some of the highly touted "clean" alternative sources–solar energy, for instance, and wind–were actually riskier to society than some conventional energy systems, including nuclear energy in particular.

This verdict was a bombshell. For years, opponents of nuclear energy had been making their most effective arguments by pointing to the seemingly obvious dangers of atomic power. Now it looked as if their consistent emphasis on health and safety might backfire on them.

After the report emerged in April, Dr. Inhaber and the Canadian Atomic Energy Control Board began to receive the expected suggestions and criticisms. Some of them were used to correct the report; as a result it has gone through two full revisions, and one section has been changed and updated four times.

But late in 1978 the tone of the debate began to change. The first version of the report came to the attention of Dr. John Holdren, a Berkeley physicist and a prominent opponent of nuclear power. Dr. Holdren did not like the Inhaber report.

The document, he contends, is chock full of inaccuracies. For example, he says, Dr. Inhaber misread a table and radically overestimated the materials requirements for a windmill, thus substantially inflating the risks of wind-driven energy production. When confronted with the error, says Dr. Holdren, Dr. Inhaber tried to manipulate his calculations so as to conceal it. In another example Dr. Inhaber made a series of errors in dealing with methanol systems, each time overestimating the risk of a methanol plant by applying assumptions that he did not impose when treating conventional energy systems.

Dr. Holdren's bill of allegations continues. Dr. Inhaber, he says, has implied that he has counted up all the risks associated with conventional energy sources including waste disposal. But the papers and articles he cites, Dr. Holdren says, contain no such comprehensive information.

Dr. Inhaber also claims he's used "upper limit" values on the risks various authors have estimated for nuclear power when in fact he has not. More generally, says Dr. Holdren, Dr. Inhaber has relied on faulty sources and committed gross arithmetic errors.

But more striking than Dr. Holdren's criticisms has been the manner in which he and like-minded colleagues have chosen to make them. Dr. Holdren first wrote to the AECB to tell them that the Inhaber report constituted a "serious embarrassment" to the board. He then wrote a general circular to "Colleagues," repeating his charges. He sent a memo to "Science Editors of Major U.S. Publications" to call his critique to wider attention.

He wrote letters which were published in Nuclear News and Science. He wrote a later letter to Science expanding on his first critique and two further letters to Nuclear News. He wrote again to the AECB, protesting their refusal to disavow Dr. Inhaber. (He has not yet written to Dr. Inhaber himself.) Sympathetic colleague of Dr. Holdren's have written to news media calling attention to the matter. Paul Ehrlich is planning an article on what he calls the "Inhaber Report Scandal" for the Mother Earth News.

Dr. Holdren's language has also been unusual. He has called Dr. Inhaber and his report "shockingly incompetent." He has termed the document "the shabbiest hodge-podge of misreadings, misrepresentations, and preposterous calculational errors I have ever seen between glossy covers." Dr. Holdren has further called the paper "by far the most incompetent technical document I have ever known to have been distributed by grown-ups." He has charged repeatedly that Dr. Inhaber's errors are beyond the realm of honest mistakes and bespeak a conscious intention to mislead. Kent Anderson, who serves as Energy Information Specialist for Dr. Holdren's Energy and Resources Program at Berkeley, derides the AECB's suggestion that the Inhaber report is "food for thought." Using the report as "food for thought," Mr. Anderson says, "is equivalent to eating garbage."

Moreover, Dr. Holdren's people know what they are doing. They hope their vehemence will persuade outsiders that their complaint is not just a "dispute among experts."

So it is not surprising that even for a field like physics, which has had its share of controversies, the Holdren attack has been egregious in its invective and threats.

Dealing With Specific Criticisms

Dr. Inhaber has tried in various documents to deal with specific criticisms the report has aroused. Some of them, such as the methanol problem, he accepts: "I never claimed that the report was without errors," he points out. "That's why we've been willing to put it through revisions." On some matters, he says, such as parts of the wind calculation, the manipulation of figures that Dr. Holdren claimed to see was no more sinister than a process of updating that went on as new information came in.

In some cases, Dr. Holdren had alleged misuse of sources when Dr. Inhaber had in fact quoted them accurately; the ambiguities in the report came from ambiguities in the original sources themselves. In some instances—such as the issue of the "upper limit" figures for nuclear risks—Dr. Inhaber says he simply never made the misleading assertions that Dr. Holdren says he did. In fact, Dr. Inhaber points out, the risk figures he chose to use were sometimes higher than the ones that Dr. Holdren had come up with in his own work on nuclear risks.

More generally, Dr. Inhaber comments, the field of risk assessment is young and filled with enormous problems in collections and comparability of data; that is why he has repeatedly said that the precise numbers attributed to any particular energy system are less illuminating than the systems' safety rankings when they are compared with one another. And after correcting for the legitimate criticisms that Dr. Holdren and others have made, he has found that the rankings have not significantly changed. Here are some of his current estimates:

	Estimated range of deaths for a specified energy output (10 GWy)
Coal	50-1600
Oil	20-1400
Wind	120-230
Solar, space heating	80-90
Uranium	2½-15
Natural gas	1-4

Certainly, Dr. Inhaber argues, there was nothing in the original document to justify the kind of personal vilification that's been visited on him.

Professor Richard Wilson, a physicist whose Harvard Energy and Environmental Policy Center sent a detailed critique to Dr. Inhaber, doubts that many of the criticisms Dr. Holdren makes would have a substantial effect on the report's overall conclusions. He also points out that Dr. Inhaber had to rely on his sources. And some of the sources on alternative energy systems are not only confusing but deliberately so.

A Discouraging Discussion

But particularly discouraging, Professor Wilson thinks, is that the incivility of the attack has distracted attention from the improvements that can and should be made now in this type of study. More attention should go to the distinction between the "first-order risks" of energy production—coal emissions, say, or nuclear accidents—and "second-order risks" like those involved in construction and transport. A similar distinction can profitably be emphasized between widely distributed risks and those concentrated on a specific segment of the population, like coal miners.

And a major step has to be taken in relating risk assessments to calculations of cost-effectiveness. When we think about the efficiency of a methanol system, for instance, we should probably be thinking in terms of how much oil it will free up for electricity production if we use the methanol to replace gasoline in running automobiles. When we think about the materials requirements for a solar energy collector, we should envision them at a level low enough to make solar energy systems economically feasible; otherwise we are calculating risks for systems that will almost surely never exist.

There is little doubt that Professor Wilson's suggestions and others like them could make energy risk assessment increasingly useful in public policy. But unless disputes like the Inhaber debate can be shorn of their hysteria, the country may have a hard time using such contributions properly or making energy decisions with a relatively clear-eyed view of the risks and benefits of the alternatives before us.

Suzanne Weaver is a member of the Journal's editorial page staff.

The Washington Post
January 28, 1979

AGAINST THE GRAIN
by William Greider

In the atomic age of rock 'n roll affluence, one of the continuing burdens of citizenship is keeping up with the latest public threats to life and limb. Here is another to add to your list: killer windmills.

I have in my possession an official government study which sounds the alarm. When windmills are spread across America generating electricity, thousands of people will be maimed and killed as a result.

Solar power will be nearly as dangerous as windmills, the report warns. Why haven't we been told this before?

Good grief. The cold statistics have an even scarier message for us. Solar power and windmills, according to this government report, will be more dangerous than nuclear power plants. This is chilling news, if you thought the world was headed toward a soft energy future of wind and sun, where nobody gets hurt.

I am being playful about killer windmills and the death-dealing side effects of solar power because I assume, no matter what dreadful statistics I offer, most people of ordinary common sense will refuse to be scared. If I add the small fact that this particular study—"Risk of Energy Production" by Herbert Inhaber—was done for the Atomic Energy Control Board of Canada, that would irrevocably tarnish its conclusions for most skeptics.

Nevertheless, the Inhaber report is an elegant body count, provocative because it perversely turns the tables on the "soft technology" futurists. With prosaic charts and tables, Inhaber insists that all of the casualties be counted, not merely the imagined masses who perish in our collective fright vision of nuclear calamity, but the real ones who die everyday, in the here and now.

Implicitly, Inhaber attacks the religious assumptions which surround this question of risk and fear. If windmills are not safe, from whence will come our salvation?

Everyone is free to choose his own fears. If the fourth freedom is freedom from fear, the fifth is freedom *to* fear, picking and choosing from the parade of potential catastrophes. If the breakfast bacon does not poison us, perhaps the hamburger at lunch will. If fire does not fall from the sky someday, then maybe a silent, odorless, irradiated wind will come around the corner and blow death up our noses.

As our attention scurries from one grave risk to another, as experts offer new entries on the list, our pursuit of fear coalesces, strangely enough, with our pursuit of entertainment. Great fires, plane crashes, nuclear holocaust, dread diseases revived from the Middle Ages—these are staple themes on TV. These disaster dramas all teach the same lesson: The stresses and strains of mass death bring out the best in people. Or reveal serious character flaws in the arrogant: The fat man dies well, but the blustery cowboy turns to jelly.

This is not very entertaining, really. It is more like sick laughter in a greasy burlesque hall. The lesson on "character" makes us feel righteous, while actually we are laughing at both the fat man and the blustery cowboy. Because they are dying and we aren't.

Ultimately, this proliferation of entertaining fears becomes a crippling condition, a confusion that deepens the modern sense of impotence. The supposed dangers of modern life—an unseen wind, a menacing cloud, a stain in the hamburger—are not only beyond our personal control, but even beyond the grasp of our ordinary com- choose reliably among the best things to worry about, even killer windmills, because we don't understand statistical probability or even low-grade chemistry. All fears may be regarded as equally frightening in the land of the free, home of the brave, if you are inclined toward worry.

Some citizens, confronted with this confusion of potential killers, choose to relax and enjoy our hamburgers. This attitude is usually denounced as dumbheaded and hedonistic. Certainly, it's hedonistic but I'm not so sure it's that dumb.

On the whole, we prefer the long, pleasurable life of modern comforts to the shorter, sometimes dreadful lifespans which preceded industrial society. Having examined the available evidence, we conclude that something or other will kill us all, someday, despite fears, despite eternal vigilance. We accept a heretical notion: Dying can't be all that bad, if everybody does it.

Other citizens, who are more conscientious, demand government action. If you wish to understand the federal government in all of its eccentricities, think of a gigantic and diverse organization dedicated to defending us against death. Study the budget—from the Pentagon to the anti-cancer research, from the Federal Trade Commission to the campaign for safe cars—and you will see many worthy and many ridiculous goals. Huge heaps of money are stacked against rather modest dangers, almost the way primitive peoples placed sheep and goats on the altars of vengeful gods.

This religious quality of modern spending—deliverance in the here-and-now, not the hereafter—explains why people feel especially bitter if the government betrays them and does something to kill people, instead of saving them. This deliverance is noble work but bound to fail, ultimately, for reasons which, alas, lie beyond Congress or the president.

In the modern realm of public fears, real risks in the here-and-now somehow sound less frightening than the limitless possibilities of future calamity. Thus, 150 or so miners will be killed every year in coal-mine accidents; another 14,000 or more will suffer disabling injuries. This is happening every year, with extraordinary regularity, despite the federal safety laws, but nobody is marching to close down the coal mines. For that matter, nobody is turning off their lights to save a coal miner.

Coal kills ordinary citizens every year, too, through pollution of the air. So does oil. This is happening right now, in our own time, but it does not seem to scare people half as much as the possibility that someday, somewhere, if something goes wrong, a nuclear power plant might melt down and spew forth radiation and kill, rather quickly, 10,000 or, who knows, 100,000 people.

So far, this calamity has not happened in nearly 25 years of operating commercial nuclear plants. Indeed, according to the U.S. Nuclear Regulatory Commission, not a single American worker has died in a commercial nuclear plant because of radiation accidents, a claim which is not vigorously disputed by the anti-nuke critics. That is a marvelous industrial-safety record compared with competing fuels, especially coal, yet it doesn't really settle any of the fears. The basic fear casts forward into future generations and defines probability curves which translate into nightmares.

For many, this is more compelling than the known record. The Inhaber report can state with absolute reliability that some people, perhaps many people, will be killed by windmills and solar power. There is no mystery about it: Both of these "soft" energy systems, regardless of their other virtues, require heavy manufacturing of hardware, much more than a nuclear power plant when compared on the basis of how much electricity each produces. That means additional industrial deaths and injuries in the steel and glass factories (and in coal mines), plus more public hazard from industrial pollution.

For example, Inhaber estimates that a solar thermal plant (or plants) requires 20 times more steel, 13 times more concrete, twice as much construction labor as a nuclear plant producing the same amount of electricity. In proportion, he adds up the known frequency of industrial deaths and accidents from steel making, from coal mined to make steel, from glass and aluminum, from contruction. Then he adds the known pollution effects of producing those materials. (Solar advocates, incidentally, consider this a virtue because the heavy hardware means more industrial jobs.)

Inhaber adds another risk to the "soft" energy sources—the work-related accidents of a coal-fired back-up plant. Neither windmills nor solar plants are constantly reliable, but most of us still want electricity on windless, cloudy days, so presumably we will insist upon a back-up system which can be easily turned on or shut down, depending on the weather. That means a coal-burning auxiliary generator.

Inhaber does similar calculations for oil and coal which demonstrate rather dramatically, how dangerous those energy sources are to human life, compared to all others. The safest source, by his calculations, is natural gas—followed closely by nuclear power.

That claim, of course, is where the arguments start. Inhaber's statistics already have been attacked for grossly overestimating the material requirements and, thus, the human risk of windmills. Anti-nuke experts in Canada and elsewhere have savaged Inhaber's claims. The National Research Council of Canada attacked both his methodology and his research as unreal. He underestimates the probabilities of death from a nuclear meltdown and low-level radiation emissions, while he grossly exaggerates the necessary industrial hazards of manufacturing the "soft" alternatives, the critics charge. Some of their complaints sound, strangely enough, like the complaints which the nuclear industry makes when its critics draw scary pictures of a nuclear holocaust in our future.

What are the probabilities of accidental disaster? The question stretches over an infinity of future generations, so that only experts can pretend to know. Inhaber points out that the western industrial nations have now had 2,000 reactor years of operating experience without a meltdown. That doesn't prove a meltdown will never happen. It does imply that the odds are at least more remote than 1 in 2,000.

The anti-nuclear argument responds: That's swell, but suppose the odds are 1-in-5,000 or 1-in-10,000? The more nuclear plants we build, the sooner we will get to the catastrophe. It is an article of faith among nuclear opponents that one catastrophe, anywhere, would close down the whole business.

Frightened citizens would march on their neighborhood nuke and insist that it be turned off.

Maybe. It depends. It depends on who gets killed and how many get killed. The nation did not close down coal mines when hundreds of miners died or stop flying on airplanes when the big jets started crashing. The level of risk might simply be absorbed by most folks, as another of life's uncertain elements. Nobody wants to die, but we also don't want to be without electric lights.

So which risk shall we fear the most? The actual ongoing deaths caused by coal and oil? The nearly certain deaths, smaller in number, which will be caused by manufacturing solar and wind power hardware? Or the possibility of truly staggering calamity from a nuclear accident which has never happened, but some day might?

A reasonable person does not wish to choose among these fears, but there's the catch of our modern condition. You must choose. Somebody is going to get killed in order to bring you electricity. The choice involves more than entertaining fears, yet the choice requires calculations which most of us are not equipped to make.

The anti-nuke experts have a wonderful phrase for this predicament. It's a mathematical expression: "the zero-infinity dilemma." The Union of Concerned Scientists explains the "zero-infinity dilemma" for us: "Zero times infinity is indeterminate. This dilemma prevents the public from being able to reasonably assess the risk: Is the risk negligible because the event may never happen or large because the consequences are so great?"

I think the "zero-infinity dilemma" aptly describes the political confusion of the modern citizen with his infinity of fears and his longing for zero risk. It is a civilizing departure for human society to look far forward into the future and worry about what this scary new technology will do to the grandchildren or the grandchildren's grandchildren. But it is also cheap entertainment—an easy fear with zero risk in the here-and-now for those of us who are not coal miners.

Atom
August 1979

The Risks of Energy Production

Dr. P.M.S. Jones of the Authority's Economics and Programmes Branch comments on a recent report by Dr. Herbert Inhaber of the Canadian Atomic Energy Control Board (Report AECB 1119, March 1978)

This report, a short version of which appeared in New Scientist (18th May, 1978) sets out to compare the risks to health and life of employees and the general public from a range of energy technologies. It breaks new ground in its attempt to apply the same criteria to conventional (including nuclear) and non-conventional technologies. The latter cover solar thermal electricity, solar voltaic, solar space heating, wind power, ocean thermal gradients, and methanol from vegetation.

The risks are calculated on a full systems basis, taking into account material acquisition and construction, emissions caused by material production, operation and maintenance, energy back-up systems, energy storage and transportation. The 'risks' associated with construction are spread over the system life. The concept is good and draws attention to the fact that the risks associated with non-conventional sources are hidden because they are associated with the extraction, production and fabrication of material in quantities that significantly exceed those needed for the centralised conventional systems.

The data and conclusions are not, however, directly transferable to the United Kingdom. The statistics for risks in mining and manufacturing are based on North American figures and the natural gas and oil data relate to land-based rather than North Sea operations. For comparison/occupational risks between fossil and nuclear sources the data produced by the Health and Safety Executive in 'The Hazards of Conventional Sources of Energy' (HMSO, 1978) are therefore preferable. They omit risks in construction and those associated with materials of construction, but these are comparatively small for conventional and nuclear electricity as the Canadian report points out. The HSE figures limit themselves to accidents and exclude occupational illness because of the 'uncertainties'.

Despite these differences the occupational impacts presented in the two studies are broadly comparable, particularly at the bottom end of the AECB range (Table 1).

The AECB calculations for the non-conventional systems are based upon simplifying assumptions about the nature and quantities of materials required for their installation and properly include allowance for back-up systems and storage, both of which tend to be neglected by their protagonists. The general *ranking* of the systems and their relationship to the conventional systems is probably reasonable though the numbers themselves could not be accepted until the calculations have been confirmed for the UK (See figs 1 and 2 reproduced from AECB 1119).

The calculations of public risk are not transferable to the UK environment. They are based on air pollution

Table 1 — Comparison of occupational deaths per GWy of electrical energy sent out.

	HSE (UK conditions)		*AECB (N. American conditions)*	
Primary Source	**Operation**	**Deaths**	**Deaths**	**Man days lost**
Coal	Extraction	1.4		
	Transport	0.2		
	Generation	0.2		
	Total	1.8	2.2-10	70,000
Oil and Gas	Extraction	0.3		
	Transport	Insignificant		
	Generation	None reported		
			oil — 0.2-2	oil 2,000-19,000
	Total		gas 0.4	gas 6,000
Nuclear	(USA) Extraction	0.1		
	Transport	Insignificant		
	Generation & reprocessing	0.15		
	Total	0.25	0.2-1.3	1800-9,000

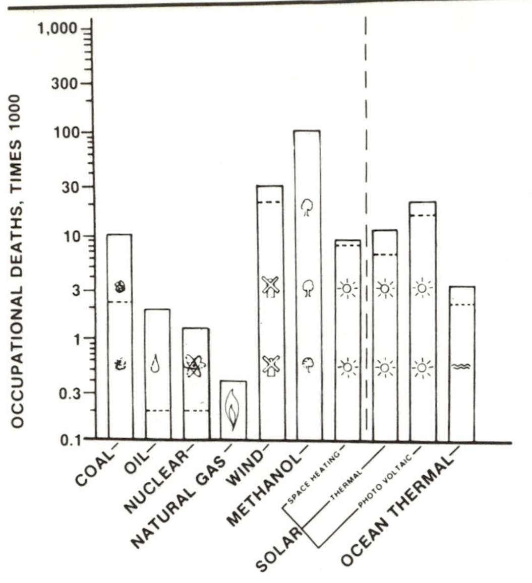

effects arising from the combustion of fossil fuels for direct electricity production or in producing the materials to manufacture non-conventional systems. To these figures are added estimated effects of potential accidents. The principal contributions are estimated to arise from emissions in material production and the production of back-up energy using coal as fuel. Nuclear risks to the public are based on the Rasmussen report and other publicised works.

Whilst there are correllations between fossil fuel combustion and morbidity and mortality the issue is a complex one with the effect of pollutants on population dependent on emission levels, temperature and height of emission, weather, topography and population density, as well as the chemical nature of the pollutant. Discharge of sulphur dioxide and particulates from a 200ft. power station chimney is very much less damaging than emission of an equivalent quantity from a domestic coal fire. (See for example P.M.S. Jones *et al* 'An Economic and Technical Appraisal of Air Pollution in the UK', HMSO, 1972).

For these reasons figures for public risk in the UK could differ radically from those estimated by AECB for N. America.

Above

Figure 1. Occupational Deaths, times 1000, per Megawatt-year, as a Function of Energy System. The values refer to one megawatt, net output, over the life of the system. For example, coal would have a maximum of 10/10,000 = 0·010 deaths per megawatt output per year over the 30-year system life. The top of the bars indicates the upper end of the range of values; the dotted lines within the bars, the lower. Where no dotted line is shown, the upper and lower ends of the range are similar. Those bars to the right of the vertical dotted lines indicate values for technologies less applicable to Canada. Most of the non-conventional technologies have higher values than the conventional systems.

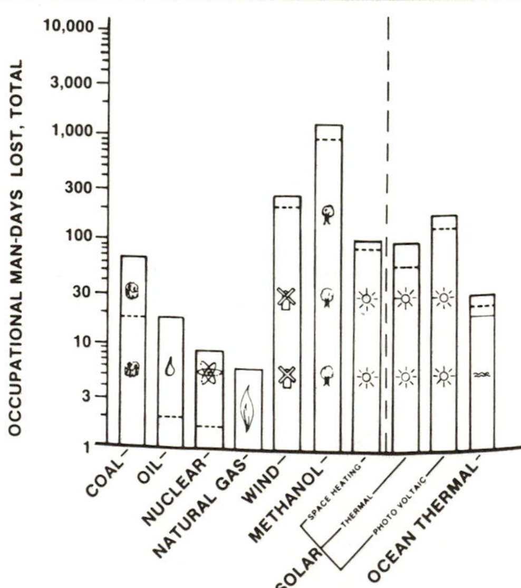

Right

Figure 2. Occupational Man-Days Lost per Megawatt-year Net Output over Lifetime of System. As in the previous graph, these values refer to the risk incurred in particular activities related to gathering and handling fuels, acquiring material and equipment, and operation and maintenance of power plants. Risk incurred by the public is not included. For calculational purposes, each death is counted as 6000 man-days lost. Methanol has by far the greatest values, a factor of about 3 greater than windpower.

The New York Times
January 2, 1980

FLAWS IN AN ARGUMENT ON THE SAFETY OF NUCLEAR POWER

To the Editor:

Theodore M. Besmann's Dec. 20 letter about the risks and benefits of nuclear energy cites a report by Herbert Inhaber of the Atomic Energy Control Board of Canada as evidence that nuclear energy entails less risk than other energy sources. The acute controversy surrounding the Inhaber Report since its publication in April 1978 has caused the Atomic Energy Control Board to deny all responsibility for the accuracy of that study and to halt its circulation.

The Inhaber Report was an attempt to compare the occupational and public health risks of 11 energy technologies. Inhaber concluded that the nonconventional technologies were far riskier than nuclear technologies. Scientists, however, upon checking Inhaber's references, data and computations, found a morass of errors that completely invalidated his conclusions.

In June 1979, the Energy and Resources Group at the University of California, from which Inhaber claimed to have taken much of his data, published a comprehensive 232-page critique demonstrating that the correction of only the most important and obvious errors so altered the results that the various solar alternatives were found to be considerably safer than the nuclear ones.

In November 1979, the Atomic Energy Control Board, responsible for publication and dissemination of the Inhaber Report, withdrew it from circulation. Thus, the main premise of Mr. Besmann's letter, that nuclear energy poses fewer risks than other energy sources, is at best unproved and at worse false.

His second premise, that nuclear energy is safer than many human activities, is misleading. Although the use of automobiles and planes may entail more risks than producing nuclear energy, the comparison is invalid. Any comparison of risks must be a comparison of those arising from alternate methods of achieving an end result, i.e., travel by railroad versus travel by car versus travel by plane, not travel by car versus production of energy by solar technologies.

In addition, individuals tend to perceive involuntary risks as greater than the same level of risk to which they voluntarily expose themselves. Similarly, the catastrophe that kills many people at once is perceived to be worse than numerous isolated incidents resulting in the same number of deaths. In other words, all risks are not created equal.

The issue of comparative risks of energy production is worthy of extensive investigation, but care must be taken to insure that one's comparisons are valid and one's data accurate.

PETER H. GLEICK
University of California
Berkeley, Calif., Dec. 22, 1979

The writer, who is associated with the Energy and Resources Group, is a coauthor of the Inhaber critique.

The New York Times
January 26, 1980

A NUCLEAR RISK STUDY THAT DIDN'T DIE

To the Editor:

Peter H. Gleick's Jan. 2 letter, "Flaws in an Argument on the Safety of Nuclear Power," states that "the acute controversy surrounding the Inhaber Report" (on relative risk of energy systems) caused the Atomic Energy Control Board of Canada to deny all responsibility for its accuracy and to halt its circulation. "In November 1979," writes Mr. Gleick, "the Atomic Energy Control Board, responsible for publication and dissemination of the Inhaber Report, withdrew it from circulation."

Neither has the report been withdrawn nor has its circulation been halted. The board has simply ceased to publish this exploratory work, now two years old, of which some 2,500 copies are still in circulation worldwide. Furthermore, all editions, from the original to the second revision, are available to the public in the A.E.C.B. library in Ottawa.

As to the question of denying all responsibility for accuracy, it should be noted that each year the board issues a number of technical and general information publications, written by staff members or persons under contract. Some of these are marked with a clear disclaimer of responsibility while others are not, but in every case the work is the author's to defend. It would take resources beyond our command to verify all material of this nature or to rise to the defense of such work at each criticism.

The A.E.C.B. published the Inhaber Report in 1978 in order to foster discussion on the subject of acceptable risk in the energy field. The study has since become a key ingredient in the stew The Wall Street Journal has dubbed "the passionate risk debate."

Perhaps we can take encouragement from the words of Emerson in 1860, "Passion, though a bad regulator, is a powerful spring," or Hegel in 1832, "Nothing great in the world has been accomplished without passion."

HUGH J.M. SPENCE
Chief, Office of Public Information
Atomic Energy Control Board
Ottawa, Jan. 16, 1980

NOTES AND REFERENCES

1. J.F. Coates, "Technology Assessment," in *McGraw-Hill Yearbook of Science and Technology,* McGraw-Hill, New York, 1974.
2. *The Energy Index 1976,* Energy Information Center, New York, 1977, p. 430.
3. A.B. Lovins, "Energy Strategy: The Road Not Taken?," *Foreign Affairs* 55, 65–96 (October 1976).
4. H. Inhaber, "Risk Accounting: A New Tool for Assessing Systems," in preparation.
5. Reference 3, p. 81.
6. R. Caputo, "An Initial Comparative Assessment of Orbital and Terrestrial Central Power Systems," Jet Propulsion Laboratory, Pasadena, California, March 1977, Report 900–780.
7. K.R. Smith, J. Weyant, and J.P. Holdren, "Evaluation of Conventional Power Plants," Energy and Resources Program, University of California, Berkeley, California, July 1975, Report ERG 75-5.
8. *Ibid*., p. 112.
9. *Ibid*., p. 113. The system described also includes lime scrubber flue gas desulfurization. However, the report admits that this is only "under development," and only operational systems are considered here.
10. "Path to Self-Sufficiency: Directors and Constraints," Bechtel Corp., San Francisco, California, 1975.
11. "Basic Estimated Capital Investment and Operating Costs for Coal Strip Mines," Bureau of Mines, Washington, D.C., 1974, Report IC-8661.
12. "Environmental Considerations in Future Energy Growth," Battelle Laboratories, Columbus, Ohio, April 1973.
13. "Analysis of Requirements and Constraints on the Transport of Energy Materials," Vol. 1, Federal Energy Administration, Washington, D.C., November 1974.
14. "Comparative Risk–Cost–Benefit Study of Alternative Sources of Electrical Energy," Atomic Energy Commission, Washington, D.C., December 1974, Report WASH-1224.
15. "Occupational Injuries and Illnesses in the United States, By Industry, 1973," Bureau of Labor Statistics, Washington, D.C., 1975.
16. Interoffice Memorandum to T.D. English from S.R. McReynolds, May 26, 1976, Jet Propulsion Laboratory, Pasadena, California.
17. Ben J. Walterberg, ed., *U.S. Fact Book,* Grosset and Dunlap, New York, 1975, p. 716.
18. Ben J. Walterberg, ed., *U.S. Fact Book,* Grosset and Dunlap, New York, 1976, p. 690.
19. L. Smith, "Manpower Requirements for Construction of Various Solar Plants," Jet Propulsion Laboratory, Pasadena, California, April 1976, Report 342-76-B-905.
20. D.J.F. van Oss, *Chemical Technology,* Barnes and Noble, New York, 1970.
21. G. Herrera, ed., "Assessment of R, D & D Resources, Health and Environmental Effects, O & M Costs and Other Social Costs for Conventional and Terrestrial Solar Electric Plants," Jet Propulsion Laboratory, Pasadena, California, January 1977, Report 900-782, Table B-3.

22. Reference 7, p. 122.

23. C.L. Comar and L.A. Sagan, "Health Effects of Energy Production and Conversion," *Annual Review of Energy* 1, 581–599 (1976).

24. Reference 6, pp. 4–15.

25. "A Study of Social Costs for Alternative Means of Electrical Power Generation for 1980 and 1990," Argonne National Laboratory, Argonne, Illinois, 1972.

26. L.D. Hamilton, Editor, "The Health and Environmental Effects of Electricity Generation–A Preliminary Report," Brookhaven National Laboratory, Upton, New York, 1974.

27. "Energy and the Environment: Electrical Power," Council on Environmental Quality, Washington, D.C., 1973.

28. L.A. Sagan, "Health Costs Associated with the Mining, Transport, and Combustion of Coal in the Steam-Electric Industry," *Nature* 250, 107–111 (1974).

29. E.W. Hall *et al.*, "Conceptual Design of Thermal Energy Storage Systems for Near Term Electric Utility Applications," General Electric Co., July 1979, p. 3–100.

30. "Air Quality and Stationary Source Emission Control," Commission on Natural Resources, National Academy of Sciences, Washington, D.C., March 1975.

31. "Environmental Impacts, Efficiency and Cost of Energy Supply and End Use," Hittman Associates, Inc., Columbia, Maryland, 1974-197 Report HIT-593.

32. Reference 7, p. 38.

33. Reference 7, p. 140.

34. "OCS Oil and Gas–An Environmental Assessment," Council on Environmental Quality, Washington, D.C., April 1974.

35. "Transport Statistics in the U.S., 1970, Part 6, Oil Pipe Lines," Interstate Commerce Commission, Washington, D.C., 1971.

36. "Cargo Handling is One-Man Job on New Esso Tanker," *Marine Engineer* 75(2), 46 (1970).

37. Reference 7, p. 148.

38. J.S. Shewchun and D.B. Curtis, "Solar Energy and the Canadian Mining Sector: A Demand Forecast," Queen's University, Kingston, Ontario, August 1978, Working Paper No. 12, p. 37.

39. F. Robson, "Technology and Economic Feasibility of Advanced Power Cycles and Methods of Producing Non-Polluting Fuels for Power Plants," Environmental Protection Agency, Washington, D.C., December 1970, Report APTD-0661.

40. "Petroleum in the Marine Environment," National Academy of Sciences, Washington, D.C., 1975.

41. P.E. Baesch, *Oil Spills and the Marine Environment,* Ballinger, Cambridge, Massachusetts, 1974.

42. "Oil: Possible Trends of Future Production," Federal Energy Agency, Project Independence Task Force, Washington, D.C., November 1974.

43. L.B. Lave and L.C. Freeburg, "Health Effects of Electricity Generation from Coal, Oil and Nuclear Fuel," *Nuclear Safety* 14, 409–428 (1973).

44. "Reactor Safety Study: An Assessment of Accident Risks in U.S. Commercial Nuclear Power Plants," Nuclear Regulatory Commission, Washington, D.C., 1975, Report WASH-1400.

45. Reference 7, p. 30.

46. Reference 7, p. 150.

47. Reference 7, p. 159.

48. L.A. Sagan, "Human Costs of Nuclear Power," *Science* 177, 487–493 (1972).

49. "Utility Staffing and Training for Nuclear Power," Atomic Energy Commission, Washington, D.C., June 1973, Report WASH-1130.

50. R.W. Dentsch, "Nuclear Manpower Crisis Ahead," *Nuclear News,* May 1974.

51. R.H. Bryan *et al.,* "Estimated Quantities of Materials Contained in a 1000 MW (e) PWR Power Plant," Oak Ridge National Laboratory, Oak Ridge, Tennessee, June 1974, Report TM-4515.

52. "Nuclear Energy," Federal Energy Administration, Project Independence Task Force, Washington, D.C., November 1974.

53. Advisory Committee on the Biological Effects of Ionizing Radiations, "The Effects on Populations of Exposures to Low Levels of Ionizing Radiation," National Academy of Sciences and National Research Council, Washington, D.C., November 1972.

54. "The Potential Radiological Implications of Nuclear Facilities in the Upper Mississippi River in the Year 2000," Atomic Energy Commission, Washington, D.C., January 1973, Report WASH-1209.

55. "Environmental Radiation Dose Commitment: An Application to the Nuclear Power Industry," Environment Protection Agency, Washington, D.C., February 1974, Report 520/4-73-002.

56. "Environmental Survey of the Uranium Fuel Cycle," Atomic Energy Commission, Washington, D.C., April 1974, Report WASH-1248.

57. Reference 21, Section 3.

58. Martin–Marietta Corporation, "Central Receiver Solar Thermal Power System, Phase I," Energy Research and Development Administration, Washington, D.C., April 1976, Report SAN/1110-76/Tl.

59. Senate Committee on Small Business, 94th Congress, First Session, May 13–14, 1975, U.S. Government Printing Office, Washington, D.C., 1975, p. 3901.

60. "BLS Reports Results of Survey of Occupational Injuries and Illnesses for 1974," Bureau of Labor Statistics, Washington, D.C., 1975, Report USDL-75-647.

61. *Statistical Yearbook 1976,* United Nations, New York, 1977, p. 320.

62. "Solar Program Assessment: Environmental Factors. Solar Thermal Electric," Energy Research and Development Administration, Washington, D.C., 1977, p. 18–19.

63. Reference 199, p. 18.

64. Reference 21, Section 6–1.

65. R. Manvi, "Terrestrial Solar Power Plant Performance," Jet Propulsion Laboratory, Pasadena, California, March 1977, Report EM 341, p. 11.

66. "Criteria for Energy Storage R & D," National Academy of Sciences, Washington, D.C., 1976, p. 74.

67. Reference 6, p. 6–12.

68. Reference 21, Table 28.

69. Reference 21, Table 5.

70. Reference 14, pp. 4–30.
71. Reference 6, p. 6–21.
72. C. Bell, "Assessment of a Solar Photovoltaic Power Conversion System for Central Station Application," Jet Propulsion Laboratory, Pasadena, California, June 1975, Report 900-702.
73. "Accidents and Unscheduled Events Associated with Non-Nuclear Energy Resources and Technology," Environmental Protection Agency, Washington, D.C., February 1977, Report EPA-600/7-77-016.
74. "Solar Program Assessment: Environmental Factors, Photovoltaics," Energy Research and Development Administration, Washington, D.C., March 1977, Report ERDA 77-47/3, Table IV-2.
75. Reference 59, p. 4235.
76. M. Davidson, D. Grether, and K. Wilcox, "Ecological Considerations of the Solar Alternative," Lawrence Berkeley Laboratory, Berkeley, Cal., Feb. 1977, Part IV.
77. M.G. Gandel, P.A. Dillard, D.R. Sears, S.M. Ko, and S.V. Bourgeois, "Assessment of Large-Scale Photovoltaic Materials Production," Environmental Protection Agency, Cincinnati, Ohio, August 1977, Report EPA-600/7-77-087, p. 4.
78. C. Daey Ouwens, "Does Solar Energy Demand More Land Surface, and More Materials or Energy Investment than Nuclear Energy or Fossil Fuels," in *International Conference on Solar Electricity,* Toulouse, France, March 1976, p. 1005–1015.
79. Reference 6, p. 4–30.
80. R.H. Turner, "Thermal Energy Storage for Solar Power Plants," Jet Propulsion Laboratory, Pasadena, California, July 1976, Report 900-754.
81. "Solar Program Assessment: Environmental Factors, Solar Heating and Cooling of Buildings," Energy Research and Development Administration, Washington, D.C., March 1977, Report ERDA 77-47/1.
82. *Ibid,* p. 29.
83. Energy and Environmental Studies Department, "An Examination of the Potential for Solar Energy Utilization in Ontario," Ontario Hydro, Toronto, 1975, Report 75092.
84. M.K. Berkowitz, "Implementing a Solar Energy Technology in Canada: The Costs, Benefits, and Role of Government," Institute of Policy Analysis, University of Toronto, Toronto, May 1977, p. 11.
85. R.M.R. Higgin, "Solar Heating for Buildings in Ontario," in *Sharing the Sun: Solar Technologies in the Seventies,* held in Winnipeg. International Solar Energy Society, Cape Canaveral, Florida, 1976, p. 213.
86. "Renewable Energy Resources in Ontario: Environmental Implications of Solar and Wind Energy Technologies," Federal Department of Fisheries and Environment, Burlington, Ontario, May 1977, p. 17.
87. Reference 59, p. 3846–3851.
88. "Solar Heating and Cooling of Buildings, Phase O," Westinghouse Electric Corporation, Baltimore, Maryland, 1974, Report NSF-RA-N-74-023D, p. U-21.
89. F.R. Kalhammer, "Energy-Storage Systems," *Scientific American* 241 (6) 56 (1979).

90. *Encyclopedia Americana,* Americana Corporation, New York, 1976, Vol. 7, p. 761.
91. "Statistical Abstract of the United States, 1976," U.S. Department of Commerce, Washington, D.C., 1976, p. 368.
92. *Ibid.,* p. 709.
93. *CBS News Almanac, 1975,* Hammond Almanac Incorporated, Maplewood, New Jersey, 1977.
94. Middleton Associates, "The Transition to Renewable Energy: Economic Opportunities for Canada," prepared for Science Council of Canada, Ottawa, 1976.
95. Reference 88, p. U-25.
96. Private communication from P.J.O. Choquette, Chief, Pollution Data Analysis Division, Environmental Protection Service, Fisheries and Environment Canada, Ottawa, June 1979.
97. J.R. Williams, *Solar Energy Technology and Applications,* Ann Arbor Science Publishers, Ann Arbor, Michigan, 1974, p. 98.
98. Based on data from a Darrieus vertical-axis wind turbine in the Ottawa, Canada, area, constructed by Dominion Aluminum Fabrication and operated by Defence Research Establishment Ottawa. Data compiled by Maryl Weatherburn.
99. Namden for Energiproduktionsforskning, "Vindenergi i Sverige: Resúltatrapport juni 1977," Stockholm, Sweden, 1977, Report NE1977: 2, pp. 80, 135.
100. Reference 15, p. 38.
101. L. Vadot, "The generation of electricity from windmills," *La Houille Blanche* (1), 15-22 (Jan-Fév. 1959).
102. Reference 59, p. 4010. The term used was "other workers," taken to mean operating and maintenance personnel. The data are for 1980.
103. Reference 59, p. 4061-4089.
104. J.L. Boot and J.G. McGowan, "Feasibility Study of a 100 Megawatt Open Cycle Thermal Difference Power Plant," University of Massachusetts, Amherst, Massachusetts, August 1974.
105. J.L. Boot and J.G. McGowan, "Preliminary Investigation of an Open Cycle Thermal Difference Power Plant," University of Massachusetts, Amherst, Massachusetts, August 1974.
106. W.G. Pollard, "The Long-Term Prospects for Solar Energy," *American Scientist* 64 (4), 424–429 (1976).
107. "Solar Program Assessment: Environment Factors, Ocean Thermal Energy Conversion," Energy Research and Development Administration, Washington, D.C., March 1977, Report ERDA 77-47/8, p. 28.
108. "Statistical Abstract of the United States, 1972," U.S. Department of Commerce, Washington, D.C., 1972, p. 723.
109. Reference 90, Vol. 1, p. 747.
110. Reference 108, p. 594.
111. *Ibid.,* p. 710.
112. Reference 59, p. 4077.
113. Reference 107, p. 26.
114. Intergroup Consulting Economists, "Economic Pre-Feasibility Study: Large Scale Methanol Fuel Production From Surplus Canadian Forest Biomass, Part 1,

Summary Report," Fisheries and Environment Canada, Ottawa, September 1976.

115. "1972 Census of Transportation, Vol. III. Commodity Transportation Survey," Bureau of the Census, U.S. Department of Commerce, Washington, D.C., June 1976, p. 8–1.

116. *Ibid.*, p. 8–9.

117. Reference 114, Part 2, Working Papers, p. II-94-95.

118. Reference 14, p. 5–27.

119. Reference 115, p. 8-84.

120. "Energy Conservation. The Data Base. The Potential for Energy Conservation in Nine Selected Industries," Federal Energy Administration, Office of Industrial Programs, Washington, D.C., 1975, p. 70.

121. Reference 61, p. 321.

122. A large part of this factor is due to the fact that the southern locations are sunnier than those of the north. A smaller part is due to latitude. These points are covered in detail in K.G.T. Hollands and J.F. Orgill, "Potential for Solar Heating in Canada," University of Waterloo, Waterloo, Ontario, February 1977, Report 77-01.

123. John Holdren, the senior author of Ref. 7, from which much of these data are taken, is a well-known critic of nuclear power. For example, in a *Bioscience* article ["Population and Panaceas," 19(12), 1065–1071 (1969)] written with Paul Ehrlich, he makes three statements: "1) (nuclear power) shares the propensity of fast-growing technology to unpleasant side effects; 2) radioactive waste problems may in the long run prove a poor trade; 3) (waste heat) has potentially disastrous effects in the local and world ecological and climatological balance." In another article ["Hazards of the Nuclear Fuel Cycle," *Bulletin of the Atomic Scientists* 30 (8), 14-23 (1974)] he makes the following statements: "1) the possibility of radioactive contamination of the environment exists at many stages of the nuclear operation; 2) routine emissions (have not) always been controlled; 3) in the late 1950's . . . some members of the public in the mining regions (received) 1.5 to 3 times the maximum permissible intake of radium in drinking water; 4) the potential for harm associated with nuclear accidents or sabotage . . . is enormous." It is clear that risk of nuclear power has not been treated lightly by Holdren and his colleagues.

124. Reference 44, p. 108.

125. Energy, Mines and Resources Canada, "An Energy Strategy for Canada," Supply and Services Canada, Ottawa, 1976, p. 51.

126. K.F. Tupper, National Research Council, Ottawa, September 1977, personal communication.

127. S.C. Morris, "Comparative Effects of Coal and Nuclear Fuel on Mortality," presented at 137th Annual Meeting, American Statistical Association, Chicago, Illinois, August 15–18, 1977.

128. A.M. Weinberg, "Energy Needs, Nuclear Power and the Environment," in *Radiation Research: Biomedical, Chemical and Physical Perspectives. Proceedings of the Fifth International Congress of Radiation Research,* held at Seattle, Washington; July 14–20, 1976, Academic Press, New York, 1975. As former head of the Oak Ridge National Laboratory, Weinberg is not noted as an advocate of solar energy.

129. G. Winstanley and B. Emmett, "Energy Requirements Associated with Selected Canadian Energy Developments," Energy, Mines and Resources Canada, Ottawa, March 1977.
130. Reference 7, p. 132.
131. Nuclear Energy Policy Study Group, *Nuclear Power: Issues and Choices*, Ballinger Publishing, Cambridge, Massachusetts, 1977, p. 195.
132. *Ibid.*, p. 229.
133. Reference 125, p. 52.
134. Reference 125, p. 68.
135. Reference 7, p. 3.
136. "Safety Statistics–1976," Safety Department, Ontario Hydro, Toronto, 1977, p. 7.
137. A.D. Jassby, "Environmental Effects of Hydroelectric Power Development," University of California at Berkeley, Lawrence Radiation Laboratory, October 1976, Report LBL–5296.
138. W.G. Morison, Director, Design and Development, Ontario Hydro, Toronto, personal communication, December 1977.
139. *Encyclopedia Brittanica*, 15th edition, Encyclopedia Brittanica Inc., New York, Vol. 4, p. 1076.
140. J.P. Albers, W.J. Bawiec, and L.F. Rooney, "Demand for Nonfuel Minerals and Materials by the United States Energy Industry, 1975–90," in *Demand and Supply of Nonfuel Minerals and Materials for the United States Energy Industry, 1975-90–A Preliminary Report*, U.S. Government Printing Office, Washington, 1976, Geological Survey Professional Paper 1006–A,B, p. A12.
141. H.H. Thomas, *The Engineering of Large Dams, Part 1*, Wiley, New York, 1976, p. 12.
142. R. Peele, editor, *Mining Engineers' Handbook*, 3rd edition, Wiley, New York, 1941, p. 3–03.
143. Bechtel Corporation, *Manpower, Materials, Equipment and Utilities Required to Operate and Maintain Energy Facilities*, National Science Foundation, Washington, 1975, p. 425.
144. Department of Interior, *Energy Perspectives II*, U.S. Government Printing Office, Washington, D.C., June 1976, p. 39.
145. *Ibid.*, p. 143.
146. Statistics Canada, "Electric Power Statistics, Vol. 2," Ottawa, 1977, Publication 57-202.
147. Federal Power Commission, "Hydroelectric Plant Construction Cost and Annual Production Expenses. 18th Annual Supplement," Washington, 1974, p. VII.
148. M. Potier, *Energie et Sécurité*, Mouton & Cie., Paris, 1969, p. 188.
149. Reference 73, p. 32.
150. A.D. Bertolett and R.J. Fox, "Accident-Rate Sample Favors Nuclear," *Electrical World*, July 15, 1974, pp. 40-41.
151. Reference 148, p. 189.
152. J. Darmstadter, P.D. Teitelbaum, and J.G. Polach, *Energy in the World Economy*, Johns Hopkins Press, Baltimore, 1971, pp. 43 and 97.
153. S. Chatterjee and A.K. Biswas, "The Human Dimensions of Dam Safety," *Water Power* 23, 446–453 (1971).

154. P.F. Gast, "Divergent Public Attitudes Toward Nuclear and Hydroelectric Plant Safety," presented at American Nuclear Society meeting, Chicago, Illinois, June, 1973.

155. Reference 148, p. 207.

156. A.O. Babb and T.W. Mermel, "Catalog of Dam Disasters, Failures, and Accidents," U.S. Department of Interior, Bureau of Reclamation, Washington, D.C., 1968.

157. N.S.G. Rao, "Failure of the Khadakwasla and Panshet Dams," *Journal of the Institution of Engineers (India)* 47 (11), 1123–1144 (1967).

158. E. Gruner, "Classification of Risk," in *Transactions of the Eleventh Congress,* International Commission on Large Dams, Madrid, Spain, 1973.

159. Associated Press, Nov. 6, 1977.

160. *World Almanac and Book of Facts, 1977,* Newspaper Enterprise Association, New York, 1977, p. 934.

161. National Science Foundation, *Large Dams of the U.S.S.R.,* U.S. Government Printing Office, Washington, D.C., 1963.

162. R.K. Mark and D.E. Stuart-Alexander, "Disasters as a Necessary Part of Benefit-Cost Analyses," *Science* 197, 1160–1162 (1977).

163. "Dam Collapses in Southeast Spur Federal Inspection Plan," *Engineering News Record* 199 (19), 13 (Nov. 10, 1977).

164. M.W. Miller and G.E. Kaufman, "High Voltage Overhead," *Environment* 20 (1), 6 (1978).

165. G.O. Robinette, *Energy and Environment,* Kendall/Hunt Publishing, Dubuque, Iowa, 1973.

166. *Financial Post,* Toronto, Feb. 18, 1978, p. 48.

167. Ontario Hydro, "Transmission–Technical. Submission of Ontario Hydro to the Royal Commission on Electric Power Planning with Respect to the Public Information Hearings," Toronto, March, 1976, p. 3.1–14.

168. Ontario Hydro, "Transmission Planning Processes. Memorandum to the Royal Commission on Electric Power Planning with Respect to the Public Information Hearings," June, 1976, p. 12.0–50.

169. *Ibid.,* p. 12.0–51.

170. Reference 167, p. 3.1–1.

171. Reference 167, p. 3.1–15.

172. Reference 167, Table 3.1.4–1.

173. Reference 167, Table 3.1.4–2.

174. Reference 168, Fig. 12–18.

175. H.A. Smith, "Electricity Supply–Generation or Degeneration," Hearn Lecture presented to Institution of Electrical Engineers, Toronto, May 2, 1973, p. 30.

176. Ontario Hydro, "Transmission–Environmental. Submission of Ontario Hydro to the Royal Commission on Electric Power Planning with Respect to the Public Information Hearings," Toronto, March 1976, p. 3.3–42.

177. Reference 18, p. 694.

178. Reference 17, p. 681.

179. Reference 18, p. 684.

180. Reference 18, p. 698.

181. Reference 17, p. 741.

182. "Energy Model Data Base: User's Manual," Brookhaven National Laboratory, Upton, N.Y., 1975, Report BNL 19200.

183. Reference 16, Table 8b.

184. Reference 16, Table 4.

185. Reference 14, p. 3–11.

186. Reference 14, p. 3–67.

187. P. Victor, G. Hathaway, and J. Lubek, "Solar Heating and Employment in Canada," Energy, Mines & Resources Canada, Ottawa, Report ER 79-1, Feb. 1979, p. 28.

188. Philip G. Hill, *Power Generation: Resources, Hazards, Technology and Costs,* MIT Press, Cambridge, Massachusetts, 1977, p. 162.

189. W.G. Morrison, Ontario Hydro Director of Design Engineering, as quoted in Canadian Press, Oct. 10, 1978.

190. Reference 117, p. I-19.

191. Reference 91, p. 553.

192. "Electric Power in Canada," Energy, Mines & Resources Canada, Ottawa, 1976, Report E1-77-5.

193. H.S. Posner, "Biohazards of Methanol in Proposed New Uses," *Journal of Toxicology & Environmental Health* 1 (1), 153–171 (1975).

194. Health and Safety Commission, "The Hazards of Conventional Sources of Energy," HMSO, London, 1978, p. 17.

195. "A Review of Large Scale Power using Windmills," Ontario Hydro, Toronto, December 1975, Report No. 75117, pp. 5–6.

196. F. Hirschfeld, "Windpower: Pipedream or Reality?" *Mechanical Engineering,* 20–28 (Sept. 1977).

197. Toronto *Globe & Mail,* July 8, 1978.

198. J.M. Noel, "French Wind Generator Systems," NSF/NASA Wind Energy Conversion System Workshop Proceedings, June 1973, Washington, D.C., p. 186.

199. "Solar Program Assessment: Environmental Factors, Wind Energy Conversion," Energy Research and Development Administration, Washington, D.C., March 1977, Report ERDA 77-47/6, p. 13.

200. D.R. Sears, D.V. Merifield, M.M. Penny, and W.G. Bradley, "Environmental Impact Statement for a Hypothetical Photovoltaic Solar–Electric Plant," Lockheed Missiles & Space Co., Inc., Huntsville, Alabama, July 1976, EPA temporary no. IERL-Ci-160.

201. "Energy Alternatives: A Comparative Analysis," Science and Public Policy Program, University of Oklahoma, Norman, Oklahoma, May 1975, p. II-17.

202. Reference 99, p. 115.

203. Reference 73, p. 141.

204. Reference 73, p. 124.

205. R.J. Leicester, V.G. Newman, and J.K. Wright, "Renewable Energy Sources and Storage," *Nature* 272 (5653), 518–21 (1978).

206. Office of Technology Assessment, "Application of Solar Technology to Today's Energy Needs," United States Congress, Washington, D.C., June 1978, p. 394.

207. "Accident Facts–1977 Edition," National Safety Council, Chicago, Ill., 1977, p. 27.

208. Reference 7, p. 146.

209. W.D. Devine, Jr., "An Energy Analysis of a Wind Energy Conversion System for Fuel Displacement," Institute for Energy Analysis, Oak Ridge, Tennessee, February 1977, Report ORAU/IEA(M)-77-2.

210. D.R. Stein, "Statistische Erfassung und Auswertung der Energieerzeugung von Windkraftwerken," *Elektrizitaetswirtschaft* 50(10), 279–285 (Oct. 11, 1951), translated as "Statistical Summary and Evaluation of Electric Power Generation from Windpower Stations," NASA Technical Translation TT-F-15651, National Aeronautics and Space Administration, Washington, D.C., June 1974.

211. J.I. Zerbe, *Science* 203(4386), 1237 (March 1979).

212. E.F. Gorzelnik, "Solar test gives some, not all answers," *Electrical World,* May 15, 1978.

213. Private communication from D. Robinson, energy system consultant, New England Electrical Systems, Boston, Mass., November 1978.

214. "Turn on the Sun," Ontario Ministry of Energy, Toronto, Canada, Sept. 1977, p. 36.

215. E.P. Cockshutt, "Solar Energy," National Research Council, Ottawa, Canada, Feb. 1977.

216. D. Rogers, "Energy Resource Requirements of a Solar Heating System," National Research Council, Ottawa, Canada, Feb. 1979, Report EPO-79-1.

217. John A. Garate, "Wind Energy Mission Analysis, Executive Summary," General Electric Space Division, Philadelphia, Pa., Feb. 1977, Report COO/2578-1/1, p. 21.

218. Private communication from W. Engle, Boeing Co., Seattle, Washington, March 1979. It was stated that eventual construction times would be about one-half that noted in the text. However, this report is concerned with present-day conditions. Also, experimental values are to be generally chosen over theoretical.

219. *Chemical Engineers Handbook,* McGraw-Hill, New York, 1973, Fifth Edition, p. 25-19.

220. "A Nationwide Inventory of Air Pollution Emissions (1974)," Air Pollution Control Directorate, Fisheries and Environment Canada, Ottawa, Report 3-AP-78-2, Dec. 1978, p. iv.

221. "Solar Energy for Space Heating & Hot Water," Division of Solar Energy, Energy Research and Development Administration, Washington, D.C., May 1976, Report SE 101, p. 5.

222. Reference 214, p. 14.

223. G.C. McKoy, "Penetration Analysis and Margin Requirements Associated with Large-Scale Utilization of Solar Power Plants", Aerospace Corp., El Segundo, California, August 1976, Report ER-198, p. 60.

224. Reference 131, p. 231.

225. B.L. Cohen, "Impacts of the Nuclear Energy Industry on Human Health and Safety," *American Scientist* 64 550 (1976).

226. F. Bove, "Here Comes the Sun: The Government Discovers Solar Energy", *Science for the People* 11 (2) 7 (1979).

227. "EPA Study Pinpoints the Real Villain in Nuclear Waste: Technetium-99," *Energy Daily* 7 (20), 4 (Jan. 29, 1979). While it is difficult to determine the estimated risk per unit energy output from this report, the study indicates that it is small.

228. W.F. Merritt, "The Leaching of Radioactivity from Highly Radioactive Glass Blocks Buried below the Water Table–Fifteen years of Results," in *Management of Radioactive Wastes from the Nuclear Fuel Cycle,* International Atomic Energy Agency, Vienna, 1976, Vol. II, pp. 27–36.

229. "BEIR Committee Issues New Estimates of Radiation Risks," (news release) National Research Council, Washington, D.C., Feb. 5, 1979.

230. *Nucleonics Week,* April 26, 1979, p. 17.

231. *Ad Hoc* Population Dose Assessment Group, "Population Dose and Health Impact of the Accident at the Three Mile Island Nuclear Station," available from the Bureau of Radiological Health, U.S. Department of Health, Education and Welfare, Rockville, Maryland, May 10, 1979.

232. Reference 14, p. 3-91 and Table 3-1.

233. T. Donakowski, "Health Risks of High-BTU Gas Pipeline and Electric Power Transmission Systems," *Energy–the International Journal* 5 (7), 609–616 (1980).

234. C. Hohenemser, R. Kasperson, and R. Kates, "The Distrust of Nuclear Power," *Science* 196, 25–34 (April 1977).

235. E.E. Pochin, "The Acceptance of Risk," *British Medical Bulletin* 31 (3), 184–190 (1975).

236. Norges Offentlige Utredninger, "Nuclear Power and Safety," State Printing Office, Oslo, Norway, June 1978, Report NOU 1978: 35C, p. 127.

237. *Ibid.,* p. 138.

238. *Ibid.,* p. 251.

239. *Ibid.,* p. 285.

240. *Ibid.,* p. 249.

241. *Ibid.,* p. 250.

242. *Ibid.,* p. 284.

243. "High-Level Radioactive Waste Management Alternatives," U.S. Atomic Energy Commission, Washington, D.C., 1974, Report WASH-1297. Cited in Ref. 236, p. 183.

244. Nuclear Energy Agency, "Objectives, Concepts, and Strategies for the Management of Radioactive Waste Arising from Nuclear Power Programmes," Organization for Economic Co-operation and Development, Paris, Sept. 1977, p. 46.

245. Reference 236, p. 169.

246. Reference 236, p. 178.

247. "Emergency Investigation of Coal Mine Embankments," NTIS Report PB 234 104/8WP, June 1974. Available from National Technical Information Service, Springfield, Virginia.

248. P. Putnam, *Power from the Wind,* Van Nostrand, New York, 1948, p. 155.

249. T.W. Reddoch and J.W. Klein, "No Ill Winds for New Mexico Utility," *IEEE Spectrum,* 57–61 March 1979.

250. "Wind Energy Systems Program Summary," Department of Energy, Washington, D.C., December 1978, Report DOE/ET-0093, p. 13.

251. W.H. Robbins and J.E. Sholes, "ERDA/NASA 200 kW–MOD OA Wind Turbine Program," in *Third Wind Energy Workshop, Vol. 1,* Department of Energy, Washington, D.C., May 1978, Report CONF–770921/1, p. 59.

252. R.J. Barchet, "MOD 1 Wind Turbine Generator Program," in Ref. 251, p. 76.
253. T.R. Kornreich and D.M. Tompkins, "An Analysis of the Economics of Current Small Wind Energy Systems," in Ref. 251, p. 156.
254. B. Wolff and H. Meyer, *Wind Energy,* Franklin Institute Press, Philadelphia, Pennsylvania, 1978, p. 57.
255. D.M. Simmons, "Wind Power," Noyes Data Corp., Park Ridge, New Jersey, pp. 217–218.
256. *Ibid.*, pp. 278, 287–288, 297.
257. *Ibid.*, p. 249.
258. *Ibid.*, p. 87.
259. *Ibid.*, p. 128.
260. *Ibid.*, p. 148.
261. W.B. Edmondson, "A 100 kW Windmill Generator," *Solar Energy Digest* 2 (4), 4 (April 1974).
262. Reference 236, p. 253.
263. Reference 7, p. 119.
264. National Safety Council, "Work Injuries and Illness Rates, 1979 edition," Chicago, Ill., 1979, p. 6.
265. J. Miettinen, I. Savolainen, P. Silvennoinen, E. Tornio, and S. Vuori, "Risk–Benefit Evaluation of Nuclear Power Plant Siting," *Journal of Nuclear Energy* 3, 489–500 (1976).
266. E.I. Vorob'ev, L.A. Ilin, V.A. Knizhnokov, D.I. Gusev, and R.M. Barkhudarov, "Atomic Energy and The Environment," *Atomnaya Energiya* 43 (5), 374–384 (1977).
267. A.B. Meinel and M.P. Meinel, *Applied Solar Energy: An Introduction,* Addison–Wesley, Reading, Massachusetts, 1977, p. 463.
268. *Ibid.*, p. 585.
269. *Ibid.*, p. 586.
270. Reference 216, p. 28.
271. S. Baron, "Solar Energy, Will It Conserve Our Non-renewable Resources?," cited in Ref. 216, p. 28.
272. Reference 216, p. 30.
273. L. Carter, "Uncontrolled SO_2 Emissions Bring Acid Rain," *Science* 204 (4398), 1179–1182 (June 1979).
274. "Energy Index 77," Environment Information Center, New York, December 1977, p. 106.
275. "Electrical Energy in Canada 1977," Energy, Mines & Resources Canada, Ottawa, 1978, p. 58.
276. *Ibid.*, p. 5.
277. *Ibid.*, p. 2.
278. *The Unfinished Agenda,* T.Y. Crowell Co., New York, 1977, p. 84.
279. J. Kronlund, "Organizing for Safety," *New Scientist* 82 (1159), 899–901 (June 14, 1979).
280. Reference 86, p. 62.
281. Reference 86, p. 31.
282. Reference 86, p. 34.
283. Reference 86, p. 29.

284. Reference 86, p. 20.

285. M.K. Reckard, "Decentralized Energy: Technology Assessment and System Description," Brookhaven National Laboratory, Upton, N.Y., June 1979, Report BNL 50987, p. 187.

286. Reference 117, pp. I-17, III-23.

287. Reference 114, p. 54.

288. Reference 117, p. I-5.

289. Reference 117, p. IV-2.

290. A. Carlisle, "The utilization of forest biomass and forest industry wastes for the production and conservation of energy," Canadian Forestry Service, Department of the Environment, Ottawa, Canada, 1976, p. 50.

291. Reference 117, p. I-13.

292. Energy Resources Conservation Board, Report 76E, Calgary, Alberta, July, 1976.

293. Synfuels Interagency Task Force Report to President's Energy Resources Council, "Synthetic Fuels Commercialization Report," Washington, D.C., Nov. 1975, vol. III.

294. Reference 117, p. II-91.

295. Reference 117, p. II-79.

296. Reference 290, p. 7.

297. Environmental Protection Agency, "Comprehensive Standards: The Power Generation Case," Washington, D.C., June 1978, Report EPA-600/9-78-013, p. 111.

298. "Nuclear Generation Costs Stable in 1978: AIF," *Nuclear News* 22 (9), 33 (1979).

299. "Comparative Risk–Cost–Benefit Study of Alternative Sources of Electrical Energy," *Nuclear Safety* 17 (2), 171–184 (1976).

300. "Environmental Aspects of Commercial Radioactive Waste Management," Document DOE/ET-0029, Department of Energy, Washington, D.C. Cited in "Comparative Risks of Electricity Generation with Uranium and Coal," Science Applications Inc., Oak Ridge, Tenn., Report SAI-OR-79-140-05, May 1979 (draft), p. 121.

301. "Proceedings of Fifth Ocean Thermal Conversion Conference," Feb. 1978, Miami Beach, Florida, p. II-284.

302. Reference 122, p. 36.

303. R.L. Loftness, *Energy Handbook*, Van Nostrand, New York, 1978, p. 403.

304. D. Lapedes, Editor, *McGraw-Hill Encyclopedia of Energy*, McGraw-Hill, New York, 1976, p. 219.

305. Department of Energy, "The Prospects for the Generation of Electricity from Wind Energy in the United Kingdom," HMSO, London, 1977, Energy Paper, No. 21, p. 26.

306. D. Mackay and R. Sutherland, "Methanol in Ontario," Ontario Ministry of Energy, Toronto, Sept. 1976, p. 19.

307. Council on Environmental Quality, "Environmental Quality–1976," Washington, D.C., Sept. 1976, p. 238.

308. Office of Energy Research and Planning, State of Oregon, *Transition*, Prometheus Unbound Books, Portland, Oregon, 1975, p. A-211.

309. C.K. Brown and R. Higgin, "Preliminary Assessment of the Potential for Large Wind Generators as Fuel Savers in Ontario," Ontario Research Foundation, Mississauga, Ontario, March 1976, p. 70.

310. "Energy Expenditures Associated with Electric Power Production by Nuclear and Fossil-fueled Plants," Atomic Energy Commission, Washington, D.C., December 1974, Report WASH-1224A, p. 4.

311. *Problems Involved in Developing an Index of Harm,* ICRP Publication No. 27, Pergamon Press, London, 1977, p. 21.

312. Reference 306, p. 71.

313. Reference 305, p. 35.

314. Reference 308, p. A-142.

315. Reference 300, p. 116.

316. Reference 301, p. II-299.

317. Reference 301, p. VII-29.

318. Reference 303, p. 358.

319. Reference 308, p. A-207.

320. Reference 219, p. 6-76.

321. Reference 195, p. 11.

322. Reference 303, p. 355.

323. Reference 195, p. 14.

324. C.H. Stone, Editor, *Producing Your Own Power,* Rodale Press, Emmaus, Pennsylvania, 1974, p. 21.

325. *Ibid.,* p. 29.

326. A.R. Patton, "Solar Energy for Heating and Cooling of Buildings," Noyes Data Corp., Park Ridge., New Jersey, 1975, p. 165.

327. Aerojet Nuclear Co., "Space Heating Systems in the Northwest–Energy Usage and Cost Analysis," Idaho National Engineering Laboratory, January 1976, Report ANCR-1276.

328. R.T. Ruegg, "Solar Heating & Cooling of Buildings: Methods of Economic Evaluation," National Bureau of Standards, Washington, D.C., July 1975, Report NBSIR 75-712, pp. 26, 28.

329. R.L. Field, "Design Manual for Solar Heating of Buildings and Domestic Hot Water," Solpub Co., Gaithersburg, Maryland, April 1977, Publication 501, p. 57.

330. R.C. Jordan and B.Y.H. Liu, Editors, "Applications of Solar Energy for Heating & Cooling of Buildings," American Society of Heating, Refrigerating and Air Conditioning Engineers Inc., New York, 1977, Report ASHRAE GRP 170, p. II-2.

331. Reference 329, p. 67.

332. Reference 328, p. 27.

333. J.W. Dickey, J.T. Beard, F.A. Iachetta, and L.U. Lilleleht, "Annual Collection and Storage of Solar-heated Water for the Heating of Buildings," presented at Conference on Solar Energy in Cold Climates, June 1976, University of Detroit, Michigan.

334. D. Lorriman, "First Annual Report, Meadowvale Solar Monitoring Programme," March 1978, submitted to National Research Council, Ottawa, p. 13.

335. *Ibid.,* pp. 1, 9.

336. "Steam-Electric Plant Air and Water Control Data," Federal Power Commission, Washington, D.C., March 1975, p. x.

337. *Ibid.,* p. xi.

338. Reference 324, p. 27.

339. Reference 324, p. 23.

340. B.J. Graham, "Feasibility of Heating Domestic Hot Water for Apartments with Solar Energy," U.S. Naval Academy, Annapolis, Maryland, March 1975, p. 9.

341. Reference 248, p. 105.

342. T.E. Lenchek, "Energy Expenditures in a Solar Heating System," *Alternate Sources of Energy* 21, 13–18 (June 1976).

343. Reference 303, p. 344.

344. "Dead of Un-natural Causes," *Miners' Voice,* 11–12 (October 1977).

345. Reference 6, p. 3-4.

346. Reference 6, p. 4-19.

347. Reference 6, p. 4-17.

348. Reference 6, p. 4-28.

349. Reference 21, Section 5.3.

350. Reference 15, p. 27.

351. W.L. Wardrop & Associates Ltd., "Net Utilization Efficiency of Oil, Gas, and Electricity as Energy Supply Systems for Buildings," Canadian Electric Association, Montreal, August 1979, p. 17.

352. *Ibid.,* p. 19.

353. *Ibid.,* p. D-3.

354. *Ibid.,* p. E-21.

355. Reference 195, p. 12.

356. M.F. Simon and H. Michel, "Solar Energy and Domestic Heating Needs," *Energy Development* (IEEE Power Engineering Society Paper), 3, 142–146 (1977).

357. J.R. Sasaki, "Solar-Heating Systems for Canadian Buildings," National Research Council, Ottawa, Canada, December 1975, p. 4.

358. J.C. Ward and G. Lof, "Evaluation of the Solar Heating System in the Lof Residence, Denver, Colorado," Colorado State University, Fort Collins, Colorado, April 1976, p. 6.

359. "Proceedings of the Workshop on Solar Heating and Cooling of Buildings, Washington, June 1974," School of Engineering, University of Virginia, Charlottesville, Virginia, Report NSF-RA-N-74-126, p. 12.

360. C.G. Justus, "Wind Energy Statistics for Large Arrays of Wind Turbines (New England and Central U.S. Regions)," Georgia Institute of Technology, Atlanta, August 1976, p. i-b.

361. Reference 6, p. 6-14.

362. Reference 359, p. 104.

363. Reference 358, p. 4.

364. Reference 6, p. 6–36.

365. S.C. Morris, K.M. Novak, and L.D. Hamilton, "Databook for the Quantitation of Health Effects from Coal Energy Systems (Preliminary Draft)," Brookhaven National Laboratory, Upton, N.Y., May 1979, p. 20.

366. H.E. Rockette, "Mortality Among Coal Miners Covered by the UMWA Health and Retirement Funds," Department of Health, Education and Welfare, Morgantown, West Virginia, NIOSH Publication 77-155, 1979.

367. Reference 365, Table II-1.

368. R.S. Bohm, J.R. Morre, and F. Schmidt-Bleek, "Benefits and Costs of Surface Coal Mine Reclamation in Appalachia," in *Energy and Man: Technological and Social Aspects of Energy,* M.G. Morgan, Editor, IEEE Press, New York, 1975.

369. Reference 365, p. 22.

370. Reference 365, p. 31.

371. Reference 365, p. 34.

372. Reference 365, p. 47.

373. Reference 6, p. 6–37.

374. J. Boulton, Editor, "Management of Radioactive Fuel Wastes: The Canadian Disposal Program," Atomic Energy of Canada Limited, Chalk River, Ontario, October 1978, Report AECL 6314, p. 24.

375. R.W. Besant and R.S. Dumont, "Comparison of 100 per cent Solar Heated Residences Using Active Solar Collection Systems," *Solar Energy* 22 (5), 451–453 (1979).

376. C.K. Brown and D.F. Warne, "An Analysis of the Potential for Wind Energy Production in Northwest Ontario," Ontario Research Foundation, Mississauga, Ontario, Nov. 1975, p. 53.

377. *Ibid.,* p. 87.

378. O. Ljungstrom, Editor, *Advanced Wind Energy Systems,* Styrelsen for Teknisk Utvekling, Stockholm, Sweden, Vol. II, 1974, p. 6-33.

379. *Ibid.,* p. 5-79.

380. H.C. Westh, "A comparison of wind turbine generators," in *Proceedings of the Second Workshop on Wind Energy Conversion Systems, June 9–11, 1975,* MITRE Corp., Report NSF-RA-N-75-050, p. 157.

381. J. Fisher, "The past and the future of wind energy in Denmark," *ibid.,* p. 162.

382. G.R. Watson and G.R. Bainbridge, "Design, Construction, and Proving of Low-cost 5 kW Wind-Powered Turbine for Isolated Applications," in *International Conference on Future Energy Concepts,* Institution of Electrical Engineers, London, 1979, p. 277.

383. P. McGeehin and J. Jensen, "Large Scale Energy Storage: Batteries for Transport and Stationary Applications," *ibid.,* p. 191.

384. T. Mensforth, "Windpower Generation on a Large Scale," *ibid.,* p. 268.

385. B. Anderson, *Solar Energy: Fundamentals in Building Design,* McGraw-Hill, New York, 1977, p. 236.

386. *Ibid.,* p. 252–254.

387. P.R. Sabady, *The Solar House,* Newnes–Butterworths, London, 1978, p. 89.

388. *Ibid.,* p. 92.

389. *Ibid.,* p. 94.

390. S.V. Szokolay, *Solar Energy and Building,* 2nd edition, Wiley, New York, 1977, p. 66.

391. *Ibid.,* p. 69.

392. *Ibid.,* p. 71.

393. *Ibid.,* p. 73.

394. *Ibid.*, p. 75.
395. *Ibid.*, p. 77.
396. *Ibid.*, p. 79.
397. *Ibid.*, p. 89.
398. *Ibid.*, p. 96.
399. *Ibid.*, p. 97.
400. Reference 376, p. 15.
401. Reference 376, p. 123.
402. Reference 85, Vol. 3, p. 4.
403. Reference 85, Vol. 3, 271.
404 G.E. Bush, in Ref. 85, "A Simple Home Solar Heating System (What can be done Now)," Vol. 10, p. 120.
405. P. Shippee, "A Solar Water Heating System," Ref. 85, Vol. 8, p. 355.
406. A. Shamo and R. Fichtenbaum, "The Feasibility of Solar House Heating: A Study in Applied Economics," Ref. 85, Vol. 9, p. 32.
407. M. Davidson, D. Grether and M. Horowitz, "Assessment of the Socio-Economic and Environmental Aspects of the Central Receiver Power Plants," Ref. 85, Vol. 9, p. 91.
408. R.D. McConnell, J.H. Beaudet, B. Piche, and E. Maille, "Use of Off-peak Electricity for Solar-Heated Homes," Ref. 85, Vol. 9, p. 128.
409. W.D. Stewart, "Wind Power Today—Its Place in our Energy Picture," in *Wind Energy Conversion Systems,* G.A. Fuller, Editor, Great Plains Research Centre, Regina, Saskatchewan, 1978, p. 9.
410. "Solar Program Assessment: Environmental Factors, Fuels from Biomass," Energy Research and Development Administration, Washington, D.C., March 1977, Report ERDA 77-47/7, p. 35.
411. *Ibid.*, p. 117.
412. *Ibid.*, p. 82.
413. D.L. Pulfrey, *Photovoltaic Power Generation,* Van Nostrand, New York, 1978, p. 4.
414. *Ibid.*, pp. 44–45.
415. *Ibid.*, p. 188.
416. "Statistical Abstract of the United States," 94th (1973), 95th (1974) and 99th (1978) editions, Department of Commerce, Washington, D.C.
417. "Mineral Yearbook 1973" and "Mineral Yearbook 1974," Bureau of Mines, Washington, D.C.
418. W.K. Gummer, Atomic Energy Control Board, private communication.
419. "Accident Facts," National Safety Council, 1969–1976 editions, Chicago, Illinois.
420. Reference 91, p. 705.
421. Science Policy Research Division, *Energy Facts II,* U.S. Government Printing Office, Washington, D.C. August 1975.
422. Interagency Task Force on Solar Energy, "Project Independence Blueprint Final Task Force Report, Solar Energy," Federal Energy Administration, Washington, D.C., November 1974.

423. J.G. McGowan, "Ocean Thermal Energy Conversion–a Significant Solar Resource," *Solar Energy, 18,* 81–92 (1976).

424. G.L. Dugger, E.J. Francis, and W.H. Avery, "Technical and Economic Feasibility of Ocean Thermal Energy Conversion," *Solar Energy* 20, 259–274 (1978).

425. R. Lidstone, Whiteshell Nuclear Research Laboratory, Pinawa, Manitoba, May and September, 1979, private communication.

426. Reference 7, p. 138.

427. R.C. Erdmann *et al.,* "Status Report on the EPRI Fuel Cycle Accident Risk Assessment," Science Applications Inc., Palo Alto, Cal., July 1979, Report NP-1128, p. 1-4.

428. Reference 275, p. 60.

429. Canadian Press, Oct. 17, 1979.

430. J. Reissland and V. Harries, "A Scale for Measuring Risks," *New Scientist,* 809-811 (Sept. 13, 1979).

431. Government of Saskatchewan, "Final Report–Cluff Lake Board of Inquiry," Regina, Saskatchewan, June 1978, Vol. 2, p. 623.

432. Reference 6, p. 6-32.

433. S.H. Schurr *et al., Energy in America's Future: The Choice Before Us,* John Hopkins University Press, Baltimore, Maryland, 1979, p. 329.

434. *Ibid.,* p. 361, 365.

435. *Ibid.,* p. 363.

436. Reference 285, pp. 126–127.

437. L.C. Hebel *et al.,* "Report to the American Physical Society by the study group on nuclear fuel cycles and waste management," *Reviews of Modern Physics* 50 (1), S86 (Jan. 1978).

438. Nuclear Regulatory Commission, "Final Generic Environmental Statement on the Use of Recycle Plutonium in Mixed-Oxide Fuel in Light-Water Cooled Reactors," NUREG-0002, Washington, D.C., August 1976, Chapters II–IV.

439. E.E. Pochin, "Estimated Population Exposure from Nuclear Power Production and other Radiation Sources," Nuclear Energy Agency, Organization for Economic Co-operation and Development, Paris, 1976.

440. Reference 437, p. S81.

441. T. Tani, S. Sawata, T. Tanaki, and S. Karaki, "Estimation of Collector and Electrical Energy Cost for STEPS in Japan," in *Sun: Mankind's Future Source of Energy, Proceedings of International Solar Energy Society Congress,* New Delhi, India, Jan. 16–21, 1978, Pergamon Press, New York, p. 175-179.

442. J.J. Iannucci, R.D. Smith, and C.J. Swet, "Energy Storage Requirements for Autonomous and Hybrid Solar Thermal Electric Power Plants," *ibid.,* p. 482–486.

443. R.J. Schwing, "New Type of Solar Power Plant in the Mojave Desert," *Proceedings of the 1979 Energy Information Forum and Workshop for Educators,* Michigan Educators Energy Forum, Lansing, 1979, p. 103-105.

444. United Nations Environmental Programme, "The Environmental Aspects of Production and Use of Energy, Part II, Nuclear Energy," UNEP Executive Director, Nairobi, Kenya, Sept. 1979, p. 33.

445. *Ibid.,* p. 35.

446. *Ibid.,* p. 42.

447. *Ibid.,* p. 108. From Ref. 438, Vol. 3, Tables IV J(E) 9-16.

448. *Ibid.*, p. 125.

449. *Ibid.*, p. 126.

450. H. Brooks and J.M. Hollander, "United States Energy Alternatives to 2010 and Beyond: The CONAES Study, *Annual Review of Energy* 4, 1–70 (1979), p. 56.

451. *Ibid.*, p. 61.

452. As reported in *Oak Ridge Review* 12 (4), 20–21 (1979). Natural gas, solar, nuclear, oil, hydroelectricity, coal, geothermal, wind, methanol, and synthetic fuels were considered.

453. T.S. Dean, *Thermal Storage*, Franklin Institute Press, Philadelphia, 1978, p. 29.

454. S.F. Gilman, E.R. McLaughlin, and M.W. Wilson, "Field Study of a Solar Energy Assisted Heat Pump Heating System," in *Solar Cooling and Heating: Architectural, Engineering and Legal Aspects,* T.N. Veziroglu, Editor, Hemisphere Publishers, Washington, D.C., 1978, Vol. II, p. 577–614.

455. T.J. McNamara, "A Solar Energy System for Domestic Hot Water," *ibid.*, p. 669–682.

456. T.V. Esbensen and V. Korsgaard, "Dimensioning of the Solar Heating System in the Zero Energy House in Denmark," in *European Solar Houses,* Royal Institution, London, 1976, p. 39–51.

457. B. Rosengren and E. Morawatz, "The Termoroc House: An Experimental Low Energy House in Sweden," *ibid.*, p. 11–27.

458. R.L. Gotchy, "Health Effects Attributable to Coal and Nuclear Fuel Cycle Alternatives," U.S. Nuclear Regulatory Commission, Washington, D.C., Sept. 1977 (draft), Report NUREG-0332, p. 17.

459. *Ibid.*, p. 10.

460. Federal Minister of Research and Technology, "The German Risk Study: Summary," Gesellschaft fur Reaktorsicherheit, Cologne, Federal Republic of Germany, August 1979, p. 39.

461. "A Methodology and Data Base for Examining the Health Risks of Electricity Generation from Uranium and Coal Fuels," Science Applications Inc., Oak Ridge, Tenn., Report SAI-OR-79-140-20, Dec. 1979 (Draft). All references to this work imply a 1000 megawatt reactor with a load factor of 0.7, unless noted otherwise.

462. *Ibid.*, p. 50.

463. *Ibid.*, p. 64.

464. *Ibid.*, p. 79.

465. *Ibid.*, p. 124.

466. *Ibid.*, p. 130.

467. *Ibid.*, p. 174.

468. *Ibid.*, p. 152.

469. *Ibid.*, p. 137.

470. *Ibid.*, pp. 40, 54, 65, 82–83 and 111.

471. *Ibid.*, p. 128.

472. *Ibid.*, pp. 152 and 183.

473. *Ibid.*, p. 187.

474. *Ibid.*, p. 182.

475. *Ibid.*, pp. 45, 56, 85, 131, and 186.

476. *Ibid.*, p. 136.

477. *Ibid.*, p. 140.

478. *Ibid.*, p. 172.

479. *Ibid.*, p. 156.

480. *Ibid.*, pp. 72, 90, and 118.

481. *Ibid.*, p. 240. Load factor of 0.65.

482. *Ibid.*, pp. 63, 76, 94, and 122.

483. "GM plans battery-powered car for commuter use by mid-1980s," Toronto *Globe & Mail,* Feb. 5, 1980, p. 82.

484. D.J. Miller, "Alternative Energy Sources," in *Nuclear or Not? Choices for Our Energy Future,* G. Foley and A. van Buren, Editors, Heinemann, London, 1978, p. 101-106.

485. F. Fagnani and C. Maccia, "L'Evaluation des risques Associés aux differentes énergies," presented at Colloque sur les risques des differentes énergies, Paris, Jan. 1980.

486. C.C. Travis *et al.,* "A Radiological Assessment of Radon-222 Released from Uranium Mills and Other Natural and Technologically-Enhanced Sources," U.S. Nuclear Regulatory Commission, Washington, D.C., Feb. 1979, p. 202.

487. Reference 427, p. 3-4.

488. R.J. Schwing, Townsend and Bottum Inc., El Monte, Calif., Nov. 1979, private communication.

489. W. Ramsay, "Small-Scale Energy Technologies," draft prepared for the National Energy Strategies Project, Resources for the Future, Washington, D.C., June 1978, p. 18-11.

490. *Ibid.*, p. 18-13.

491. *Ibid.*, p. 18-92.

492. *Ibid.*, p. 18-94.

493. *Encyclopedia Americana,* Americana Corp., New York, 1970, Vol. 1, p. 642.

494. G.E. Mueller, "Energy Self-sufficiency: The Need is Now," System Development Corp., Santa Monica, Cal., Oct. 26, 1979, p. 14.

495. Reference 194, p. 9.

496. Reference 194, p. 28.

497. G.D. Bell, "The Calculated Risk–A Safety Criterion," in *Nuclear Reactor Safety,* F.R. Farmer, Editor, Academic Press, New York, 1977, pp. 49–72.

498. Reference 107, pp. 30, 31, 36, 40.

499. Reference 14, p. 4-34.

500. Karn-Bransle-Sakerhet, "Handling of Spent Nuclear Fuel and Final Storage of Vitrified High Level Reprocessing Waste," AB Teleplan, Solna, Sweden, 1978, Vol. 1, p. 115–119.

501. UK Section of the International Solar Energy Society, *Solar Energy: a U.K. Assessment,* Royal Institution, London, May 1976, pp. 134–135.

502. *Ibid.*, p. 130.

503. S. Karaki, G.O.G. Lof, and P.R. Armstrong, "Space Heating with Solar All-air Systems–CSU Solar House II," in Ref. 441, *op. cit.,* Vol. 3, p. 1398.

504. J.G.F. Littler and R.B. Thomas, "Wind Generation of Electricity for a Novel Dwelling Independent of Servicing Networks," in Ref. 441, *op. cit.,* p. 1822.

505. A.A.M. Sayigh, "Effect of Dust on a Flat Plate Collector," in Ref. 441, *op. cit.*, Vol. 2, p. 960.

506. "Plant Engineers Solar Energy Handbook," Lawrence Livermore Laboratory, University of California, Livermore, Cal., March 1978, report LLL/M-087, p. 5-10.

507. D.J. de Renzo, editor, "Windpower–Recent Developments," Noyes Data Corp., Park Ridge, New Jersey, 1979, p. 41.

508. *Ibid.*, p. 183.

509. *Ibid.*, p. 188.

510. *Ibid.*, p. 187.

511. *Ibid.*, p. 211.

512. *Ibid.*, p. 231.

513. *Ibid.*, p. 275.

514. E.M. Noll, *Wind/Solar Energy*, H.W. Sams & Co., Indianapolis, Indiana, 1975, p. 170.

515. S.T. DiNovo *et al.*, "Preliminary Environmental Assessment of Biomass Conversion to Synthetic Fuels," U.S. Environmental Protection Agency, Cincinnati, Ohio, Oct. 1978, Report EPA-600/7-78-204, p. 201.

516. R. Rayment, "Wind Energy in the UK," *Building Services Engineer* 44(3), 63 (1976).

517. "The Solar Survey," National Center for Appropriate Technology, Butte, Montana, 1979, pp. 6, 11–13, 15, 17–18, and 20.

518. "Summary Report on Design and Installation of Solar Heating Systems in Demonstration House at Thunder Bay, Ontario," National Research Council, Solar Energy Project, Ottawa, Ontario, report DEMO-4, p. 12.

519. "Design of Solar Heating System in Demonstration House–A. Penney House," *ibid.*, report DEMO-18.

520. "Design and Installation of Solar Heating System in Demonstration House–C.J.M. Ives House. Final Report," *ibid.*, report DEMO-17.

521. A.D. Darlington, P.T. Davies, and J.L.J. Rosenfeld, "Estimation of Size requirements for Solar Photovoltaic Systems," in *Storage in Solar Energy Systems*, U.K. Section of International Solar Energy Society, London, 1978, p. 59.

522. R.A. Shaw, "Residential Solar Space Heating in New Zealand," University of Auckland, Auckland, New Zealand, Sept. 1978, p. 29.

523. Reference 15, 1977 edition, Jan. 1980, Bulletin 2047, p. 25.

524. *Ibid.*, p. 65.

525. *Ibid.*, p. 62.

526. "Design and Test Program for Transportable Solar Laboratory Program," Honeywell Inc., Minneapolis, Minn., Oct. 1974, p. 3-35.

527. *Ibid.*, p. 3-1.

528. Toronto *Globe & Mail*, "Windmill Grows in Stature as Outlook Brightens for Electricity Generation," April 14, 1980, p. B6.

529. Reference 526, p. 3-4.

530. D. Wolf, A. Tamir, and A. Kudish, "A Central Solar Domestic Hot Water System Performance and Economic Analysis," *Energy* 5, 191–205 (1980). The "over" 300,000 collectors in Israel produce "almost" a third of Israeli requirements for domestic hot water. If the pre-1967-borders population of Israel is about 2.2 million, this corresponds to about 0.5 million families. The proportion of hot water supplied for those families which have collectors is then $0.5/3 \times 0.3 = 0.56$. This is probably somewhat optimistic, since it is noted that there are more

than 300,000 collectors and the proportion of total domestic hot water supplied is less than one third. This proportion is higher than the value of 0.46 for an experiment described in the above-noted paper. Each of the apartments shared by couples in the experiment which had electrical hot water heating used 895 kWh for water heating per year. It is felt that this arrangement is more typical of society than the four singles who shared other apartments. Typical Israeli collectors have an area of 2–3 m^2. If the value of 895 kWh/year is representative of heat demand per household, the energy collected per unit area is 0.56 × 895/(2–3) = 167–250 kWh/m^2. If the average Israeli insolation on a horizontal surface is about 250 W/m^2, the average efficiency is about 1000 (167–250)/250 × 8760 = 0.08–0.12.

531. International Commission on Large Dams, "World Register of Dams 1973," p. 671.

532. K.T. Segerson, "Economic, Environmental and Social Factors Influencing the Cost of Wood-Generated Electricity in Vermont," in *1978 National Conference on Technology for Energy Conservation, January 1978, Albuquerque*, Information Transfer Inc., Rockville, Maryland, 1978, p. 49.

533. J. Lopez-Cotarelo, Sener S.A., Madrid, Spain, Aug. 1978, private communication.

534. *Statesman* (India), Aug. 19, 1979, p. 7.

535. *New York Times*, Aug. 26, 1979, p. 15.

536. M.E. Wrenn, "A Comparison of Occupational Human Health Costs of Energy Production: Coal and Nuclear Electric Generation," *Energy and Health*, presented at Society for Industrial and Applied Mathematics Institute for Mathematics and Society Conference on Energy and Health, June 1978, Alta, Utah.

537. F. Kotrappa *et al.*, "Radiation Protection Aspects in Decommissioning of a Fuel Reprocessing Plant," Fifth International Congress of the International Radiation Protection Association, Jerusalem, March, 1980, Vol. 1, p. 11.

538. Y.G. Gonen, "Risk Estimates of Stochastic Effects due to exposure to Radiation–A Stochastic Harm Index," *ibid.*, p. 149.

539. P.E. Metcalf and B.C. Winkler, "Risk Ratios for Use in Establishing Dose Limits for Occupational exposure to Radiation," *ibid.*, p. 211.

540. S. Vignes *et al.*, "Evaluation Quantitative et Comparative des Risques Liés aux Centrales Nucléaires," *ibid.*, p. 219.

541. L. Battist and H.T. Peterson, "Radiological Consequences of the Three Mile Island Accident," *ibid.*, Vol. 2, p. 263.

542. G. Herrmann, "Radiation Exposure of Personnel in a Reprocessing Plant," *ibid.*, Vol. 3, p. 209.

543. Bureau of Labor Statistics, "Number of Occupational Injury and Illness Fatalities, Private Sector, by industry, United States, 1976," Washington, 1978.

544. A-M Ericsson, "Transportation of Radioactive Materials in Sweden," Kemakta Konsult AB for Swedish Nuclear Power Inspectorate, Sept. 1979, p. 1.

545. L.J. Habegger, J.R. Gasper, and C.D. Brown, "Comparative Health and Safety Assessment of Alternative Future Electrical Generation Systems," presented at I.A.S.T.E.D. Energy Symposium, Montreal, May, 1980.

546. Easton (Pennsylvania) *Express*, Oct. 28, 1979, p. F-13.

547. G.E. Kouba, W.P. Moran, and B.V. Ketcham, "Energy Evaluation and Component Characterization Data in a Residential Solar Monitoring Program," presented at 5th Annual UMR-DNR Conference on Energy, Rolla, Missouri, Oct. 1978, p. 502-509.

548. M.F. Merriam, "Wind Energy for Human Needs," *Technology Review* 79(3), 28–39 (1977).

549. "Going with the Wind," *EPRI Journal* 5(2), 6 (1980).

550. Reference 15, 1977 edition, Washington, D.C., Jan. 1980, Bulletin 2047, p. 18.

551. L. Stang, "Avoiding Future Health Problems Related to Photovoltaic Technology," Brookhaven National Laboratory, Upton, New York, report BNL 27334.

552. A. Oudiz and J. Lochard, "Comparaison des couts marginaux de protection dans les centrales thermiques classiques et nucleaires," Ref. 537, *op. cit.*, p. 215.

553. Reference 29, pp. 5–9.

554. Reference 207, p. 23.

555. J.C. Mears, J.M. Nash, and J.T. Smok, "Comparison of Predicted and Measured Solar Energy System Performance," presented at Winter Annual Meeting, American Society of Mechanical Engineers, New York, Dec. 1979, paper 79-WA/Sol-39.

556. S.J. Flaim *et al.*, "Economic Feasibility and Market Readiness of Solar Technologies, Draft Final Report," Solar Energy Research Institute, Golden, Colorado, Sept. 1978, report SERI-TR-52-055, p. 250

557. *Ibid.*, p. 84.

558. *Ibid.*, p. 186.

559. *Ibid.*, p. 188.

560. *Ibid.*, p. 238.

561. *Ibid.*, p. 239.

562. *Ibid.*, p. 245.

563. *Ibid.*, p. 158. Some of the data in this reference may be the same as in Ref. 316 and 317.

564. S.J. Flaim *et al.*, "Economic Feasibility and Market Readiness of Eight Solar Technologies. Interim Draft Report," Solar Energy Research Institute, Golden, Colorado, June 1978, report SERI-34, pp. 9, 17.

565. *Ibid.*, p. 160.

566. *Ibid.*, p. 161.

567. *Ibid.*, p. VII-A-4. This includes costs necessary to obtain and prepare a site, transport the components to the site, assemble and erect the system, and connect it to the utility grid. It is assumed here that land costs are not included (see footnote 3 of this reference).

568. Rocket Research Company, "Chemical Energy Storage for Solar Thermal Conversion," Redmond, Washington, April 1979, report RRC-80-R-678/SAND79-8198, abstract.

569. *Ibid.*, p. 2-12.

570. *Ibid.*, p. 12.

571. National Research Council, "NRC Solar Information Series. 1. Solar Heated Homes in Canada," Ottawa, 1980, p. 22.

572. R.W. Besant *et al.*, "The Passive Performance of the Saskatchewan Conservation House," in *Proceedings of the 3rd National Passive Solar Conference,* San Jose, Cal., Jan. 1979, International Solar Energy Society, Newark, Delaware, 1979, Vol. 3, p. 713.

573. "Passive Solar Buildings," Sandia Laboratories, Albuquerque, New Mexico, July 1979, report SAND 79-0824, p. 143.

574. F.W. Heuser and K. Kotthoff, "Results of the German Risk Study–Activity Releases and Consequences," in *Proceedings of the American Nuclear Society /European Nuclear Society Topical Meeting on Thermal Reactor Safety,* April 1980, Knoxville, Tennessee, June 1980, report CONF-800403, Vol. 1, p. 348.

575. Reference 460, p. 41.

576. "Application of Solar Technology to Today's Energy Needs," Office of Technology Assessment, U.S. Congress, Washington, D.C., Vol. II, p. 689.

577. A. Lovins, "Soft Energy Paths," in *Proceedings of the Fourth Annual Conference of the Solar Energy Society of Canada,* London, Ontario, August 1978, Vol. 2, p. 23.

578. J. Jensen, *Energy Storage,* Newnes–Butterworths, London, 1980, p. 72.

579. W.E. Carscallen and B.E. Sibbitt, "First year performance data and lessons learned in the NRC 14 house solar demonstration program," Ref. 577, *op. cit.,* paper 3-1-5.

580. J.M. Nash and G.W. Cunningham, "Thermal Performance of DHW Solar Energy Systems in the National Solar Data Network during the 1978–1979 Heating Season," in *Preconference Proceedings–Solar Heating and Cooling Systems Operational Results,* Colorado Springs, Colo. Nov. 1979, Solar Energy Research Institute, Golden, Colo., p. 7. The gross area of collectors was used in the calculations since the net area was not given.

581. H.E. Taylor, "The Performance of a Daystar HW2/F-B Solar Water Heater in Southern New Jersey," *ibid.,* p. 13.

582. J.A. Tipton and H.A. Rockefeller, "Thermal Performance of a Solar Heated Hot Water System," *ibid.,* p. 19. The system is described as "operating extremely well."

583. A.H. Fanney and S.T. Liu, "Performance of Six Domestic Hot Water Systems in the Mid-Atlantic Region," *ibid.,* p. 25.

584. W.K. Aungst, "Preliminary Results: Solar Domestic Hot Water Monitoring in Pennsylvania," *ibid.,* p. 35.

585. *Ibid.,* p. 36.

586. R.S. Skidmore and J.T. Smok, "Thermal Performances of Solar Energy Space Heating Subsystems," *ibid.,* p. 89.

587. S. Karaki *et al.,* "Thermal Performance of a Commercially Available State-of-the-art Solar Air-Heating System," *ibid.,* p. 95.

588. H.S. Murray, "Solar Heating Results for the Nambe Community Center," *ibid.,* p. 127.

589. G. Darkazalli *et al.,* "Combined Solar Thermal/Photovoltaic System–Cooling Season," *ibid.,* p. 155.

590. J.L. Cain *et al.,* "Economics of Heating Maryland Broiler Houses with Solar Energy," *ibid.,* p. 175. A total of $14.50 in solar maintenance (assumed to be mostly electric fan operation) was spent per test chamber. If the cost was about 4¢/kWh, this is 1450/4 = 363 kWh. A total of 41,600 MJ of solar energy was available to 34 test chambers, or 41,600/34 × 3.6 = 340 kWh per test chamber. The ratio of fan to solar energy is then 363/340 = 1.07.

591. D.S. Ward, "Introduction," in *Conference Proceedings–Solar Heating and Cooling Operational Results,* Colorado Springs, Colo. Nov. 1978, Solar Energy Research Institute, Golden, Colo., report SERI/TP-49-063, p. 1.

592. W. Freeborne, "Performance of HUD residential Solar Systems," *ibid.,* p. 3.

593. D.L. Nemetz, "Thermal Performance Evaluation of the A-Frame Industries Solar Energy Hot Water System," *ibid.*, p. 25. Since the solar energy supplied was greater than the load, only the latter is considered in calculations.

594. W.H. McCumber, "Thermal Performance Evaluation of the Facilities Development Solar Energy Hot Water System," *ibid.*, p. 43.

595. S. Karaki, "Solar Air-heating and heat pump cooling in CSU Solar House II (1977–78)," *ibid.*, p. 61.

596. J.C. Hedstrom *et al.*, "Performance of Los Alamos Solar Mobile/Modular Home Unit No. 1," *ibid.*, p. 67.

597. D.J. Leverenz and D.M. Joncich, "The Performance of a Residential Scale Solar Heating System," *ibid.*, p. 91.

598. A. Eden and J.T. Tinsley, "Performance of the USAF Academy Solar Test House–A Retrofit Application," *ibid.*, p. 99.

599. A.F.G. Bedinger and J.F. Bailey, "Performance Results and Operating Experience of the UT-TVA Solar House (1976-78)," *ibid.*, p. 131.

600. R.N. Jensen, "The Langley Systems Engineering Building: Solar Heating and Cooling Experience and Performance," *ibid.*, p. 163.

601. H.L. Armstrong, "Thermal Performance Evaluation of the Aratex Services Inc. Solar Energy Hot Water System," *ibid.*, p. 177.

602. J.W. Crum, "Performance Evaluation of the Trinity University Solar Energy System," *ibid.*, p. 195.

603. H.S. Murray *et al.*, "Solar Heating Results for the Nambe Community Center," *ibid.*, p. 227.

604. J.C. Ward, "Electricity and Gas Consumption of 24 Solar Homes Compared with 26 Conventional Homes Having Identical Heating Loads," *ibid.*, p. 385.

605. H.E. Remmers and A.L. Tonelli, "Operation of Six Different Solar Retrofit Systems in Single Family Residences," *ibid.*, p. 405.

606. D.R. Boleyn, "Operating Results from Solar Homes in the Pacific Northwest," *ibid.*, p. 425.

607. "Mini-OTEC in Hawaii," *Sunworld* 3 (1), 22 (1979).

608. Reference 460, p. 5.

609. B.F. Williams, "Prospects of Low Cost Photovoltaic Supplies," in *Tutorials of the American Section of the International Solar Energy Society Inc.*, August 1978, Denver, Colorado, p. 91.

610. M.D. Fraser, "The Present Status and the Future Potential of Solar Industrial Process Heat," *ibid.*, p. 109.

611. E.J. Carnegie and J. Pohl, "Construction of an Agricultural Solar Drying System," in *Proceedings of the 1978 Annual Meeting of the American Section of the International Solar Energy Society*, August 1978, Denver, Colorado, Vol. 2.1, p. 25.

612. H.J. Hale and L.J. Murphy, "Solar Energy Performance Results from the National Solar Data Program," *ibid.*, p. 537.

613. R. Bruno *et al.*, "The Phillips Experimental House: A System's Performance Study," *ibid.*, p. 552.

614. R. Bruno and W.S. Duff, "Solar Heating, Cooling and Hot Water Production: A Critical Look at CCMS Installations," *ibid.*, p. 764–776, Table IX. A system with a tubular collector was not included, since this is not considered in the calculations of this report. Some of the installations in this reference from

which the average efficiency was calculated may also be mentioned elsewhere in Appendix G.

615. A.C. Myers III and L.L. Vant-Hull, "The Net Energy Analysis of the 100 MW (e) Commercial Solar Tower," *ibid.*, p. 786.
616. J.L. Watkins, "Concentrating Photovoltaic Array Testing," *ibid.*, Vol. 2.2, p. 316.
617. D.P. Grimmer, "Solar Energy Breeders," *ibid.*, p. 558.
618. K.A. Lawrence, "A Review of the Environmental Effects of Three Solar Energy Technologies," *ibid.*, p. 592.
619. S. Baron, "Solar Energy–Will it conserve Our Non-renewable Resources?," *ibid.*, p. 617.
620. J.G. McGowan *et al.*, "Operational Evaluation of a Wind Powered Heating System," *ibid.*, p. 746.
621. "Georgia Looks to Wood Wastes," *Mechanical Engineering* 102 (8), 23 (1980).
622. "Ecological and Biological Effects of Effluents from Near-Term Electric Vehicle Storage Battery Cycles," Argonne National Laboratory, Argonne, Ill., May 1980, Report ANL/ES-90, p. 8.
623. U.S. Bureau of the Census, "Statistical Abstract of the United States 1977," Washington, D.C., 1977, p. 73.
624. United States Congress, Office of Technology Assessment, "Energy From Biological Processes," Washington, D.C., July 1980, Vol. 1, pp. 83–85.
625. "Solar Heating and Hot Water System installed at St. Louis, Missouri. Final Report Tao (William) and Associates," April 1980, report NASA CR 161420.
626. Central Electricity Generating Board," Report and accounts for the year ending 31 March 1978," London 1978.
627. R.A.D. Ferguson, "Comparative Risks of the Electricity Generating Fuel Systems in the U.K., Revised Draft Report," Energy Centre, University of Newcastle upon Tyne, September 1980, pp. A-2, A-3.
628. *Ibid.*, p. A-5.
629. *Ibid.*, p. A-7.
630. *Ibid.*, p. A-11.
631. *Ibid.*, p. D-21.
632. *Ibid.*, p. C-1.
633. *Ibid.*, p. C-2.
634. *Ibid.*, p. C-13.
635. *Ibid.*, p. C-30.
636. *Ibid.*, p. C-33.
637. *Ibid.*, p. C-32.
638. *Ibid.*, p. C-39.
639. *Ibid.*, p. C-43.
640. *Ibid.*, p. C-57.
641. *Ibid.*, p. D-9.

642. *Ibid.*, p. D-17.

643. *Ibid.*, p. D-25.

644. *Ibid.*, p. D-28.

645. *Ibid.*, p. D-31.

646. *Ibid.*, p. E-9.

647. *Ibid.*, p. E-23.

648. *Ibid.*, p. E-27.

649. *Ibid.*, p. E-29.

650. *Ibid.*, p. E-30.

651. *Ibid.*, p. E-32.

652. P.D. Moskowitz *et al.*, "Photovoltaic Energy Technologies: Health and Environmental Effects Document (draft)," Brookhaven National Laboratory, Upton, New York, Sept. 1980, p. 12.

653. *Ibid.*, p. 15.

654. *Ibid.*, p. 46.

655. *Ibid.*, p. 47.

656. *Ibid.*, p. 51.

657. IBM Corp., "Solar Energy System Performance Evaluation. Seasonal Report for Decade 80 House, Tucson, Arizona," Huntsville, Alabama, May 1980, pp. 2, 47.

658. *Ibid.*, p. 70.

659. *Ibid.*, p. 59.

660. B.J. Brinkworth, "Results of Solar Heating Experiments," *Philos. Trans. R. Soc. Lond.* 295, 361 (1980).

661. H. Tabor, *Philos. Trans. R. Soc. Lond. A* 295, 376 (1980).

662. S.J. Leach, "Research at the Building Research Establishment into the applications of solar collectors for space and water heating in buildings," *Philos. Trans. R. Soc. Lond. A* 295, 403 (1980).

663. S.C. Black and F. Niehaus, "Comparisons of Risks and Benefits among Different Energy Systems," presented at International Workshop on Energy/Climate Interactions, Munster, Federal Republic of Germany, March 1980.

664. Assistant Secretary for Environment, "Technology Characterizations," U.S. Department of Energy, Washington, D.C., Jan. 1980, report DOE/EV-0072, p. 215.

665. Battelle Institute, "Comparison of Health Effects of Various Electricity Supply Technologies and First Attempt at Classifying Risks of Nuclear Energy," Frankfurt, Federal Republic of Germany, 1976, report 200/1.

666. Reference 664, p. 105.

667. Reference 664, p. 125.

668. L. Stang *et al.,* editors, "Health Effects of Photovoltaic Technology," Brookhaven National Laboratory, Upton, New York, June 1980, report BNL 51118, pp. ix, 29.

669. D.B. Bickler, "Low Cost Solar Array," *ibid.,* p. 32.

Index

D

E

F

G

H

I

J

K

L

S

T